The Megatectonics of Continents and Oceans

The Megatectonics of Continents and Oceans

edited by

Helgi Johnson and Bennett L. Smith

RUTGERS UNIVERSITY PRESS
New Brunswick, New Jersey

Library of Congress Catalog Card Number: 69-13555
SBN: 8135-0625-5

Manufactured in the United States of America
by Quinn & Boden Company, Inc., Rahway, N.J.

Contents

Foreword *Bennett L. Smith* vii

1. Definition of Regional Structures by Magnetics
James Affleck 3

2. Paleomagnetism and Continental Drift
David W. Collinson 12

3. Continents and Ocean Basins *Robert S. Dietz* 24

4. Crustal Deformation in the Western United States
James Gilluly 47

5. Tectonics and Geophysics of Eastern North America
Philip B. King 74

6. Developments in Seismology and Georheology
Leon Knopoff 113

7. Continental and Oceanic Geophysics *Paul R. Lyons* 147

8. The Mediterranean, Ophiolites, and Continental Drift
John C. Maxwell 167

9. The Axial Valley: A Steady-State Feature of the Terrain
Kenneth S. Deffeyes 194

10. The Large-Scale Tectonic Stress Field in the Earth
Adrian E. Scheidegger 223

11. Biostratigraphy, Magnetic Stratigraphy, and Sea Floor Spreading *Richard K. Olsson* 241

12. Science of the Earth *J. Tuzo Wilson* 253

Index 269

Foreword

"If this theory proves acceptable, as seems to be the case, a revolution will have occurred in earth science . . . ," writes J. Tuzo Wilson in one of the papers in this volume. The theory to which he refers is one of ocean floor spreading due to shallow advection in the upper mantle with new oceanic crust being generated along the mid-ocean ridges and being reabsorbed again into the deep ocean trenches off mountain and island arcs. "Vast, rigid, crustal plates, about 50 km thick, are moving about, with continents embedded in some of them." Much that is found in general and historical geology textbooks, he concludes, may be out of date and many ideas about petrogenesis and economic geology may need to be revised.

After continental drift had fallen into disgrace even as a subject of conversation in the forties, it suddenly became respectable again in the mid-fifties with the availability of paleomagnetic information. The papers in this book, based mostly on lectures given at a symposium on matters relating to continental drift at Rutgers University in the summer of 1966, collectively provide a comprehensive account of the development of modern global tectonics since that paleomagnetic breakthrough. The "paleomagnetic method" is described by David Collinson. Wilson and R. F. Dietz each outline aspects of the development of the theory of sea floor spreading and some of its implications. Philip B. King, John C. Maxwell and James Gilluly have studied for years the geology of parts of the world that are of particular interest in the continental drift controversy; their three papers review those areas in the context of the drift controversy.

Leon Knopoff explains, from a seismologist's point of view, some of the difficulties of invoking convection as a mechanism for drift. Adrian Scheidegger examines the tectonic stress field of the crust. James Affleck and Paul Lyons point to evidence from basement magnetic anomalies which they find difficult to reconcile with continental drift. Finally, in two papers which were invited more recently, Kenneth Deffeyes discusses a proposed model for the area of generation of new crust and R. K. Olsson outlines the impact of paleomagnetic dating and sea floor spreading on advances in biostratigraphy. The original symposium was arranged by Helgi Johnson and William Gilliland as one of the events celebrating the bicentennial of the founding of Rutgers.

The particular contribution of paleomagnetic studies has been to make it possible to compile a record of former apparent positions of magnetic poles and when this is done for rocks of differing ages from the same continent, it generally is found that there has been a systematic movement of the pole relative to the continent. Variations in pole positions for a single land mass might be reconciled with wandering of the poles but divergent polar tracks for different land masses provides powerful evidence for drift. Collinson notes that impressive correlations have accumulated between paleoclimatic data and magnetically determined paleolatitudes; for example, most of the redbeds appear to have been formed within 30 degrees of the equator.

The recent rapid advances in geophysics that have provided so much new data so suddenly are traced by Wilson. He points out that the theory of ocean floor spreading in its modern form might have remained one of many general suggestions had it not been related to observations "which give it a precision quite foreign to the rest." Most important has been the suggestion that reversals of the magnetic field are magnetically imprinted on materials of the ocean floor as they spread, leading to a quantitative basis for ocean floor spreading theory. The concept has been supported by paleontologic evidence of increasing age of cores away from mid-ocean ridges. It seems that the oceans are all young and there are fundamental differences in the mechanics of behavior of ocean basins and continents. A proto-Atlantic Ocean may have closed during the Paleozoic era, then reopened along a slightly different line; this could account for Cambrian and Ordovician faunas of European type in Eastern Canada and New England.

R. F. Dietz, in reviewing theories as to the origin of the continents and ocean basins focuses on the discontinuous distribution of sial on the surface of the earth. The ocean basins have been swept clean of sial by the process of sea floor spreading which has "opened and maintained vast windows into the earth's upper mantle." The prime mover in this process is thought to be mantle convection, offering a mechanism by which sialic materials, differentiated from the mantle, are skimmed from the ocean basins leading to the formation and growth of the continents by accretion. Detrital materials eroded from the continents would tend to be returned to them by the same mechanism. He favors Wilson's suggestion concerning the closing and reopening of the Atlantic.

Philip B. King, in reviewing the concept of continental accretion, suggests that beginning with the older Canadian Shield rocks of the Lake Superior area, one might consider four significant tectonic belts, progressively younger in age: Grenville, Avalonian, Appalachian and a fourth represented in a part of the Continental Shelf. The major problem for a supporter of the accretion hypothesis is that of explaining what happens to the Appalachian belt northeasterly beyond Newfoundland. He illustrates some of the possibilities of matching structures on either side of the North Atlantic Ocean and concludes that there is a plausible continuity between the Appalachian fold belt in Newfoundland and the Paleozoic fold belts in the British Isles. He points out however that there are formidable problems such as the difficulty of explaining what happens to the Grenville belt in Europe and what to do about Spain and Alpine structures in general.

The Mediterranean Sea area is of particular interest because, as John C. Maxwell points out, various working hypotheses for the drifting of rigid continental blocks have required large movements of Africa with respect to Eurasia, with concomitant closing of an ancient and extensive oceanic area. He concludes that the "geologic fit" between Europe and Africa is so striking that if they were not now close together we would certainly postulate that they once had been and had drifted apart. Orogenic systems in Eurasia characteristically occur within and across continental masses, in contrast to American experience. There is convincing evidence that the entire Mediterranean was a sialic, continental area after the Hercynian episode. Ancient Tethys was epicontinental, not oceanic. A kind of "oceanization" took place in the Mesozoic with the emplacement of enormous masses of mantle material, accompanied by the generation of a dynamic ridge and basin environment. The present pattern of Mediterranean basins largely reflects subsidence since the late Miocene.

In a review of the western United States, James Gilluly concludes that tectonic activity of a region must be connected with crust-mantle interaction within the region and cannot be the result of world-wide stress fields operating through the crust and mantle alone. The crust deforms passively under the drag of subcrustal motion in the mantle for it is too inhomogeneous and too weak to transmit stress over wide regions. He finds no evidence of major rotation in the geology of the western States that might be related to continental drift.

Leon Knopoff reviews geophysical information which he interprets as indicating that the crust cannot any longer be considered as independent of the mantle, floating in it in isostatic equilibrium, "as an iceberg might do in a sea, the properties of which are independent of the type of iceberg above it." He points out that recent seismologic investigations have demonstrated considerable structural complexity in the crust, mantle and core; moreover the lower mantle is significantly different from the upper one in seismic velocity and viscosity. Because of the highly viscous nature of the lower mantle, thermal convection is unlikely, but if such convection is confined to the upper mantle he concludes that it cannot have enough lateral extent to have driven the continents as far apart as they now are.

There has been a long controversy over the nature of the stress field in the earth. One of the main questions, according to Adrian Scheidegger, is whether anisotropic stresses are active only during periods of tectonic activity. He concludes that there is a tectonic stress field which cannot be explained merely by the action of gravity, that it presently is uniform over large areas but differs fundamentally from one large region to another. It seems to have remained fairly static in character for the last 60 million years except perhaps in Europe.

James Affleck is impressed by evidence from large-scale aeromagnetic surveys of a strong statistical bias toward dominance of anomaly trends in directions that are almost exactly north-south, east-west, northeast-southwest, and northwest-southeast and by evidence interpreted as indicating a Precambrian origin of anomalies with these trends. Although conceding that the paleomagnetic evidence for drift is impressive, he suggests that the large-scale aeromagnetic evidence indicates permanency of the continents and a constant relationship since Precambrian times between the crust and the spin axis of the earth.

Paul L. Lyons is also impressed by dominant east-west magnetic trends in basement rocks in an article which is concerned mainly with the controversial Conrad discontinuity, regarded by some as marking a boundary between overlying granitic and underlying gabbroic layers. He postulates that the Conrad surface may be equivalent to the profound unconformity separating Archean rocks from those of younger age; thus there are places where the Conrad discontinuity comes to the surface, as for example, around Lake Superior. The Conrad layer, according to Lyons, may be universally associated with east-west folding and faulting and he suggests that observed shifts of the crust could have taken place along these ancient lines of weakness.

In one of the two articles recently invited for this volume, Deffeyes presents a quantitative model to account for various parameters of the valley characteristic of a large part of the mid-ocean ridge at the axis of sea floor spreading. He postulates that the presence or absence of an axial valley can be the simple consequence of whether new basaltic materials are supplied over a width greater or narrower than the "zone of extension." Based in part on an analogy with exposures at Cyprus, he suggests that the proposed model accounts for the existence of a steady state terrain, the shape of the valley, the patterns of normal fault steps on either side, the magnetic record and other factors.

In the second of the recently invited papers, R. K. Olsson reviews evidence from micropaleontologic studies of deep sea cores that tends to support the concept of sea floor spreading referred to indirectly in Wilson's paper. He also describes the evidence of an apparent relationship between times of magnetic reversals and of extinction of species; such times of extinction appear to provide datums which are highly reliable over very broad areas and which apparently are more precise than datums based on evolutionary changes or times of appearance of species at given localities. Correlations have been made between extinction datums and the time scale based upon geomagnetic reversals making it possible to convert paleontologic age determinations to absolute time.

Since the mid-fifties the main continental drift controversy has shifted and broadened to include the behavior of the ocean floors and the crust as a whole. The important contribution of this book is that of bringing together such a wide variety of information from people who have contributed in one way or another to the new "plate tectonics."

Bennett L. Smith

New Brunswick, New Jersey
March, 1970

The Megatectonics of Continents and Oceans

1

Definition of Regional Structures by Magnetics *

JAMES AFFLECK

Gulf Research & Development Company,
Pittsburgh, Pennsylvania

Closely controlled aeromagnetic surveys, used in exploration for petroleum, also provide data for research in basement tectonics. Three results of such studies, which bear on continental drift and polar wander, are described here, without elaboration.

Persistence of Precambrian Stress Patterns

The total magnetic intensity data for part of southwestern Pennsylvania are illustrated in Figure 1. Only occasionally do a magnetic anomaly axis and a structural axis coincide; elsewhere, there is no simple correlation. Furthermore, the magnetic relief is far too great to be explained by relief of the basement alone. The anomalies are caused almost wholly by magnetization contrasts within the basement. However, there is remarkable parallelism of anomaly and structural trends, including the eastward bending in the northeast part. Basement contacts, interpreted from magnetics (Fig. 1, dashed lines), notably coincide with many of the structural axes.

These sedimentary structures in the Appalachian geosyncline are thought to be formed by compression (Rodgers, 1963). They are expressed as folds and thickening in the upper formations, and by brittle, overriding fracture in deeper formations. Generally, it is believed that basement relief is not involved in these structures. In my opinion, the basement is involved, but most probably as small fault movements of preexisting zones of weakness at points of contact. Per-

* I am indebted to the management of Gulf Oil Corporation for permission to publish this paper.

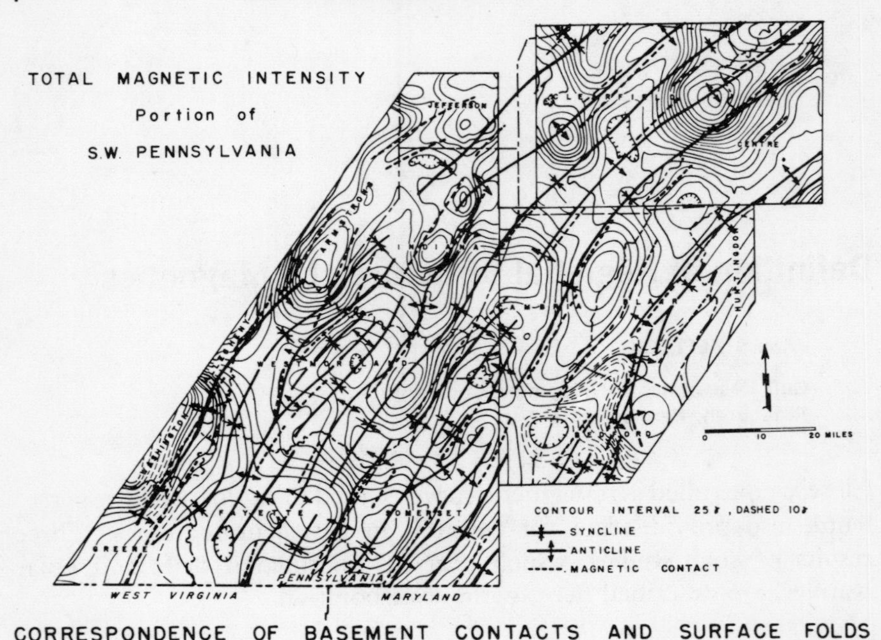

CORRESPONDENCE OF BASEMENT CONTACTS AND SURFACE FOLDS

Fig. 1. Data for Clearfield and Centre Counties from *Penn. State College Bulletin*, vol. 42, Nov. 10, 1948; for Cambria, Blair, and Bedford Counties from Aero Service Corporation, Appalachia Basin (1962); for remaining counties U.S. Geological Survey Map GP-445; surface expression of sedimentary features abstracted from *Geologic Map of Pennsylvania*, Pennsylvania Geological Survey, 4th ser. (1960).

haps the small displacements triggered the structural development at specific locales.

The basement contacts, magnetization contrasts, and associated anomalies are likely to be Precambrian developments, whereas the compression folds are related to the Late Paleozoic Appalachian orogeny. The reasoning leading to this conclusion is as follows: (1) Had the anomalies been formed during the later orogeny, any preexisting magnetic patterns would have been obliterated, since no others are present. (2) Obliteration would require heating of the basement beyond the Curie temperature of 500 C to 600 C, so that Cambrian and perhaps younger sediments would be strongly metamorphosed. But since such metamorphism is not known, the conclusion that the anomalies are related to intrabasement contrasts of Precambrian age is justified.

Similar trends of the Appalachian type are seen in all other available magnetic data within the Appalachian geosyncline, in parts of Alabama, Mississippi, Tennessee, Virginia, and elsewhere in Pennsylvania. It is therefore postulated that a Precambrian Appalachian welt was much wider than the Late Paleozoic one. Persistence of stress patterns from Precambrian times is not uncommon, as other examples, including one in Wyoming, show; there, the Laramie development of basins and mountain blocks was closely controlled by Precambrian stress patterns, which are thought to be over 3,000 million years old.

East-West Fractures

Magnetic data in the East Pacific Ocean define major fractures. The Mendocino fracture is probably related to a wide E-W magnetic interruption; correlation of dominant north-trending anomalies across the Mendocino suggests left-lateral displacement of the order of 1,000 km (Mason, 1958; Vacquier and co-workers, 1961).

What is a magnetic interruption? It is a line or zone across which a dominant magnetic anomaly pattern is broken. One form is illustrated in Figure 2; another is an alignment of anomalies suggesting intrusions; still another is an abrupt change of magnetic character along a line. Some interruptions are narrow; others may be 30 miles wide. A survey of the Pacific Ocean shows clearly the interruption of north-trending anomalies by east-trending ones in the East Pacific (Fig. 3). On the mainland, a similar interruption is found near the coast and in the Sacramento Valley. These data fail to suggest or rule out lateral displacement.

Other continental magnetic data disclose interruptions at the latitude of the Mendocino (Fig. 4). The subtle breaks in Ohio, Illinois, eastern Pennsylvania, and New Jersey are possibly speculative. Nevertheless, the interruptions appear everywhere for which data are available. These magnetic interruptions are described in detail by Fuller (1964), and Gilliland (1962) found geologic evidence along this path. Bathymetric, structural, and magnetic data on Atlantic coastal waters suggest a 100 mile right-lateral displacement (Drake and co-workers, 1963). The continental magnetic interruptions are generally clear, but there is no acceptable evidence that lateral displacement is greater than a few miles. In Fuller's (1964) opinion, the magnetic

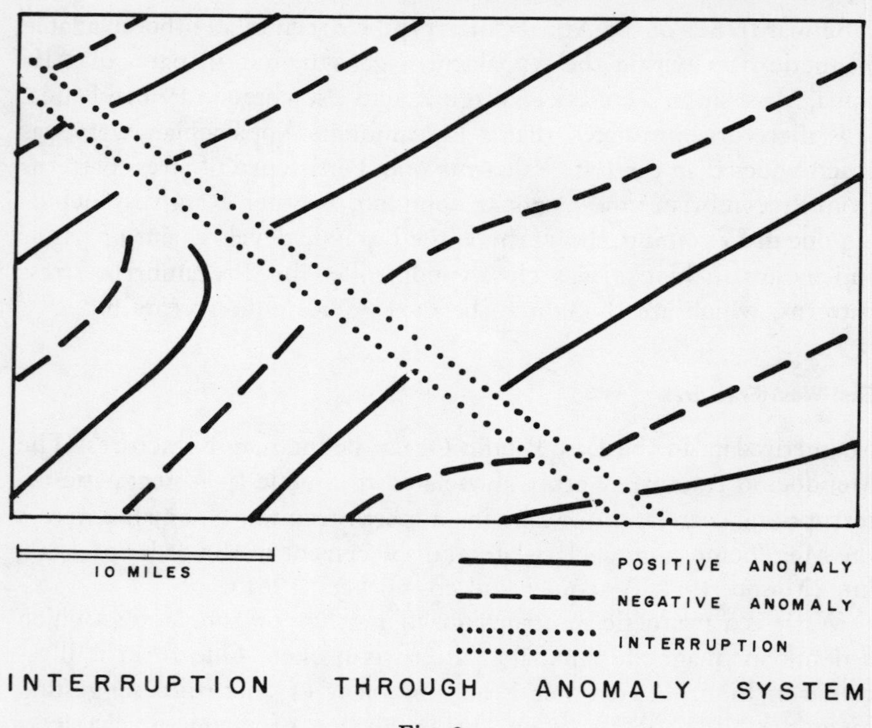

Fig. 2.

evidence suggests a fracture at great depth (below the Curie temperature level). Such a fracture might be accompanied by a major lateral displacement. If the upper basement is poorly coupled, secondary adjustment, as revealed by magnetic data, may be the only evidence of the fracture.

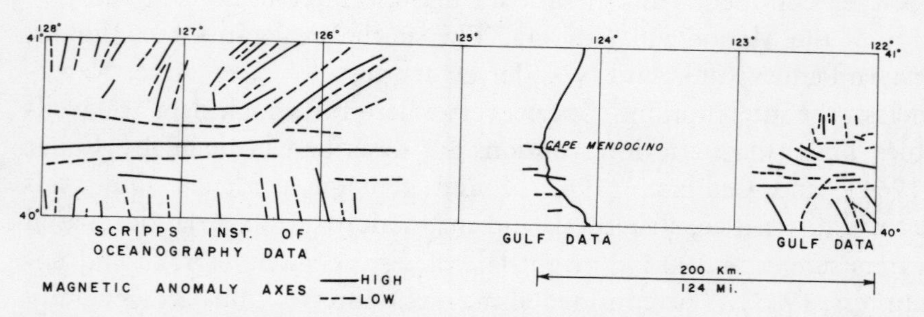

Fig. 3.

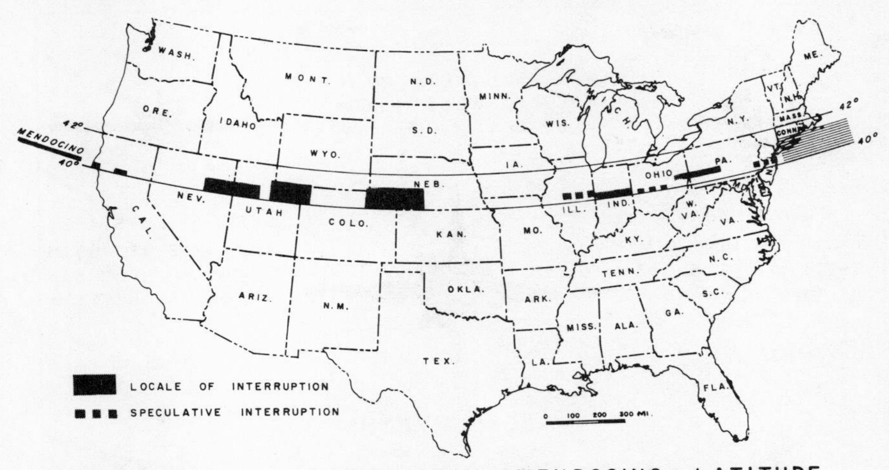

MAGNETIC E-W INTERRUPTION, MENDOCINO LATITUDE

Fig. 4.

I have found a number of similar and long E-W interruptions on magnetic surveys of the Americas and in every other available large survey. Lyons (1964) deduced, on the basis of gravity data, a similar feature in Oklahoma; this is also present on our magnetic maps. Magnetic interruptions in other directions are common, but the E-W interruptions are unique in character and persistence. Some of these features have been studied for geologic evidence of faulting or other structural distortion. It is significant that the range in the age of these structures is wide, and includes the Precambrian. The causes of fracturing in the E-W direction due to stress must therefore be considered on a world scale.

Regional Anomaly Trends

In a study of anomaly trend systems in 1963, I found a statistical bias toward the prime directions E-W, N-S, N-W, and N-E (Fig. 5). For convenience in analyzing the trends, second derivative results were used, rather than the magnetic intensity. Some investigators who have discussed these results with me have suggested that flight direction and grid bias might have affected my results. I am unable to concur with this suggestion, the results having undergone elaborate testing and checking.

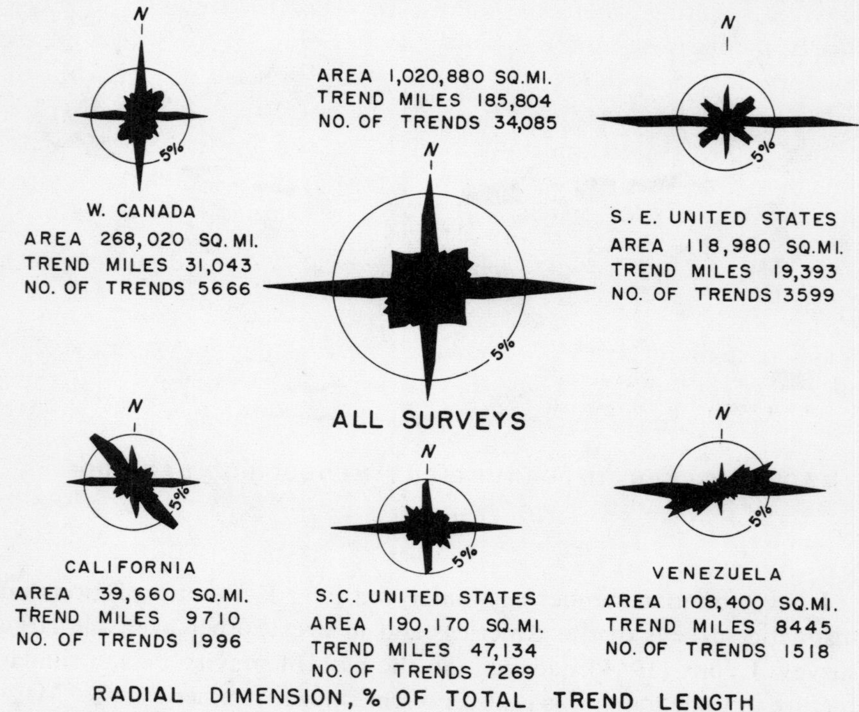

Fig. 5. Trend rosettes for five large surveys and for all surveys. (From Affleck, *Geophysics* 28:379–395, 1963)

In order to demonstrate the reality of these trends, I have recently examined extensive aeromagnetic surveys of the North Sea, Denmark, and the United Kingdom; the open control-ground magnetics from Sweden, Germany, and Holland did not influence the results greatly. For each anomaly, I selected the lineament which best defined the trend. In most cases, it was a line along the center of the steeper flank of an anomaly; in some, it was the axis. Well-defined interruptions, similar to those shown in Figure 2, were also used. The lengths of these lineaments and the geographic directions were measured and placed in 5° groups. The sums of the lengths in each group were plotted as a percentage of the total length of all the lineaments (Fig.

6). The resulting rosette is hollow, in order to demonstrate more clearly the smaller statistical peaks. As in Figure 5, the prime directions are distinct as peaks. The smaller peaks indicate the Armorican, Caledonian, and Rhine Graben structural systems; they were derived largely from magnetic data within these provinces. By reasoning similar to that used in the western Pennsylvania example, I believe that the dominant systems are of Precambrian origin.

Discussion

A sphere spinning on an axis is distorted, as if under axial compression, by the effects of centrifugal force. Axial compression would then create stress patterns leading to fracturing normal to the spin axis and

AREA 1,075,000 SQ. KM. %Σ LENGTHS-5° GROUPS
Σ LINEAMENTS 39,733 KM 100% ON SEMI-CIRCLE
NO. OF LINEAMENTS 611

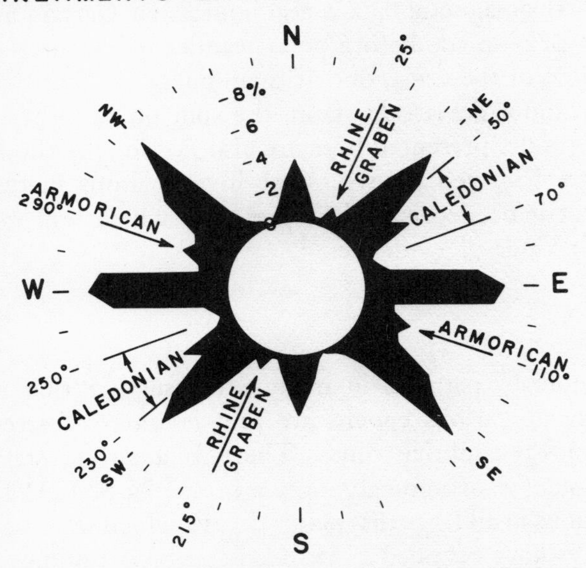

ROSETTE, MAGNETIC LINEAMENTS
NORTH SEA, UNITED KINGDOM, DENMARK
GERMANY, HOLLAND, SWEDEN

Fig. 6.

in the two directions at 45° to the axis. The axis of least stress is the spin axis. However, secondary effects caused by concentrations of stress in the other three directions can lead to fracturing along the spin axis (Anderson, 1951; Moody and Hill, 1956).

These are exactly the patterns seen in the anomaly rosettes, if the spin axis was N-S at the time the stresses were effective. Other criteria applicable to anomaly characteristics suggest that spin axes in the other prime directions are far less likely, for the following reasons: (1) a persistence of stress patterns from Precambrian time, subject to rejuvenation in later tectonic disturbances; (2) existence of E-W magnetic interruptions in all large surveys; and (3) statistical dominance on a broad scale of anomaly trends in E-W, N-S, N-E, and N-W directions and evidence of their Precambrian origins. These results obviously provide evidence which does not support the continental drift theory, or of a skidding of the crust with respect to the spin axis.

Nevertheless, evidence of polar wander and continental drift derived from paleomagnetic data is also impressive. Certain assumptions, however, must be made before polar wander is interpreted as a relative movement of the crust, one of them being the fact that the magnetic poles cannot deviate far from the spin poles. Other interpretations of the results presented here are also possible. Although impressive, they are less well known, since investigations in this field are scanty. What is needed is an interpretation which will satisfactorily explain both kinds of evidence.

Summary

Magnetic anomaly patterns in many areas suggest that major orogenetic welts of various epochs are rejuvenations of stress patterns developed in Precambrian times. There is a strong statistical bias toward dominance of anomaly trends in the N-S, E-W, N-E, and N-W directions in all large magnetic surveys which have been studied. These surveys have revealed E-W magnetic interruptions. The Mendocino fracture of the East Pacific, for example, is thought to cross under the continent into the Atlantic. These patterns all suggest the permanency of the continents, and that the relation of the earth's spin axis to the crust has been constant since Precambrian time.

References

Affleck, J.: Magnetic anomaly trend and spacing patterns. *Geophysics* 28: 379–395, 1963.

Anderson, E. M.: *The Dynamics of Faulting and Dyke Formations with Applications to Britain.* Edinburgh, Oliver & Boyd, 1951.

Drake, C. L., J. Heirtzler, and J. Hirshman: Magnetic anomalies off eastern North America. *J. Geophys. Res.* 68:5259–5275, 1963.

Fuller, M. D.: Expression of E-W fractures in magnetic surveys in parts of the U.S.A. *Geophysics* 29:602–622, 1964.

Gilliland, W. N.: Possible continental continuation of the Mendocino fracture zone. *Science* 137:685–686, 1962.

Lyons, P. L.: *Bouguer Gravity-Anomaly Map of Oklahoma.* Oklahoma Geological Survey Map GM-7, 1964.

Mason, R. G.: A magnetic survey off the west coast of the United States between latitudes 32° and 36°N, longitudes 121° and 128°W. *Geophys. J. Roy. Ast. Soc.* 1:320–329, 1958.

Moody, J. D., and M. J. Hill: Wrench-fault tectonics. *Bull. Geol. Soc. Am.* 67:1207–1246, 1956.

Rodgers, J.: Mechanics of Appalachian foreland folding in Pennsylvania and West Virginia. *Bull. Am. Ass. Petrol. Geologists* 47:1527–1536, 1963.

Vacquier, V., A. R. Raff, and R. E. Warren: Horizontal displacements in the floor of the northeastern Pacific Ocean. *Bull. Geol. Soc. Am.* 72: 1251–1258, 1961.

2

Paleomagnetism and Continental Drift

DAVID W. COLLINSON

School of Physics, University of
Newcastle-upon-Tyne, England

At the 1950 meeting of the British Association for the Advancement of Science there was a lively discussion on continental drift, chiefly between biologists and geologists. This discussion was later described by Professor P. M. S. Blackett, in his introductory remarks at the 1964 Royal Society Symposium on Continental Drift, held in London, as "the last great pitched battle between drifters and antidrifters." Whether one agrees or not, there is no doubt that the mid-1950s marked a turning point in discussions on continental drift, as a result of development of paleomagnetism. The evidence which began to accumulate appealed to many scientists because for the first time it was possible to introduce quantitative arguments into the subject. Here I shall outline the paleomagnetic method and interpretation of results, and also briefly consider the validity of the basic assumptions inherent in these interpretations.

The Paleomagnetic Method

The basic measurement in paleomagnetism is of the directions of permanent magnetization (natural remanent magnetization [NRM]) found in many types of rock, the NRM being carried by the constituent iron oxide minerals. After suitable treatment to remove unwanted components of magnetization which may be present, the direction of the "primary" NRM is often found to be consistent over a wide area within a rock formation. This direction is believed to coincide with that of the geomagnetic field at the site at the time of the rock's formation. There is thus a record of the direction of the geomagnetic field at a point on the earth's surface for an interval of geologic time, on the assumption that rock formation and magnetization are contemporaneous.

When such measurements of NRM directions are made on sediments and lavas of younger than Middle Tertiary age, it is found that the mean NRM direction (and therefore the field direction) recorded at a site corresponds closely to that which would be observed were the origin of the field a geocentric dipole aligned along the geographic axis. At present, the field corresponds rather closely to that of a dipole aligned at an angle of about 11° to the rotation axis. The evidence of paleomagnetism, therefore, is that the average geomagnetic field taken over an interval of the order of thousands of years approximates closely that of an axial dipole, with the magnetic and geographic poles coinciding. Another way of representing paleomagnetic results is to use the inclination and declination of the NRM to determine the position of the magnetic pole which will produce the observed direction of NRM at the site, on the assumption of a geocentric dipole as the origin of the field. Thus, measurement of Middle Tertiary and younger rocks shows poles clustering around the present North Pole (Fig. 1).

For the purposes of paleomagnetic interpretations and the testing of these against geologic and climatic evidence, it is assumed that the geomagnetic field has been axially dipolar throughout geologic time. A simple relation then exists between the magnetic inclination measured in rocks at a site and the ancient latitude of that site; the magnetic declination in the rock determines the attitude of the site relative to the pole. Ancient longitudes cannot be directly determined. Paleomagnetic measurements on rocks of different ages from the same continent may be represented by successive pole positions (polar wandering, see Fig. 4) or by successive continental latitudes and orientations relative to a fixed pole. This constitutes the basic data for interpretation.

Paleomagnetic Interpretation

One of the simplest applications of paleomagnetic results concerns rotation of landmasses (Fig. 2). Rocks of Mesozoic age from the N-E and S-W trending arms of Japan show mean magnetic declinations of N28°W and N30°E, respectively, whereas Cenozoic rocks from these

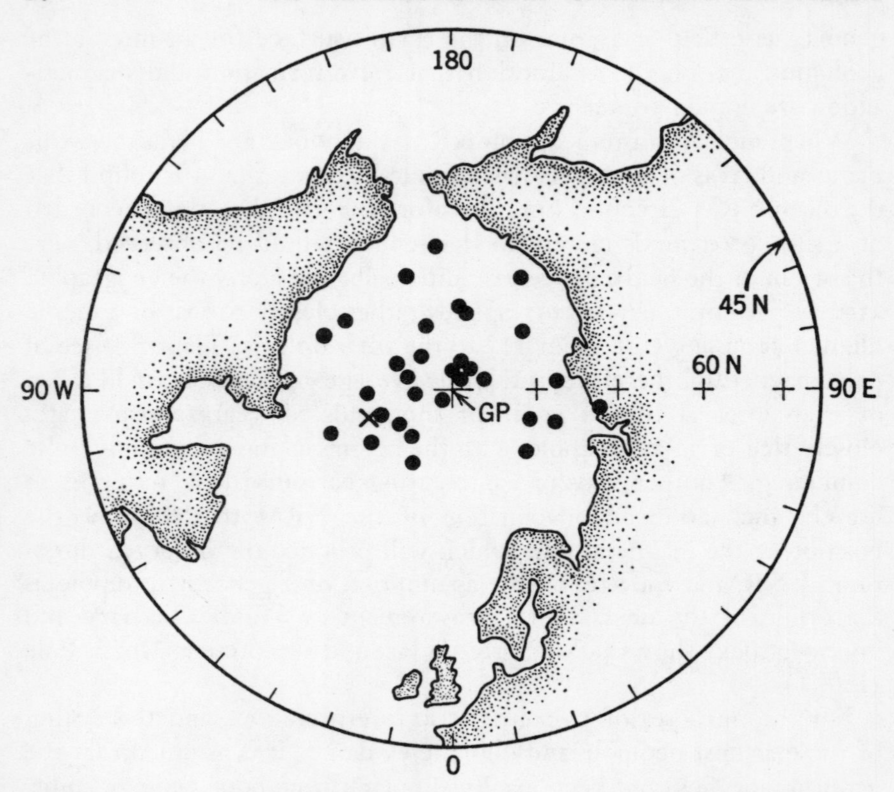

Fig. 1. Paleomagnetic pole positions derived from rocks of Plio-Pleistocene age. GP, present geographic pole; X, present geomagnetic pole. After Irving (1964).

areas show a uniform declination of about 10°E. The discrepancy between the directions in the older rocks is believed to be due to rotation of the two arms of Japan about the fossa magna in post-Mesozoic times (Kawai and co-workers, 1961). Other examples of such movement suggested by paleomagnetism are the opening of the Red Sea (Irving and Tarling, 1961) and the rotation of Spain into the Bay of Biscay (Irving, 1964).

By using both magnetic inclination and declination of NRM from rocks of different ages from the same continent, a mean pole position can be determined appropriate to each age, and a polar path constructed relative to the continent. When this is done, a generally systematic movement of the pole relative to each continent becomes

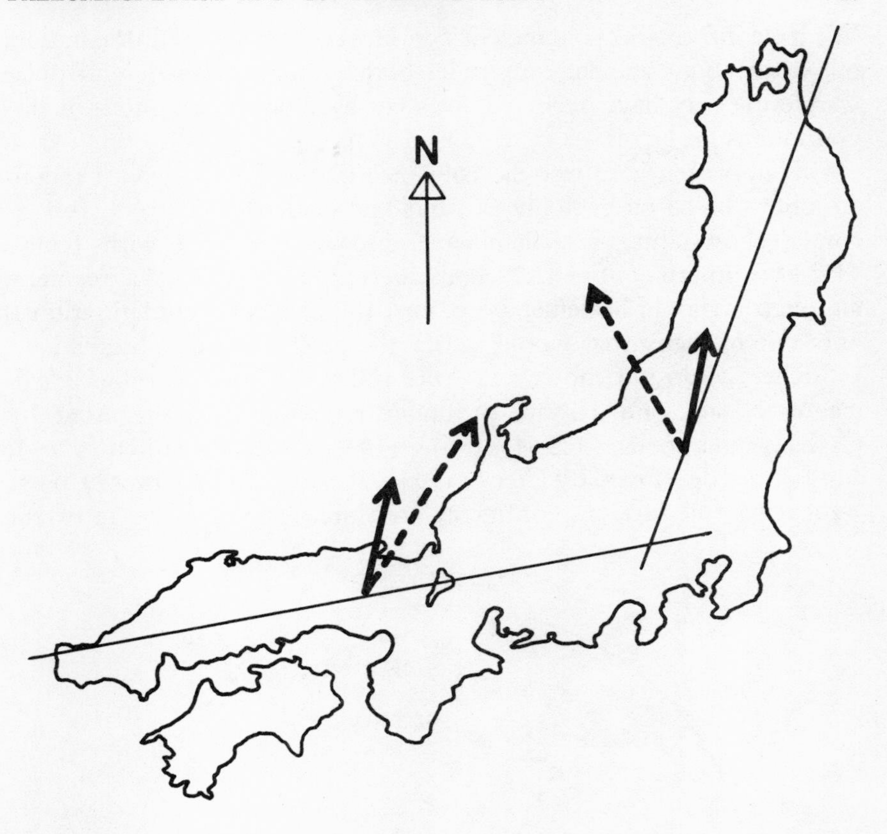

Fig. 2. Example of landmass rotation. Mean declination recorded in Cenozoic and Mesozoic rocks from N-E and S-W trending arms of Japan. Solid line (———), Cenozoic rocks; dashed line (-----), Mesozoic rocks.

apparent. For each continent this is an interesting and straightforward result; however, these polar paths do not coincide, and indeed are widely separated. If the geomagnetic field has been dipolar during geologic time, polar paths (and pole positions for a given age) should be approximately coincident if the continents have always occupied their present positions, due allowance being made for possible dating and paleomagnetic inaccuracies. This disparity between pole positions from rocks of similar age from different continents is the basic evidence favoring continental drift.

The problem, then, is the reconstruction of ancient continental positions in a way consistent with the available paleomagnetic data.

We have information on ancient continental latitudes and orientation, but none about ancient longitude; both continental drift and polar wandering may have occurred and may have been continuous or discontinuous processes.

One approach is to test the paleomagnetic evidence against reconstructions based on geologic and biologic evidence. Figure 3, for example, shows Jurassic paleomagnetic data compared with King's (1961) southern continental reconstruction. There is good agreement, although it should be remembered that the particular configuration is only one of many that would satisfy the paleomagnetic observations.

Such tests do not, however, make full use of the available paleomagnetic data, and a more profitable approach was suggested by Graham and associates (1964), Irving (1964), and Van Hilten (1964), and further developed by Creer (1964). If parts of the curves of polar wandering for different continents are similar in shape and in extent

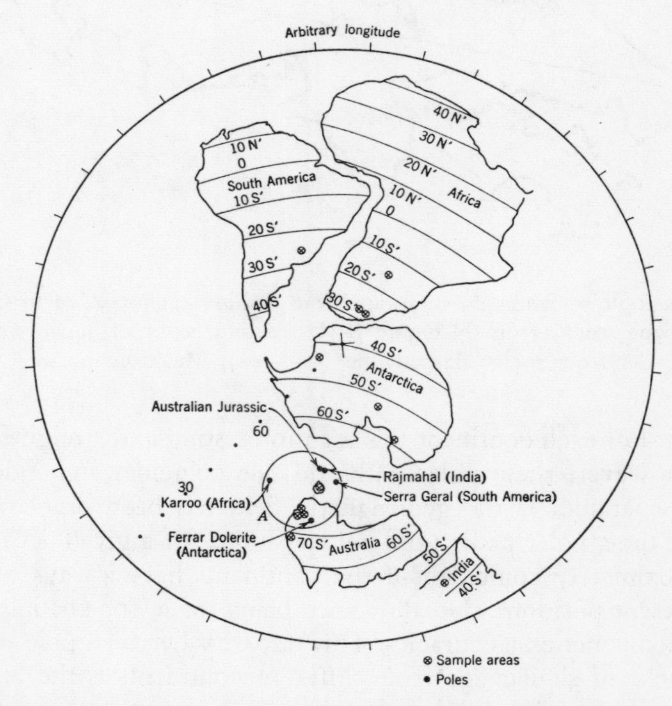

Fig. 3. Comparison of Gondwanaland reconstruction (King, 1961) with paleomagnetic data from Jurassic rocks. Cross-continental lines, paleolatitudes, after Irving (1964).

of time, a period of polar wandering without relative movement of the continents can be postulated. If this superposition can be achieved, a more restricted reconstruction will result, and information will be available regarding the time of onset of drift. Figures 4 and 5 illustrate the procedure applied to South America and Africa. The sequence suggested is polar wandering without relative drift from Silurian to Permian times, when drift began, as indicated by the divergence of the polar paths. Although based on rather scanty data at present, the fit of Africa and South America is confirmed in a striking way by paleomagnetism. Creer (1964) has extended this procedure to other continents for which suitable data are available; so far, the

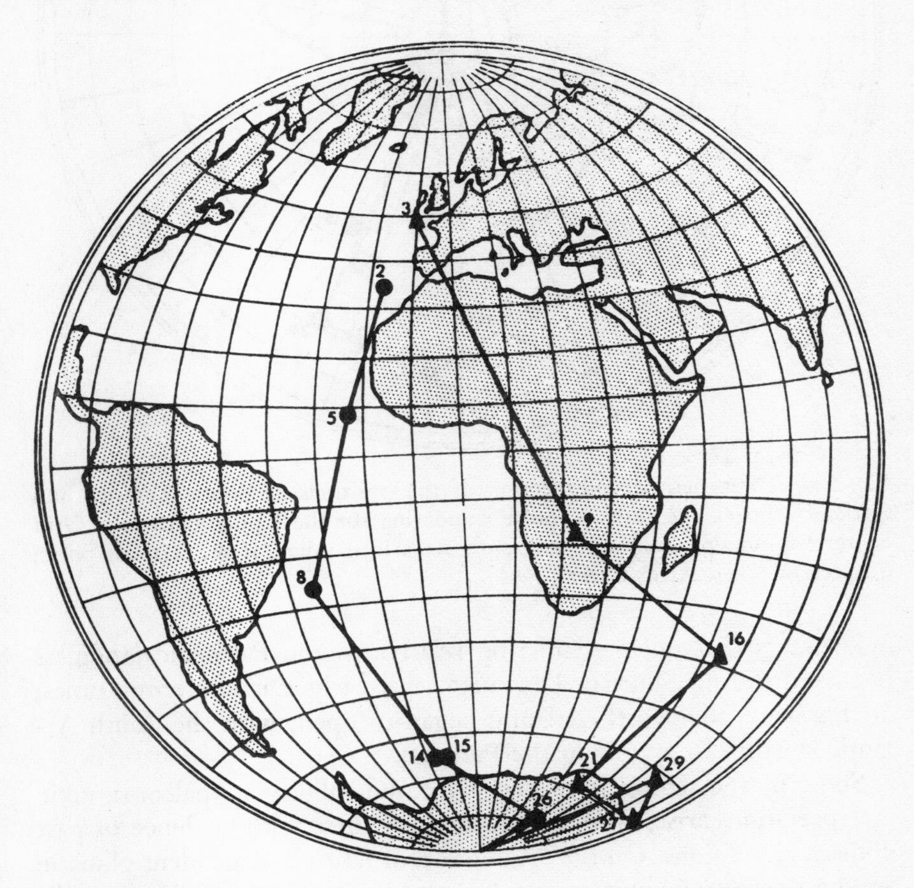

Fig. 4. Paleomagnetic south poles for South America (●) and Africa (▲). Grid represents present latitude-longitude position. After Creer (1964).

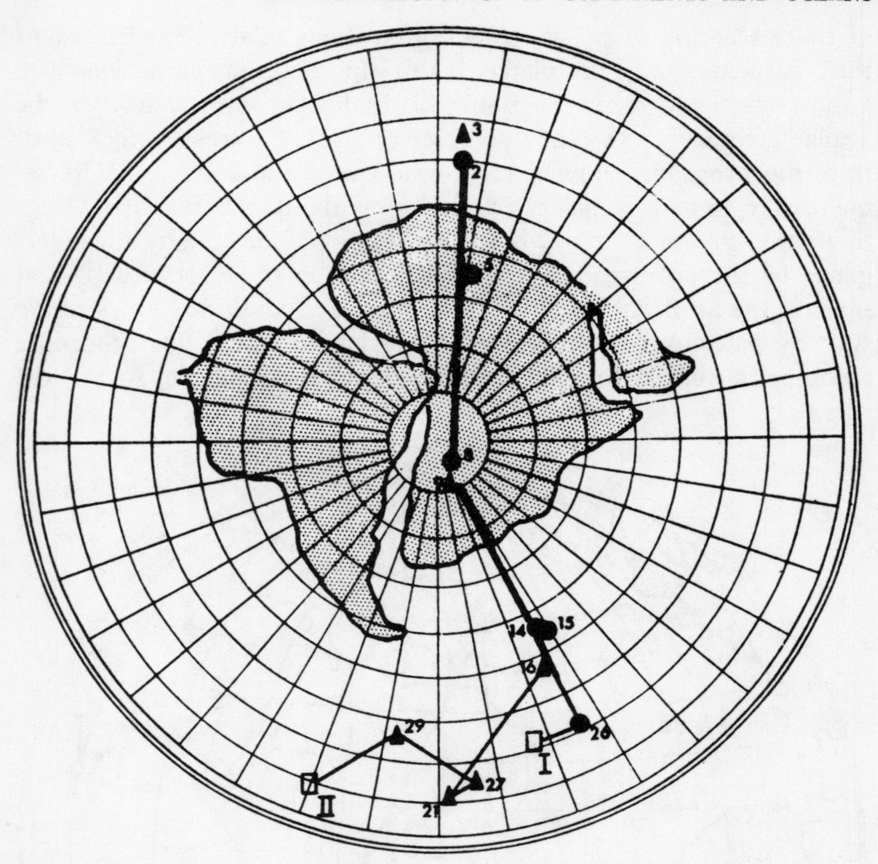

Fig. 5. Paleomagnetic reconstruction of the positions of South America and Africa by fitting the paths of polar wandering for the Upper Paleozoic. Grid center has no significance. I ●, South American poles; II ▲, African poles; □, present poles. After Creer (1964).

evidence is that polar wandering relative to the Paleozoic landmass occurred during Silurian, Devonian, and early Carboniferous times; at this stage, the North Atlantic began to open, with the South Atlantic starting to appear in the Permian.

Some of the strongest evidence for the validity of paleomagnetic interpretations arises from comparisons of these with evidence of past climatic conditions. On the assumption of latitude-dependent climatic zoning during geologic time, paleolatitudes of geologic climatic indicators, deduced from paleomagnetic measurements, should be appropriate to the climate associated with the indicator. Figure 6 shows a

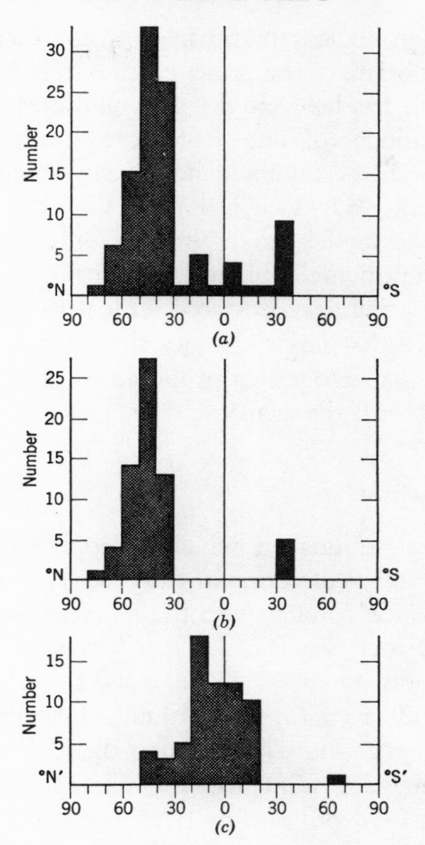

Fig. 6. Comparison of present and ancient latitudes of red beds. *Top,* latitudes of representative red beds at present; *middle,* present latitudes of beds for which paleomagnetic data are available; *bottom,* paleolatitudes of these beds. After Briden and Irving (1964).

comparison of present and ancient latitudes of red beds. These rocks are generally believed to be associated with a warm, tropical or semi-tropical environment during formation; as may be seen, most of the paleolatitudes of the red beds studied lie between 30°N and 30°S. Other geologic features studied in this way include corals, evaporites, certain fossil forms, and carbonates (Blackett, 1961; Briden and Irving, 1964); overall consistency is apparent between paleomagnetic and paleoclimatic data. Finally, earth expansion, a topic about which discussion has recently been revived, must be mentioned. This can be tested paleomagnetically by measuring the rate of change of magnetic inclination along an ancient meridian in a continent and estimating whether it is appropriate to the present radius of the earth. An alternative method which has been used is to compute pole positions from rocks from a continent of one geologic age, and choose the earth radius which gives the best coincidence of these poles (Cox and Doell, 1961; Van Hilten, 1963; Ward, 1963). Unfortunately, even quite large changes in the earth's radius produce only moderate changes in inclination at a continental site, and significant changes may be obscured by the uncertainties associated with paleomagnetic data. Present results indicate that large changes since the Cambrian have not taken place, but smaller ones, as proposed by Egyed (1957) cannot as yet be resolved with the available data.

Basis of Paleomagnetism

What are the assumptions on which paleomagnetic interpretations rest? The first is that the direction of the primary component of NRM in a rock formation is parallel with the ambient geomagnetic field at the time when the rock was laid down. There is now little doubt that magnetism is the only agent which can cause the consistency of NRM directions over wide areas of, and within, a formation. Igneous rocks acquire their magnetization when cooling through the Curie temperature of their iron oxide minerals; in sediments, magnetization may occur at the time of deposition or later, according to the particular process; and with the determination of more accurate pole positions better understanding of the mode and time of magnetization of these rocks will become important.

Various factors are recognized that can cause deviation from the direction of the aligning field of the magnetic vector in rocks. These include pressure, anisotropy of magnetic properties, heat, and chemical processes. However, it is unlikely that their effect would be the same at widely scattered sites in a formation; if they are important, it will be apparent if such sampling is carried out. Mean directions of magnetization found in the Chugwater formation, a red sandstone of Triassic age, collected at 10 sites in Wyoming are shown in Figure 7 (Collinson and Runcorn, 1960). It is possible that such effects cause some of the small but significant scatter in the mean directions, but the mean axis is well defined.

The presence of secondary magnetizations acquired by some rocks,

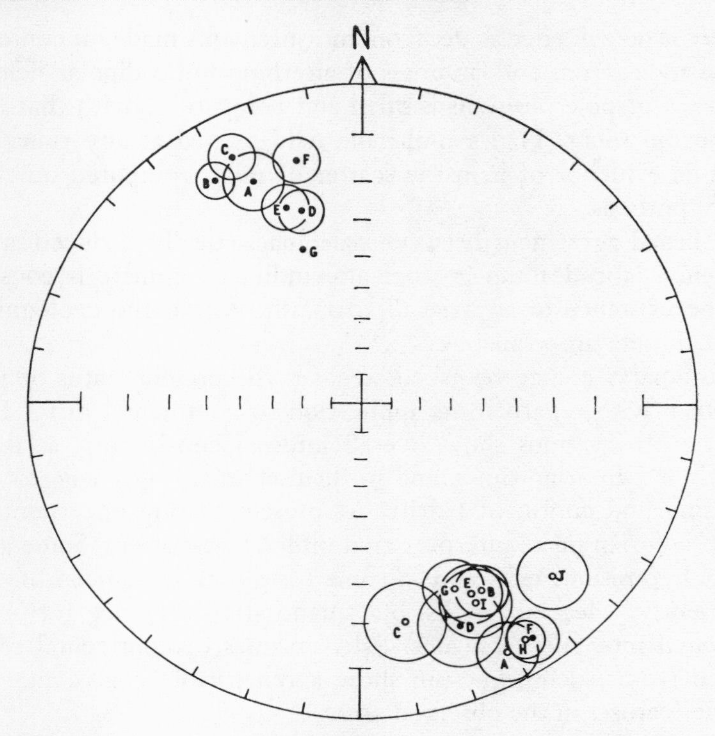

Fig. 7. Mean directions of magnetization of rocks from the Chugwater formation from 10 sites in Wyoming. Circles, 95% confidence about the mean at each site; full (open) circles, downward (upward) inclinations. After Collinson and Runcorn (1960).

often through the influence of the geomagnetic field during the last million years or so, can give rise to spurious mean directions of magnetization. The development of techniques for testing and removing these, now a routine part of the paleomagnetic method, makes reliable data more readily available.

The second basic assumption in paleomagnetic interpretation is that of the axial dipolar nature of the earth's magnetic field throughout geologic time. There is direct evidence of an axial dipolar field from paleomagnetic measurements on rocks of Middle Tertiary age or less. Theoretical work to date suggests that the field originates in motions in the earth's core, and would be expected to be axial, with the secular variation observed arising from eddies near the core surface possessing decay times of a few thousand years.

There is no evidence as yet from measurements made on contemporaneous rocks from one landmass of anything but a dipolar field, i.e., the scatter of pole positions is small and comparable with that found in Cenozoic rocks. Had a multipole field existed at any time, there should be evidence of it in the scatter of poles computed on the dipolar hypothesis.

The broad agreement between paleomagnetically deduced ancient continental latitudes and geologically indicated climate is consistent with the existence of an axial dipolar field during the geologic ages for which data are available.

In summary, it may be asked: what is the present status of paleomagnetism with regard to its application to continental drift? Paleomagnetic observations show overall internal consistency, at least as far back as Cambrian times, and in themselves strongly suggest polar wandering and continental drift. At present, dating uncertainties in the Precambrian make interpretation difficult in that era. Some anomalous pole positions exist and in some cases data have been based on insufficiently "cleaned" rocks, but quantitative evidence for ancient continental movement remains. Paleolatitudes of continental regions deduced from paleomagnetism show a remarkable consistency with geologic features in the observed areas.

Although there is now strong evidence for continental drift, there is still much scope for argument about actual continental reconstructions during a given geologic time interval. Geophysical, biologic, and geologic data must be combined to provide a mutually consistent arrangement of landmasses at any time; such a consistent pattern is now beginning to emerge. It is to be hoped that with the accumulation of reliable paleomagnetic data, more precise dating of rocks and their time of magnetization, and the development of improved procedures and new techniques, the relative positions of the continents in the past will be more accurately, if not precisely determined.

References

Blackett, P. M. S.: Comparisons of ancient climate with the ancient latitude deduced from rock magnetic measurements. *Proc. Roy. Soc.* (London), ser. A, 263:1–30, 1961.

Briden, J. C., and E. Irving: Palaeolatitude Spectra of Sedimentary Palaeoclimatic Indicators. In *Problems in Palaeoclimatology*, A. E. M. Nairn, ed. New York, John Wiley & Sons, Inc., 1964, pp. 199–224.

Collinson, D. W., and S. K. Runcorn: Polar wandering and continental drift: Evidence from paleomagnetic observations in the United States. *Bull. Geol. Soc. Am.* 71:915–958, 1960.

Cox, A., and R. R. Doell: Palaeomagnetic evidence relevant to a change in the earth's radius. *Nature* 190:36–37, 1961.

Creer, K. M.: A reconstruction of the continents for the Upper Palaeozoic from palaeomagnetic data. *Nature* 203:1115–1120, 1964.

Egyed, L.: A new dynamic conception of the internal constitution of the earth. *Geol. Rundschau* 46:101–127, 1957.

Graham, K. W. T., C. E. Helsley, and A. L. Hales: Determination of the relative positions of continents from paleomagnetic data. *J. Geophys. Res.* 69:3895–3900, 1964.

Irving, E.: *Paleomagnetism and Its Application to Geological and Geophysical Problems.* New York, John Wiley & Sons, Inc., 1964.

Irving, E., and D. H. Tarling: The paleomagnetism of the Aden volcanics. *J. Geophys. Res.* 66:549–556, 1961.

Kawai, N., H. Ito, and S. Kume: Deformation of the Japanese islands as inferred from rock magnetism. *Geophys. J.* 6:124–140, 1961.

King, L. C.: The Palaeoclimatology of Gondwanaland during Palaeozoic and Mesozoic Eras. In *Descriptive Palaeoclimatology*, A. E. M. Nairn, ed. (division of John Wiley & Sons), New York: Interscience Publishers, Inc., 1961, pp. 307–331.

Van Hilten, D.: Palaeomagnetic indications of an increase in the earth's radius. *Nature* 200:1277–1279, 1963.

Van Hilten, D.: Interpretations of the wandering paths of ancient magnetic poles. *Geol. Mijnbouw* 43:209–221, 1964 [English summary].

Ward, M. A.: On detecting changes in the earth's radius. *Geophys. J.* 8:217–225, 1963.

3

Continents and Ocean Basins

ROBERT S. DIETZ *

Environmental Science Services
Administration, Atlantic Oceanographic and Meteorological
Laboratories, Miami, Florida

When I was asked to participate in this conference by discussing the geologic relations of continents and ocean basins, I found that the facts bearing on so broad a subject were both sparse and not firmly established. Any views I shall express can therefore only be matters of personal geologic philosophy. While fully aware of these limitations and of the compounded uncertainty introduced by pyramiding assumptions, I shall attempt to touch upon some aspects of this fascinating geologic problem which seems to set Earth apart from other terrestrial planets.

I shall begin by a few general comments. To one trained in marine geology, and contrary to expectations, the chief lesson of marine geology seems to be that the continents and ocean basins are contrasting realms, almost as unlike one another as both are in turn from the Moon's surface. Our ocean basins generally owe their existence to isostasy, being devoid of any layer of crust (i.e., sialic crust). Indeed, they are covered only by a rind or an "exomantle," so that to a first approximation they are windows into the mantle. I think we have also learned that the mantle is tectonically alive, while the crust is largely dead. The dynamics of tectonism may be primarily ascribed to convection of the earth's mantle which is expressed in features outcropping on the ocean floor—the midocean rises and the trenches. Even the generation of magmas is now relegated largely to the upper mantle and not to the lower crust. The moon's so-called ocean basins,

* It is a pleasure to acknowledge scientific assistance by Walter Sproll, clerical assistance by Rose Hudson, and illustrating by John Holden.

the maria, are not genetically akin to our ocean basins (Dietz and Holden, 1964). The maria are probably explosion basins blasted into the surface of a thick lunar crust (perhaps the moon is entirely undifferentiated), and are therefore not exposures of an inner mantle. Unlike our ocean .basins, the maria are at different levels rather than at a constant depth.

Origin of Ocean Basins

The theories on the origin of ocean basins can be divided into two categories: catastrophic and evolutionary.

CATASTROPHIC ORIGIN

Best known among these theories is that of Darwin and Fisher, who held that the moon was extracted from the earth by tidal resonance leaving the Pacific basin as a scar. This concept has fallen into disrepute and perhaps the last geologist to advocate it was Escher (1939). However, geologic hypotheses, like cats, have nine lives and this one has been recently revived by an astronomer (Alfven, 1963).

Another suggestion, which may be regarded as the inverse of moon extraction, is that the oceanic depressions are explosion basins blasted out by the impact of astroids in the early history of the earth. Gilvarry (1961) and Harrison (1960) have subscribed to this view. I also once proposed (1958) such a possibility, but without interpreting the ocean basins as explosion scars, for they can hardly be explosion basins in their modern form. Instead, I suggested that the pristine, world-encircling sialic crust may initially have been fragmented by giant, crust-piercing impacts, creating the terrestrial equivalent of lunar maria. This may have occurred in very early geologic time, thereafter the ocean basins evolving by endogenic processes into their modern form. In short, impact might account for the existence of ocean basins by sialic fragmentation, but not for the modern configuration or geographic disposition. This speculation on early catastrophism strikes me as no longer appropriate. It now seems, beyond reasonable doubt, that the earth accreted as a cold body and never developed a pristine, world-encircling, granitic crust such as would be the logical consequence of the classic molten-earth hypothesis. In that case, it is meaningless to discuss its fragmentation, even in terms of the lunar model.

Certainly the known astroblemes (ancient meteorite impact scars) are mere pinpricks on the continental surfaces.

It is also pertinent to mention here the suggestion that the continents may have been cosmically derived. Berlage (1962), for example, has suggested that the continents may be composed of lunar rock which fell upon the earth when the moon was accreted. Donn and co-workers (1965), not finding a sialic source for the very earliest geosynclinal accumulations, similarly suggested the infall of sialic cosmic bodies as a source of continental rock. This belief in cosmically derived sialic substance which powdered the entire earth in early geologic time has recently been further developed by Van Bemmelen (1966). However, the mantle alone seems quite capable of providing, by gravitational differentiation, the small amount of granitic crust (0.82 percent of the earth's volume) without resorting to gratuitous sources.

EVOLUTIONARY ORIGIN

It seems evident to me, and I am sure to nearly all geologists, that the ocean basins must have evolved over aeons rather than have been formed by catastrophies. Clearly, there is some sort of balance, or "symbiotic" relation, between continents and ocean basins. For example, persistence of continents through geologic time despite offsetting erosion requires this. Any satisfactory theory must be both cyclic and vectored—the essence of evolution. The ocean basins must continuously accommodate the 1.350×10^6 cubic kilometers of water on earth, a volume which can only have grown slowly at a rate of possibly less than a meter per million years as juvenile water was squeezed out of the mantle. Geomorphically, the continental slopes are young, and cannot be considered pristine features. They are steep and incised by canyons like a young mountain front (Shepard and Dill, 1966); this cutting can hardly be of great geologic antiquity. The usual belief of geologists that continents are living and growing is not an arbitrary one.

The evolution of ocean basins may be outlined as follows: By the Urey model, the earth first accreted as a cold body from a dust cloud about 4,500 million years ago. It then remained a cold body without internal "life" for a billion years or so, much as the moon still appears today. This would account for the "lost aeon"—the time gap between the radiometric age of the earth and the date of the earliest sialic rocks. By about 3.5 billion years ago sufficient internal heat had been generated by the decay of radioactive elements to impose a thermal convection regime upon the earth such as ever since has been the *primum mobile* of diastrophism. The primitive substance with the composition of stony meteorites (chondritic) would then slowly undergo differentiation producing a highly dense nickel-iron core at the center of the earth and a low-density sialic crust of granitoid rocks rich in silica, alkalies, calcium, etc., at the surface. This cleansing of the mantle may have been somewhat more rapid at first, but it seems to have continued through geologic time to the present day.

The next question is how the sialic substances become isolated in the rafts we recognize as the continents today. But first it must be emphasized that the ocean basins appear to be windows into the earth's mantle, exposed like a vast Mohole. The ocean basins are commonly described as having a crust; but the term rind is probably more appropriate, as the oceanic "crustal" section overlying the Moho is wholly devoid of granitoid rock, so that the ocean basins do not have a sialic lining, although the Atlantic and Indian Oceans were once thought to be so lined. The nature of the oceanic layer (layer 3) is not known, but if it is partially serpentinized peridotite (Hess, 1955), i.e., simply hydrated mantle rock, it may be regarded as a sort of exomantle. Should layer 3 turn out to be basalt, its composition is still quite unlike that of the continental sial; to regard it as crust therefore leads at least to semantic confusion. Many reasons have been advanced to account for the discontinuous distribution of sial on the earth's surface, but to me it seems most likely that the ocean basins have been swept clean of sial by the process of sea-floor spreading (Dietz, 1961; Hess, 1962).

Briefly, this process causes the outcropping of mantle convection cells thus entraining the ocean floor to the top of the mantle. The midocean-rise system, with its central rift, may be regarded as the line of divergence between rising cells, while the deep-sea trenches mark the zones of downwelling. Thus, there is a conveyor-beltlike translation, bounded by fracture zones, between the rises and the trenches. Extensive horizontal creeping of the earth's surface at a rate of 1 to 5 cm per year would occur, including even continental drift, with the sialic blocks being carried passively on the backs of the spreading segments of the earth's upper mantle.

Whether this concept is valid remains unproved. However, in recent years there has been a growing inclination to believe that new sea floor is being generated along midocean rifts (Laughton, 1965; Vine, 1966; Pitman and Heirtzler, 1966; Sykes, 1967). Whether the trenches represent the take-up portion of such conveyor-belt translation has also not been established. For example, Plafker (1965) finds that underthrusting along the Aleutian Trench axis, implied by this concept, is an acceptable explanation for the deformation associated with the March 27, 1964, earthquake in southern Alaska. On the other hand, Ludwig and co-workers (1966) found no evidence for such underthrusting of the sea floor in their study by subbottom acoustic profiling of the Japanese Trench.

Assuming, for the sake of argument, that there is actual sea-floor spreading, the process can account for the ocean basins by offering an evolutionary method of skimming the sialic substance differentiated from the mantle into the discrete "rafts of slags" which we call the continents. Detritus shed into the ocean basins from the continents, or juvenile sialic material born in the ocean basins, would be returned to the continental segments. This would explain the persistence of continents through geologic time.

Of the many questions left unanswered, I shall briefly discuss one: How is sial created? The concept of sea-floor spreading suggests a two-stage evolution. First, basalts of the ocean-floor variety (tholeiites), are differentiated from the earth's periodotitic upper mantle by partial fusion, mainly in the region of the midocean rift, at the high temperatures for divergence of mantle convection cells. This basalt is then a sort of sial-sima, intermediate in density and composition between mantle rock and continental granodiorite. Sial, itself, would evolve only as a result of a further overmelting and processing in the trench realm of island-arc mountain ranges, such as the Andes. This would occur in the cool, or convergent, part of convection-cell overturn by the squeezing out of the hyperfusibles to form andesites and granodiorites.

A significant question is: What is the role of island arcs in this scheme? Perhaps the Andean foldbelt and island arcs such as Japan or the Aleutians, are basically the same type of structure. The locus of descent of convection cells may be controlled by thermal equilibrium within the mantle and may be largely independent of continent position. By sea-floor spreading, continents should tend to move to the trenches as a limiting position of their drift. The Andes then would be an "island arc" or an accretionary belt directly welded to the continent, the formation of this mountain chain being modified by the existence of a continental plate through which any magma from the upper mantle must rise. By continued horizontal shifting of the crust, other island arcs may eventually become welded to continental plates as well. From sea-floor spreading and the associated underriding of the sea floor beneath an island arc, it may be predicted that the arc would consist of two main parts: (1) an oceanward side consisting of a folded, thick accumulation of flysch (incorporating ultrabasics and other elements of the Steinmann trinity) with regional metamorphism of the low-temperature, high-pressure type; and (2) an inner belt of acidic or intermediate magmatism of high-heat flow and block movement of the crust as new buoyancy is added at depths. Matsuda (1964) has recently noted such a relation in Japan.

Continental Growth by Accretion

As already noted, sea-floor spreading suggests that continents grow in some manner by accretion, but how does this happen? [It also, in effect, refers to the whole problem of the origin of ocean basins and the growth of the continents.] The ocean basins are not really basins or depressions in a terrestrial crust but a worldwide and continent-encircling hypsometric level which must be regarded as the essentially normal level of the earth. Above the ocean basin, the continents rise like island plateaus as anomalous highs. This is unlike the moon, where the maria are truly basins, without interconnection and without a common level, in the lunar terra or "uplands." The level of the ocean basin must be subjected only to small increase in depth as more sial is squeezed out of the mantle, for the amount of sialic substance which can be sweated out of the earth's interior mantle is less than 1 percent of its volume. Clearly, to account for ocean basins, the proper question to be posed is: How did the continents originate?

Geosynclines, and especially the eugeosynclines, appear to be the fundamental building blocks of continents. But the usual theories hold that geosynclines are all ensialic, i.e., laid down on sial as intracontinental tectogenes or as ensialic, marginal, tectonic borderlands; such theories do not clearly account for the growth of continents. Any compression and folding of the geosyncline into a mountain would then result in a reduction of the continental area, not in an enlargement, if continents are to grow with time (Fig. 1).

I have elsewhere (1963, 1966) suggested a concept of geosynclines, folded mountains, and continent building in which the continental rise is equated to a eugeosynclinal prism which has collapsed to form an orogen or foldbelt. This concept regards continental rises as great ensimatic sedimentary prisms which are widespread in modern oceans and are laid down on the ocean floor at the base of the continental slope. When the prisms collapse, they are welded to the continental plate as a new accretionary belt, thereby enlarging the continental area. Figure 2 illustrates this concept in the form of a time-sequence diagram.

The accretion of island arcs offers another mode of continental growth if continents are not rooted in place but permitted the freedom of drift. But the collapse of continental rises and accretion of arcs may in fact be similar processes. Of course, with the collapse of a continental rise, the sediment returned to a continent was derived from it in the first place. Obviously, this results in a process which prevents a continent from losing substance rather than acquiring any new substance; to achieve growth, juvenile sial must be added. This new sial may evolve by the secondary processing of sea-floor igneous rocks, mainly ocean-floor basalts newly derived from the mantle. The

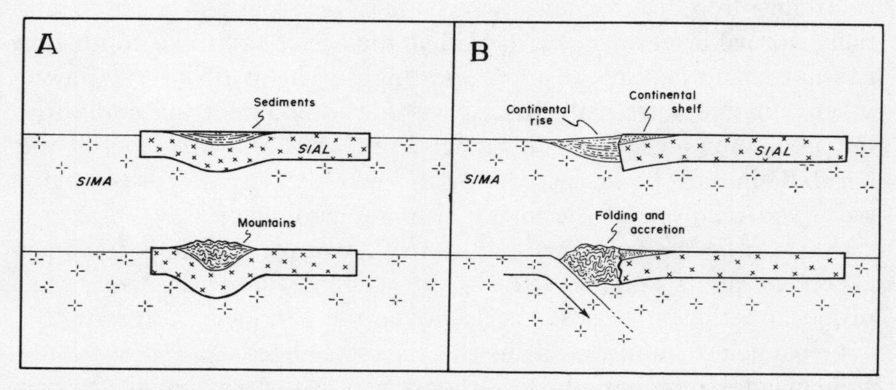

Fig. 1. *A*. By the conventional tectogene view, orogeny (folded mountain making) causes continental shortening. *B*. By the author's view, mountain making is simply a symptom of the more fundamental process of continent building. The continent grows larger by accretion.

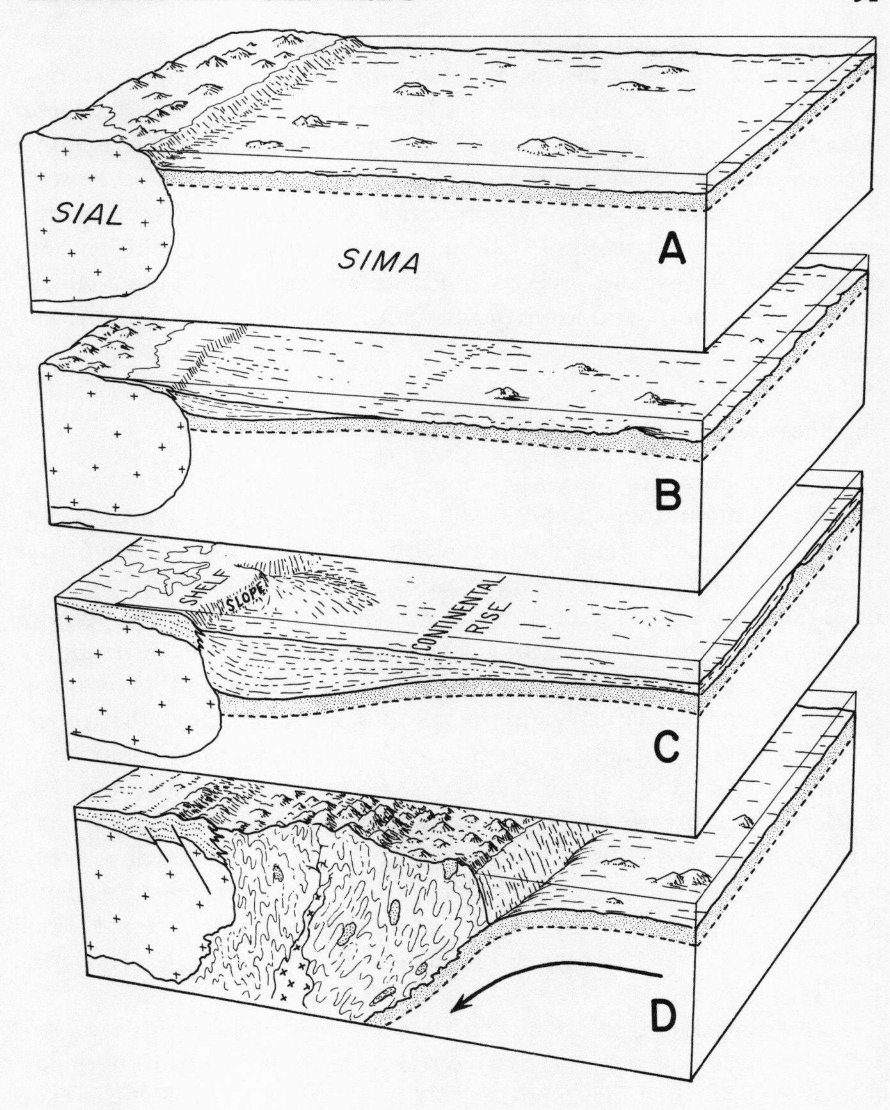

Fig. 2. Ensimatic geosynclines, mountains, and continental accretion according to the writer's actualistic concept (Dietz, 1963). Modern continental terrace wedges are equated with miogeosynclines, and the continental rise prisms are the precursors of eugeosynclines.

Andes, a modern example of a collapsed continental rise in my opinion, also may be regarded as an island arc which developed contiguous to a continent, and thus is fundamentally similar to the Japanese Islands arc which happens not to be contiguous to Asia. It is conceivable that by subsequent drift the Sea of Japan could be closed. Thus, an ancient orogen or foldbelt may be a collapsed continental rise or an accreted island arc. Which a specific one is would have to be decided by special criteria; for example, a foldbelt dominated by graywackes would tend to favor the interpretation of a collapsed continental rise.

The Supercontinents of Gondwana and Laurasia

Since continental drift is an inevitable consequence of sea-floor spreading, the configuration and geographic position of continents have not been even approximately retained over geologic time. It is noteworthy that continental drift retains some aspects of permanency: the volume of ocean basins and the total area of the continents and of the oceans. According to drift theory, the present configuration and distribution of the oceans are related to the mid-Mesozoic rifting and breakup of earlier landmasses. What were the ocean basins like before that time? I shall repeat a speculation here which I trust is more bold than bizarre (Dietz and Sproll, 1966). The continents of the modern world seem more or less randomly spread over the earth's surface, with little apparent symmetry in the disposition of these sialic (granitoid) plateaus. For example, in the Northern Hemisphere the land totals 100×10^6 square kilometers, while that in the Southern Hemisphere totals 48×10^6 square kilometers, a ratio of 2:1 (Kosinna, 1921).

Many theories have attempted to account for this disparity. An early example is Green's (1875) tetrahedral hypothesis of continent distribution, commonly found in old geology textbooks (Gregory, 1901). According to Green, the earth contracted as it cooled and as it grew smaller it tended to assume the shape of an inverted tetrahedron. The continents formed at the points and edges of the tetrahedron, so that Antarctica occupied one point, while Asia and Australia, the Americas, and Europe and Africa occupied the three edges. Green's concept can now be dismissed with a smile; we no longer believe that the earth is contracting. Furthermore, the continents are not elevated portions of the mantle, but plateaus of sialic rock resting isostatically on the mantle. But we still remain without a satisfactory theory which would explain the distribution of continents around the globe.

Perhaps a main difficulty is the common adherence to the belief in continental fixity. Instead, if the theory of continental drift is accepted, an earlier and more "logical" distribution can be envisaged. The latter theory assumes that in the Paleozoic all the continents were contained in two supercontinents: Laurasia (North America and Eurasia) and Gondwana (South America, Africa, Australia, Antarctica, India, Malagasy, and some other microcontinental fragments). The Gondwana-Laurasia concept simplifies the problem of continental distribution in several respects: (1) Originally there were only two supercontinents rather than the present six major ones plus other sialic pieces. (2) Gondwana and Laurasia were crudely circular in shape, and the former may have been entirely surrounded by a continuous mountain range (King, 1962). (3) Gondwana was located in the Southern Hemisphere, and Laurasia, at least largely, in the Northern Hemisphere. It would seem that the geography and geology of landmasses may have been somewhat simpler than at present.

An inspection of the globe reveals that the areas of Gondwana and Laurasia might have been quite similar, providing still another aspect of the simplicity of the concept. Dietz and Sproll (1966) have analyzed their areas. The present continental areas, as given in the usual atlases, were unsuitable, for their geologic limits are not delineated by the shorelines but by the 1,000-fathom isobath which marks one-half of the isostatic freeboard of continents (Carey, 1958). Such a mensuration includes extensive continental shelf and upper-slope area. We expected to find a rough similarity in the areas of the two supercontinents, but did not anticipate how nearly equal they were. The area of Laurasia was 100.7×10^6 square kilometers, that of Gondwana 100.6×10^6, a difference of only 100,000 km^2 (about the area of Iceland), or only 0.1 percent (Table 1). This close agreement must be fortuitous, as there is a considerable margin for error in our computation, including such factors as the exact position of the selected isobath, assumptions about the composition of the supercontinents, and accretions to continents after the mid-Mesozoic. However, an agreement in size to about 1 percent would be remarkable.

The total area of the earth is 510×10^6 square kilometers; nearly

Table 1. Areas of Laurasia and Gondwana *

Laurasia †	Area	Gondwana	Area
Eurasia	70.49	Africa ‡	37.80
North America	35.39	Australia and New Guinea	13.31
	105.88	New Zealand	2.59
India	−5.19	Antarctica	16.91
	100.69	South America	22.36
			92.97
		India	+5.19
		Himalayan Overlap §	+2.44
			100.60

* Areas include continental shelves and upper continental slopes to 1,000 fathom isobath. All values in million square kilometers.

† Includes Greenland. Does not include the Philippine Islands and southern Central America, an area of 1.35×10^6 square kilometers. Possible accretion to continents by orogeny after the mid-Mesozoic is not taken into account.

‡ Includes Arabia, Malagasy, and the northern portion of the Seychelles Plateau.

§ That part of the Indian subcontinent presumably underlying the Tibetan Plateau, which must be two continents thick to satisfy isostasy.

After Dietz and Sproll (1966).

40 percent, therefore, is occupied by continental crust, as revealed by our measurements. In round numbers, the earth's area is thus about 500×10^6 square kilometers, and that of Laurasia and Gondwana is about 100×10^6 each; both supercontinents thus occupied 40 percent of the earth. Menard (1964) calculated that the continental crust amounted to 41 percent. The explanation of the difference may lie in the fact that he used the 2,000-meter isobath as delimiting the continents, whereas we used the shallower 1,000-fathom isobath. Furthermore, we have not included oceanic highs (fully detached from continents) except for Malagasy and the northern Seychelles, New Zealand, and South Georgia, these clearly being sialic blocks.

Our computation of the areas of Gondwana and Laurasia was based on the National Geographic Society's 16-inch diameter globe of 1963 (scale: 1:31, 363, 200) which shows water areas and depths as well as landmasses. The area grid on their plastic quasi-hemispheric measuring cap, plus quadrilateral areas of longitude and latitude of continental interiors, were used for measuring. Although the method is not highly precise, we thought it accurate enough in view of the uncertainties in the basic assumptions. The principal microcontinent used in the Laurasia measurement is Rockall Bank (Bullard and co-workers, 1965). The Tibetan Plateau was regarded as being two continents thick (70 km), consisting of an Asian top plate and an Indian underplate; its area was therefore included as a part of the Indian subcontinent, as well as a part of Eurasia. Arabia, of course, was included as a part of Africa, which it is geologically. The contact between Gondwana and Laurasia was taken as a line along the axis of the Mediterranean Sea.

Probably additional microcontinental fragments will eventually be identified, especially in the Indian Ocean, which would increase the area of Gondwana. These additional areas, if discovered, might be partly offset by finding that the continental area around New Zealand is smaller than indicated by the 1,000-fathom isobath, which now includes an undersea area more than threefold of the landmass itself. However, the closely equal areas of the supercontinents would not be altered, except in detail, by any new facts regarding crustal extent. The principle that the two supercontinents, as conceived by the proponents of the continental drift theory, were almost identical in size seems established.

This result must be meaningful and invites some speculation. Is the Laurasia-Gondwana version of continental drift correct, or were the continents in the Paleozoic part of one landmass, like Wegener's Pangaea, as many modern geologists suppose (Carey, 1958; Wilson, 1963; Bullard and co-workers, 1965)? The equal size of Gondwana and Laurasia would seem to argue for the two-continent version and against the single landmass of Pangaea. A reasonable process whereby Pangaea were split into two equal halves is difficult to conceive; it would be fortuitous and hence unlikely.

Advocates of continental drift would also like to know whether the postulated mid-Mesozoic breakup and dispersal of the continents was a unique event in geologic history or whether, from time to time, continents have split and drifted, reconfiguring their outlines. The equal areas of the two supercontinents argue for a simplicity of drift history or even possibly that the mid-Mesozoic breakup was unique in the earth's history.

The following speculation might reasonably explain the equal areas of the supercontinents. Runcorn (1962, 1965) suggests that orogenic revolutions and continental drift may have been controlled by the growth of the earth's core, which in turn caused the latitudinal toroidal convection cells in the earth's mantle to increase from a one-cell to a five-cell pattern. However, this progression would not explain the equal areas of Gondwana and Laurasia, whereas a simpler scheme of the Runcorn type would. If the convection in the Paleozoic and earlier eras was $n = 2$, there were just two toroidal latitudinal cells (Fig. 3). The convection cells would rise at the equator, move poleward, and descend at the poles. As is now generally assumed, the

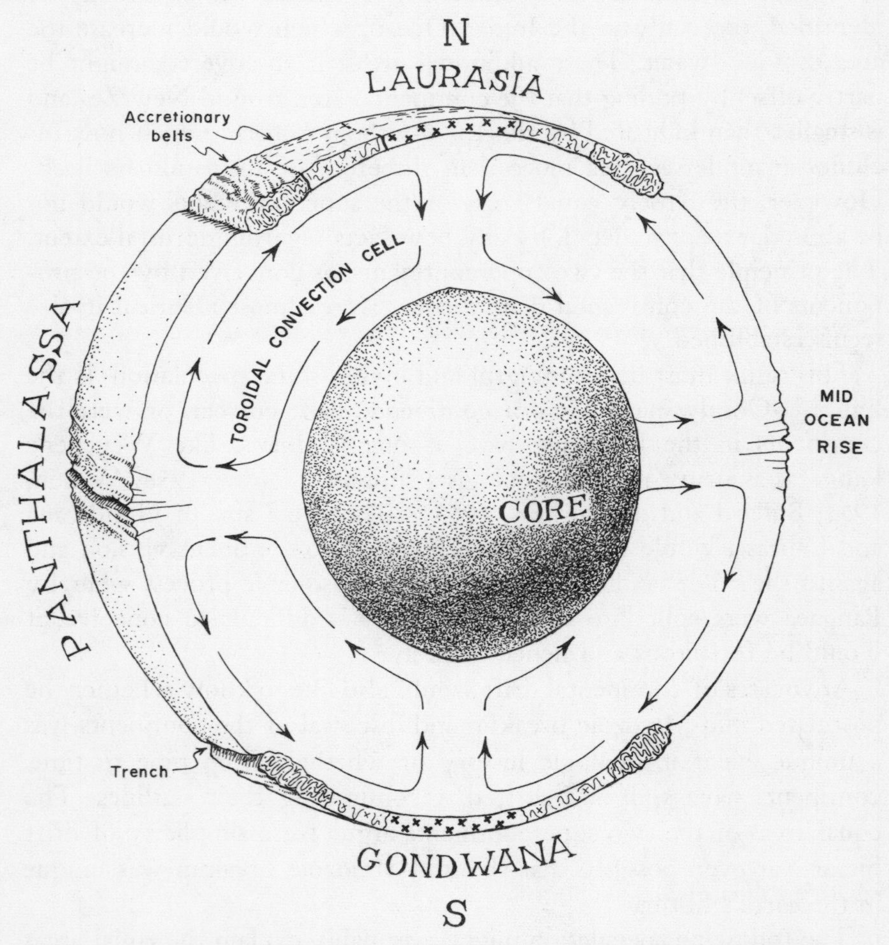

Fig. 3. Accretionary development of two supercontinents (Laurasia and Gondwana) under a two-mantle, convection-cell pattern producing two supercontinents of equal area, surrounded by the equatorial universal ocean of Panthalassa.

earth originally accreted as a cold, ultramafic body and gradually warmed up; continental or sialic rock would then have been slowly generated by differentiation from the earth's mantle. By assuming that the composition of the pristine earth was originally uniform, an equal volume of mantle would generate an equal volume of sial. In the two-convection cell scheme, the Southern Hemisphere would supply all the sial for Gondwana, the Northern Hemisphere all the sial for Laurasia. Accretion of the continents may then have occurred as a result of sea-floor spreading, with its conveyor-belt tectonics accompanied by collapsing continental rises (Dietz, 1963, 1966).

The reason for the present distribution of continents is still unresolved. Possibly, the answer lies in the new and more complex mantle convection-cell patterns which began in the mid-Mesozoic. The process is still active; the cells may have undergone recent changes, such as the possible switch of the Pacific tail of the world-wide midocean rise from beneath the Darwin Rise to beneath western North America (Menard, 1964). Girdler (1963) attempted to identify the pattern of convection cells in the modern world and concluded that it was a five-cell one. Although the validity of his conclusion is questionable, it would seem to be the correct sort of approach. When this problem is finally solved, the modern distribution of continents could be explained, using the simple Gondwana-Laurasia configuration as the initial situation. Actually, the midocean-rise system coils around the earth in a sinuous pattern, described as being like the seam on a tennis ball. If the minor offshoots are ignored, it forms a continuous swell of the earth's mantle; if there were an extension from the Gulf of Alaska to the Arctic Ocean off Siberia, it would connect unto itself, again forming an endless distorted loop. A reasonable suggestion might then be that the number of convection cells has not changed with geologic time but that their position has become distorted from more or less equatorial before the mid-Mesozoic to that of a sinuous line.

Stability of Ancient Supercontinents

One aspect of earth's history may shed some further light on our speculation: the apparent stability of ancient landmasses which were invariably peneplained and even ultrapeneplained rather than deeply dissected and cut by deep rivers, or "Grand Canyons." Such ancient filled channels as are found in the geology generally measure only a few meters. The expectation of finding ancient "Grand Canyons" may be somewhat naive, for their preservation would require unlikely geologic events; but their former presence should at least be revealed by the extensive stripping of cratonic cover formations which in my opinion are remarkably intact over the world as a whole. The stability and calm of the earth's history is much more impressive than the occasional orogenies. This was recently once more brought to my attention when I saw the 400-meter-thick carbonate series in south-central Brazil now covering an area of 200,000 mi^2. It is laid down on a peneplain of basement complex rocks, the usual relief of which is probably only several meters and a maximum of about 100 to 150 meters, as indicated by aerial geophysical survey. It is labeled as Silurian on the current geologic map of Brazil (1961), based on the doubtful identification of *Favosites* by Derbye many years ago. But this fossil is now regarded as being probably cone-in-cone structure, and therefore of inorganic origin. On the other hand, algal structures of the stromatolite type are found. Most probably, this extensive cratonic limestone deposit, which has undergone only the lowest degree of metamorphism and is largely not deformed, is Precambrian, thus indicating remarkably prolonged cratonic stability. Another example is found in the central United States, where the Pennsylvanian (Upper Carboniferous) was apparently ultrapeneplained and extremely stable, so that a sea-level rise of only several feet, or at most of a few tens of feet, sufficed to cause transgression of epicontinental platform seas of as much as 500 miles (Wanless and co-workers, 1963).

With regard to the ocean basins, their first-order features are the so-called midocean rises which trend continuously through the world's oceans. The term is rather imprecise, for rises occupy a midocean position only in the Atlantic and Indian oceans, but not in the Pacific. The reason for this would seem to be that the two former oceans are rift oceans, driven apart by the development of the midocean rise, so that this linear bulge naturally assumed a midocean position as the continents moved apart. But it seems likely that these rises represent mantle bulges somehow related to mantle tectonism and independent of the earth's sialic crust. In general, a rise may develop anywhere on earth—beneath continents or in any part of an ocean basin. The continents must then adjust themselves to these rises, even by continental drift, if necessary. In fact, it is now generally recognized that portions of the midocean-rise system do indeed lie beneath continents, although this may be largely a Neogene development. Examples are the epeirogenic plateau uplift of western North America, the high plateaus of Africa, and possibly of central Asia.

The concept just outlined implies the former existence of two supercontinents and a two-convection cell pattern; in consequence, Gondwana and Laurasia would have been in stable positions with respect to mantle overturn. The locus of convection-cell divergence and associated sea-floor spreading would be an equatorial girdle. The Paleozoic midocean rise, marking this divergence, would be located in the middle of ancient Panthalassa, and the convergence, marked by trenches, would be marginal to the supercontinents. Whenever mantle convection occurred (whether continuous, periodic, or occasional), new mountain belts would have accreted marginally to the continents, but the latter would not have been subjected to rifting and their interiors would have remained largely undisturbed.

This simple picture of the early history of the earth is perhaps erroneous, but there is some evidence to support it. The world today is out of equilibrium. The stripping of cratonic cover rocks over the continental basement complex is occurring at a much faster rate than the laying down of new cratonic formations in modern seas. The regional denudation of continents is progressing at the alarming rate of 200 meters per million years (Judson and Ritter, 1964). The "everlasting mountains" will be reduced to sea level and the continents peneplained in a mere 10 million years, a rate sufficient to have leveled North America six times during the Cenozoic period alone. The deep sea floor is receiving sediment at a much faster rate than that typical of the geologic past (Kuenen, 1946). At the modern rate of wastage, the continents could not have persisted over geologic time. This must mean that they are now unusually high due to positive epeirogenesis or plateau uplift (Fig. 4).

As Suess (1904) proposed, two main types of diastrophism have been recognized: (1) orogenesis, i.e., the creation of folded mountains; and (2) epeirogenesis, i.e., the broad, regional uplift commonly known as plateau uplift. The results of recent investigations, however, modify this concept of diastrophism. If folded mountains are generally created by the collapse of continental rises, orogenesis is simply the result of the more basic process of epeirogenesis. And if the re-

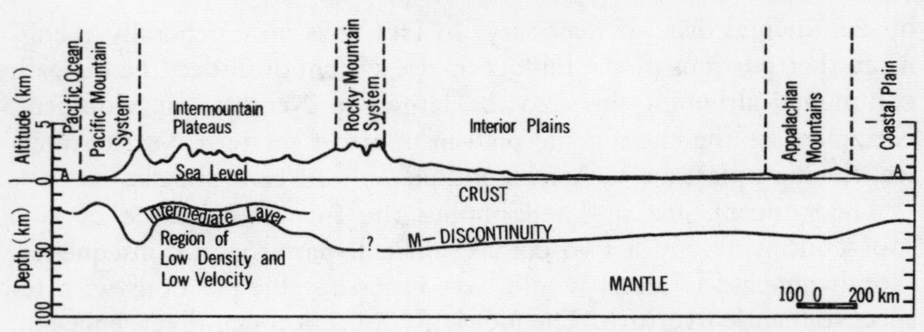

Fig. 4. Crustal thickness section across the United States (adapted from Pakiser). The Colorado Plateau (Intermountain Plateau) is shown not to be underlain by a thickened crust which supports the belief that the East Pacific Rise under-rides the western region of North America, causing the epeirogenic uplift.

sults of deep-crustal studies of continents, such as Pakiser's (1963) as a guide, epeirogenesis in the long run does not create continents but destroys them. For example, the plateau uplift of the western United States is not associated with a thickened sialic crust but with a crust which is more or less only normally thick (Figs. 4, 5). Thus, the cause of this plateau uplift is not an isostatic response to sialic thickening, but is apparently a tumescence of the mantle. This uparching may possibly be due to the rather recent positioning (Miocene?) of the East Pacific Rise under western North America. Following this line of reasoning, the uplift of the high plateaus of Africa and of central Asia (excluding the Tibet Plateau, which is most probably due to the impingement of India on Asia as a result of continental drift) also may be caused by "misplaced" midocean rises. These uplifts, about 2 km high and 2,000 km wide, are in agreement with the relief of midocean rises. Since sialic thickening is not involved, the overall effect of plateau uplift is therefore an enormous continental wastage.

In the western United States, there is a mountainous, disturbed region of crust 1,000 miles wide which familiarity leads us to accept as normal. South America, however, may be a better model of the geologic past. There, the marginal and narrow Andes are a true orogenic mountain range indicating intense tectonism. But to their east the craton lies almost wholly undisturbed in the plains of Argentina and the Amazon valley; even parts of Peru and Ecuador are already in this great valley. The juxtaposition of these two realms, without

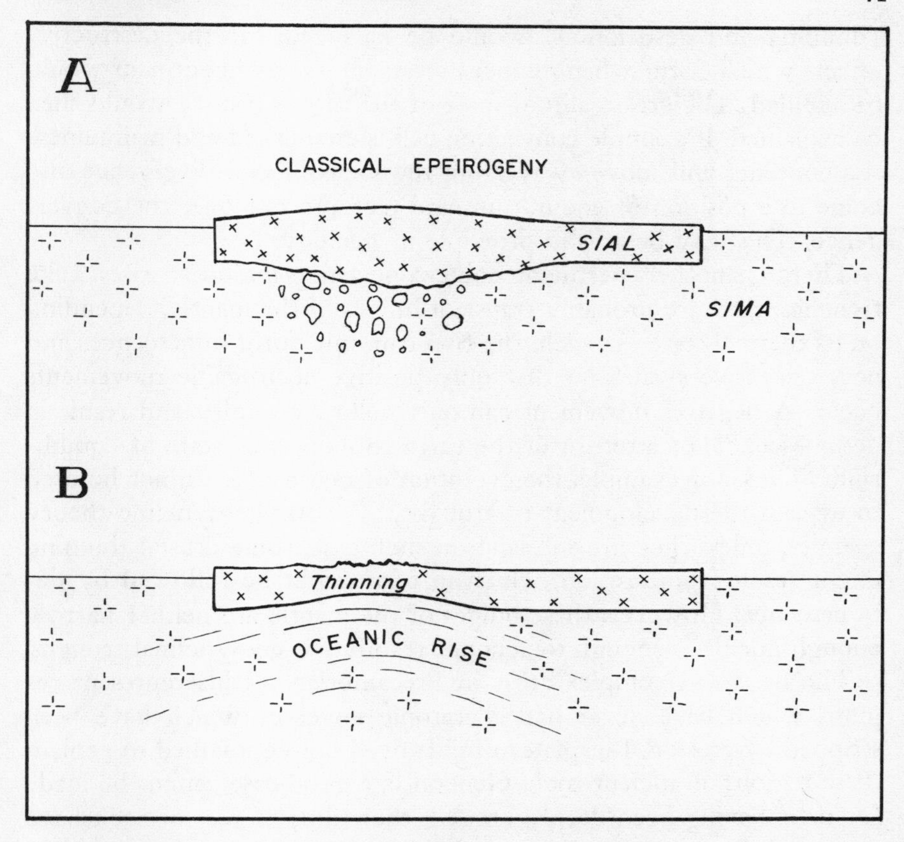

Fig. 5. *A*. By the classical view, epeirogeny (from Gk., *continent-making*) builds continents by thickening the continental plate. *B*. Recent evidence suggests that typical epeirogenic uplift (plateau uplift) results from mantle swelling and that the continental plate is not thickened. The result is plateau uplift and erosional thinning of the continental plate causing continent destruction rather than continent making. Hence the term *epeirogeny* seems a misnomer.

any transition and with vast amounts of detritus (molasse) being transported onto the craton, is a startling sight for a North American.

The obvious existence of continents in a rather healthy state of repair indicates that continent-building processes, in the balance, have more than compensated for the destructive ones. On the other hand, continental destruction is active in the modern world. Had the continents been stably positioned away from the realm of convection-cell divergence in the Paleozoic and earlier times, the two processes (building and destruction) would be balanced, for the destructive effects which occur when midocean rises underrun a continent would be avoided. The great calm of most of the earth's history would then be explained. If a simple convection cell is established and maintained, the contents will move away from the locus of cell divergence and come to a position of equilibrium and tectonic rest over the convergence. This may be a valid principle of geology.

There is another, pertinent point concerning midocean rises. Like trenches, they are probably transient bulges of the mantle, depending on its thermal state. As such, the rises can only form tumescences and never negative swales, so that only positive epeirogenic movements occur. A negative movement can only follow an uplift and result in detumescence, or a return of the earth to its normal state of equilibrium. Thus, for example, the evolution of ocean rises cannot be used to explain the development of troughs, as classic geosyncline theory assumes, unless they are subsialic; in such case, some crustal thinning might result from erosion, which in turn might be followed by detumescence. However, these bulges of the mantle are neither narrow enough nor deep enough to account readily for geosynclinal troughs.

The basement-complex areas in Precambrian shields represent regions which have never had a cratonic cover or which have been stripped by erosion. But plateau uplift need not be invoked to explain these regions of ancient rock. Generally, a good case cannot be made for their having been deeply eroded; thousands of feet of rock have commonly been removed, but probably not strata several miles thick. Structures of the Canadian Shield, such as the Sudbury lopolith which is 1.7 million years old, can be more reasonably interpreted only in terms of modest deleveling. According to Ambrose (1965), the Canadian Shield is an exhumed paleoplain established in the Precambrian and has been diastrophically stable and essentially at sea level ever since. To be sure, in some rocks evidence is sometimes found of great-load metamorphism, as with the Jurassic-Cretaceous Franciscan formation of California; but tectonic uplift rather than deleveling seems to account for their arrival at the surface.

Summary

The ocean basins exist because of sea-floor spreading, which has opened and maintained vast windows into the earth's upper mantle. The prime mover behind this spreading is mantle convection, which has led to the formation of continents and their growth by accretion.

The ocean basins approximate the primitive surface of the earth, but the rocks of the ocean floor are new—Mesozoic and Cenozoic. It is the continents which are the anomalous features of the earth.

Continental drift is a reality. The mid-Mesozoic breakup was perhaps unique in geologic history, and the resultant drift has continued, probably intermittently, to the present time. Earlier, there were two supercontinents, more or less in opposite hemispheres and separated by the universal ocean of Panthalassa. Of course, a universal ocean exists today, but the Paleozoic ocean was less interrupted. The Pacific Ocean may be regarded as the remnant of Panthalassa, with the Atlantic, Arctic, and Indian Oceans being mostly rift oceans. Although I prefer the former two-supercontinent version of drift theory, the question whether the supercontinents or a simple universal continent of Pangaea existed remains a most pressing problem to be solved.

The configuration and distribution of modern ocean basins and continents must be related to their Paleozoic location, the prebreakup position of continents, and as further complicated by the subsequent drift effect controlled by the new pattern of mantle convection marked by midocean rises and ocean trenches. There now seems to be only one midocean rise system, suggesting that there is only one sinuous, worldwide convection-cell couplet. Before the mid-Mesozoic, there may also have been one midocean rise, but in the form of an approximately equatorial girdle.

The relative area of continents (to the 1,000-fathom line), as compared to that of the ocean basins, is 2:3. As Hess (1955) noted, the depth of the ocean basins must, for reasons of isostasy, directly control the thickness of continents (about 35 km). This means that the total continental area (202 million km^2) must be related to the sial-water ratio, which by volume is 4.7:1—a fundamental ratio for understanding the first-order surface features of the earth.

The continents and oceans are in a "symbiotic" relation, so that for many purposes the hydrosphere must be considered as a geologic formation. For example, sea level imposes a base level below which continents cannot be reduced by erosion. But how did continents rise above sea level in the first place?

Not until the mideighteenth century did natural philosophers seriously speculate about the depth of the oceans. One astronomer, basing his calculations on the speed of the tide wave, stated that the oceans were 23 miles deep! Laplace more conservatively computed the depth to be 12 miles. But Maury relates that among those who were not content to regard the oceans as bottomless, the prevalent view, based merely on the "fitness of things," was that the oceans are as deep as the mountains are high. This has proved to be a fairly accurate prediction. But as with so many things about the continents and ocean basins, it remains unclear just why.

Since preparation of this manuscript in mid-1966, I have come to favor the continental drift option of Wilson (*Nature* 211:676–681, 1966), who has suggested that Laurasia and Gondwana collided in the mid-Paleozoic and then rifted apart again in the early Mesozoic essentially along the same geosuture. This temporarily juxtaposed the bulge of Africa against the eastern bight of North America. This fit becomes surprisingly good when the Bahama salient is removed from North America on the probably valid assumption that it is a post-drift excrescence laid down on oceanic crust. It still seems likely, however, that the two supercontinents were born as separate units and have maintained separate geologic identity.

The Paleozoic closing of the Atlantic Basin, of course, implies Paleozoic drift but without the complexity of the Mesozoic breakup of the continents. It also implies the existence of a trench within the primitive Atlantic toward which the opposing cratons moved and eventually collided. This trench would then later have to become the site of a spreading rift ("mid-ocean ridge"). Such a transformation probably is, in fact, feasible in terms of sea floor spreading and plate tectonics.

References

Alfven, H.: The early history of the moon and the earth. *Icarus* 1:357–363, 1963.

Ambrose, J. W.: Exhumed paleoplains of the Precambrian Shield of North America. *Am. J. Sci.* 262:817–857, 1964.

Berlage, H. P.: The basic scheme of any planetary or satellite system corrected and reanalyzed, 1 and 2. *Proc. Kon. Ned. Akad. Wetensch., Amsterdam,* Ser. B, no. 1, 62:63–83, 1959.

Bullard, E., J. E. Everett, and A. G. Smith: The Fit of the Continents Around the Atlantic. In *A Symposium on Continental Drift,* P. M. S. Blackett, E. Bullard, and S. K. Runcorn, eds. Phil. Trans. Roy. Soc. London, S.A., 258, 1965, pp. 228–251.

Carey, S. W.: The Tectonic Approach to Continental Drift. In *Continental Drift: A Symposium . . . , March, 1956*. University of Tasmania, Geology Dept., Hobart, 1958, pp. 177–355.

Dietz, R. S.: Asteroidal Impact as the Origin of Oceanic Basins. In *Symposium No. 83*, Centre National de la Recherche Scientifique, 1958, pp. 265–275 (in French).

Dietz, R. S.: Continent and ocean basin evolution by spreading of the sea floor. *Nature* 190:854–857, 1961.

Dietz, R. S.: Collapsing continental rises: An actualistic concept of geosynclines and mountain building. *J. Geol.* 71:314–333, 1963.

Dietz, R. S.: Passive continents, spreading sea floors, and collapsing continental rises. *Am. J. Sci.* 264:177–193, 1966.

Dietz, R. S., and J. Holden: Earth and moon: Contrasting tectonic realms. *Am. N.Y. Acad. Sci.* 123:631–640, 1965.

Dietz, R. S., and W. P. Sproll: Equal areas of Gondwana and Laurasia (ancient supercontinents). *Nature* 212:1196–1198, 1966.

Donn, W. L., B. D. Donn, and W. G. Valentine: On the early history of the earth. *Bull. Geol. Soc. Am.* 76:287–306, 1965.

Escher, B. G.: Moon and earth. *Proc. Kon. Ned. Akad. Wetensch., Amsterdam* 42, no. 2:127–138, 1939.

Gilvarry, J. J.: The origin of ocean basins and continents. *Nature* 188: 886–890, 1961.

Girdler, E. W.: Rift valleys, continental drift and convection in the earth's mantle. *Nature* 198:1037–1039, 1963.

Green, L.: *Vestiges of the Molten Globe as Exhibited in the Figure of the Earth's Volcanic Action and Physiography*. London, 1875.

Gregory, J. W.: The plan of the earth and its causes. *Am. Geologist* 27: 100–119, 1901.

Harrison, E.: Origin of the Pacific Basin: A meteorite impact hypothesis. *Nature* 188:1065–1067, 1960.

Hess, H. H.: The oceanic crust. *J. Marine Res.* 14:423–439, 1955.

Hess, H.: History of Ocean Basins. In *Petrologic Studies: A Volume to Honor A. F. Buddington*. Geol. Soc. Am., 1962.

Judson, S., and D. Ritter: Rates of regional denudation in the United States. *J. Geophys. Res.* 69:3395–3401, 1964.

King, L. C.: *Morphology of the Earth*, 2d ed. Edinburgh, Oliver & Boyd, 1967; New York, Hafner Publishing Company, 1967.

Kossinna, E.: *Die Tiefen des Weltmeeres*. Berlin, Veroff Inst. f. Meereskunde d. Univ. Berlin, 1921, N. F. Reih. A., no. 9, pp. 35–37.

Kuenen, P.: Rate and mass of deep-sea sedimentation. *Am. J. Sci.* 244:563–572, 1946.

Laughton, A. S.: The Gulf of Aden. *Roy. Soc. London Phil. Trans.* s. A, 259:150–171, 1966.

Ludwig, W. J., J. I. Ewing, M. Ewing, S. Murauchi, N. Den, S. Asano, H. Hotta, M. Hayakawa, T. Asanuma, K. Ichikawa, and I. Noguchi: Sediments and structure of the Japan Trench. *J. Geophys. Res.* 71:2121–2122, 1966.

Matsuda, T.: Island arc features and the Japanese Islands. *Ja. Geol. Mag.* 73:271–278, 1964 (in Japanese).

Menard, H. W.: *Marine Geology of the Pacific.* New York, McGraw-Hill Book Company, Inc., 1964.

Pakiser, L. C.: Structure of the crust and upper mantle in the western United States. *J. Geophys. Res.* 68:5747–5756, 1963.

Plafker, G.: Tectonic deformation associated with the 1964 Alaska earthquake. *Science* 148:1675–1687, 1965.

Pitman, W. C., III, and J. R. Heirtzler: Magnetic anomalies over the Pacific-Antarctic Ridge. *Science* 154:1164–1171, 1966.

Runcorn, S. K.: Towards a theory of continental drift. *Nature* 193:311–314, 1962.

Runcorn, S. K.: Changes in convection pattern in the earth's mantle and continental drift: Evidence for a cold origin of the earth. In *A Symposium on Continental Drift*, P. M. S. Blackett, E. Bullard, and S. K. Runcorn, eds. Phil. Trans. Roy. Soc. London, s. A, 1965, pp. 228–251.

Shepard, F. P., and R. F. Dill: *Submarine Canyons and Other Sea Valleys.* Chicago, Rand McNally & Co., 1966.

Suess, L.: Das Antlitz der Erde 1, 1885.

Sykes, L. R.: Mechanism of earthquakes and nature of faulting in mid-oceanic ridges. *J. Geophys. Res.* 72:2131–2153, 1967.

Van Bemmelen, R.: On mega-undations: A new model for the earth's evolution. *Tectonophysics* 3:83–127, 1966.

Vine, R. J.: Spreading of the ocean floor: New evidence. *Science* 154:1405–1415, 1966.

Wanless, H., J. Tubb, D. Gednetz, and J. Weiner: Mapping sedimentary environments of Pennsylvanian cycles. *Bull. Geol. Soc. Am.* 74:437–486, 1963.

Wilson, J. T.: Hypothesis of the earth's behaviour. *Nature* 198:925–929, 1963.

4

Crustal Deformation in the Western United States *

JAMES GILLULY

Formerly Geologist, U.S.
Geological Survey, Denver, Colo.

This conference in celebration of the bicentenary of Rutgers University has as its objective the discussion of new geologic and geophysical evidence on the existence, possible causes, and mechanics of fracturing, drifting, and rotating of continents and ocean basins. Although another session is specifically devoted to geophysical considerations, the new data from investigations of gravity, magnetics, and seismology seem to me so pertinent in guiding our thinking that they must be considered even in this primarily geologic discussion.

Isostasy

First, there is now most compelling evidence for the general existence of isostasy in the southwestern United States. The raw Bouguer anomalies shown on the Woollard and Joesting (1964) compilation reveal a truly amazing correlation with the generalized topography of the same area that I prepared for this study (Fig. 1). Note the precise coincidence in position of the greatest negative anomalies and the topographic highs of the San Juan Mountains and the Mount Elbert mass in Colorado. Note how the Front Range topographic trend parallels the Bouguer gravity trend along the Range and thence west around the end of the Sangre de Cristo Range. True, the topographic high of the Sierra Madre in New Mexico has no corresponding anomaly low, but topographic and anomaly gradients are remarkably con-

* The kindness of Paul Averitt and Don Mabey in reading an earlier draft of this paper is deeply appreciated. I remain responsible, of course, for all statements and interpretations.

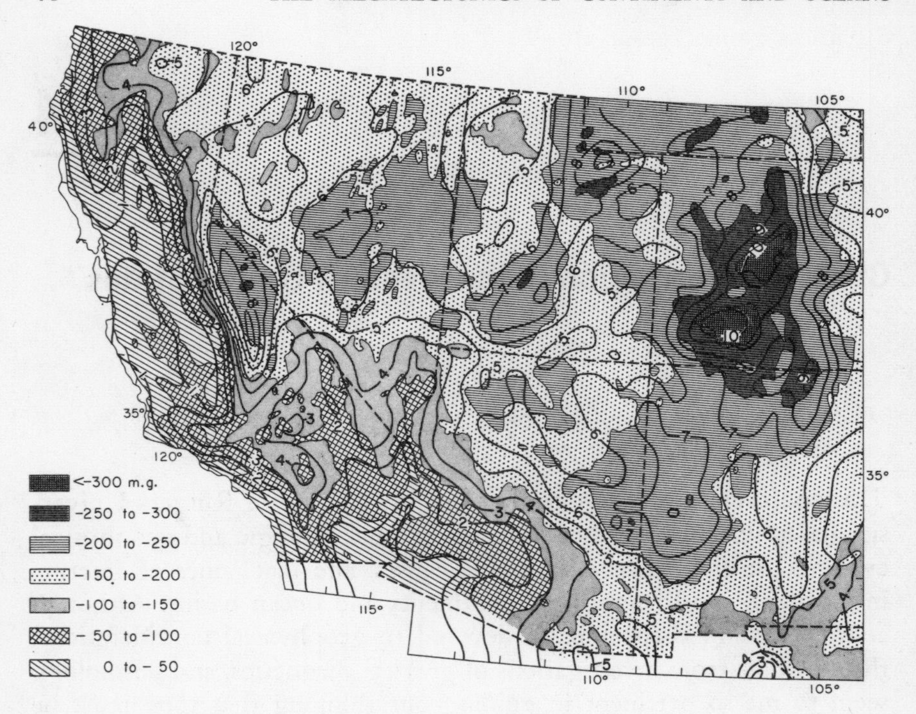

Fig. 1. Generalized topography of the southwestern United States, superposed upon the Bouguer anomaly map of Woollard and Joesting (1964).

cordant around the southern and western sides of the Mogollon and Colorado Plateaus. The parallelism in this area would have been even more marked had smaller areas been averaged for the topographic map. The low topography in southwestern Arizona is reflected in correspondingly smaller negative anomalies; the correlation is so faithful that even the Colorado River and the Death Valley lowlands are reflected.

The Sierra Nevada shows a less perfect correlation, for although the Mount Whitney massif corresponds with a large negative anomaly, as it should, the highest negative values in the region are in the calderas of Long Valley and Mono Lake, both filled with rhyolite ash. This is one of the few places where the local lithology is clearly reflected in the anomalies. Elsewhere, the large areas over which the topography has been averaged serve to mask the local lithologic variations. The Great Valley and the Klamath Mountains are reflected in the gravity patterns, as are the high area in east-central Nevada, the relative lowland of the Bonneville depression, the highland of the Wasatch front, and, most emphatically, the High Plateaus of southern Utah.

The conclusion that, at least in the Southwestern States, areas as large as about 1,500 square miles are in close isostatic adjustment is thus justified. Some of my geophysical friends tell me that had terrain corrections been applied to the raw Bouguer data the correlation would perhaps not be quite so good as it appears, but they also admit that the picture would be only slightly modified. It must therefore be assumed that there is virtually complete isostatic adjustment over areas of 1,500 square miles or more, putting decisive limits on speculations and structural models.

The recent outstanding study by Crittenden (1963a and b) of the rebound of the Bonneville Basin after unloading of its water load not only confirms the earlier work of Gilbert (1890), but provides evidence of a mantle viscosity nearly an order of magnitude smaller than that deduced from the rebound of Fennoscandia—a very weak mantle, indeed. The upper mantle beneath the Bonneville Basin and of several other places in the region has extremely low velocities for compressional waves; it is the crust-mantle mix of Cook (1962). Its strength must be notably less than the average for the region, and much less than the average for the continent.

Crustal Thickness

Explosion seismology conducted during the last 6 years under the Vela Uniform program has revealed remarkable regional variations both in the thickness of the crust and in the elastic properties of the crust and upper mantle in the western United States. Much of this information, summarized on maps by Pakiser and Steinhart (1964), is shown in Figure 2.

The most striking facts are that the crust is as thin as 20 km or even less beneath parts of the Great Valley, less than 30 km beneath much of the Great Basin province, and over 40 km thick underneath the High Plateaus of Utah. A root deeper than 50 km has long been recognized beneath the Sierra Nevada; a somewhat shallower root extends north beneath the southern Cascades, but a pronounced root has not been found beneath the northern Cascades or the Olympics.

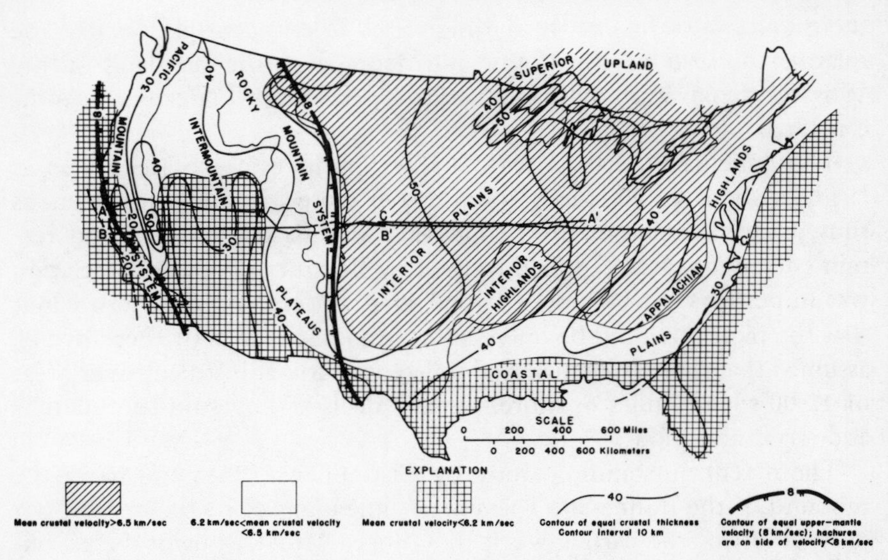

Fig. 2. Mean crustal velocity of compressional seismic waves, mean crustal thickness, and distribution of mantle in which P_n velocities are respectively greater than and less than 8 km per second. After Pakiser and Steinhart (1964).

Nor does the 50 km crustal thickness under the southern Rockies constitute an individualized root, for such or even greater thickness is also present beneath much of the High Plains.

Crustal Velocities

In areas of thin crust, the mean crustal velocity of compressional waves is generally less than 6.2 km per second, and little of the area west of the Rocky Mountain front has a mean crustal velocity of over 6.3 km, such as characterizes most of the plains and interior uplands to the east. Some seismologists recognize a Conrad discontinuity within the crust of much of the Great Basin; the presumably denser material underlying it is thinner than normal. The mean crustal velocity is therefore also lower than normal, reflecting the relatively more abundant low-density material at shallower levels. A geologist, of course, finds it difficult to accept the reality of a distinct Conrad discontinuity; in the many areas where erosion has exposed materials once buried deeper than 20 km, nothing suggests the possible existence of an abrupt discontinuity at a nearly uniform depth in the crust. If a Conrad discontinuity really exists in the Basin and Range province, it must be displaced vertically by many faults. The greater normal faults, such as those along the front of the Sierra Nevada and the Grand Wash Cliffs, even offset the Mohorovičić discontinuity (Moho) as much as 5 km, and the Moho itself slopes as much as 6° in some places (Roller and co-workers, 1964).

The mean crustal velocity is notably higher in the crust of the Sierra Nevada and the eastern side of the Great Valley than in the Great Basin to the east or along the seaboard of California and Oregon. This clearly implies that much of the Sierra Nevada root is composed of rocks considerably denser than the granites exposed in the mountains. This inference is confirmed by the heat-flow data, which show lower values than would be expected were the root composed of rock as radioactive as the exposed granites. The thin crust beneath the eastern side of the Great Valley must be considerably denser than that beneath the western side. The well-known magnetic high of the northern Great Valley is doubtless caused by dense mafic volcanics. These probably extend southward beneath the east side of the valley and thereby account for the greater mean density of the crust which must be inferred from the higher mean crustal velocities there, as suggested by Thompson and Talwani (1964).

The mean crustal velocity beneath the High Plateaus of Utah, Arizona, and New Mexico is notably higher than that under the Basin and Range province to the west. This seems to imply that the increase in crustal thickness at the western boundary of the Colorado Plateaus province is caused by a thickening of the denser material making up the lower part of the crust. The Colorado Plateaus were formerly much lower than the Basin and Range province to the south and west, yet they are now thousands of feet higher. Relatively, therefore, the Basin and Range province has dropped, while the Colorado Plateaus province has been rising. Is it possible that these vertical movements are responses to reciprocal changes in crustal thickness caused by lateral transfer of deep crustal material from the Basin and Range province to the Colorado Plateaus province?

The Upper Mantle

Compressional velocities in the upper mantle (P_n) also vary widely in the western United States (Figure 3). In general the areas of abnormally low P_n trend roughly parallel to the Pacific shore and to the structural trends of the exposed rocks, but, as shown by Herrin and Taggart (1962), this is not true in detail. Their map shows an east-northeast-trending belt almost parallel to and just south of the Uinta Range and extending well into Nevada across the Bonneville Basin, beneath which the velocity of P_n is as low as 7.5 km/sec. This is in the area of Cook's (1962) "crust-mantle mix." This belt of low-velocity P_n is almost paralleled on the south by another belt of higher P_n—as high as 8.2 km/sec—and still farther south, beneath the Colorado Plateaus by another belt of low-velocity P_n. Farther south, in southeastern Arizona, is still another region of 8.2 velocity. Except for the near parallelism of the most northerly of these belts with the Uintas, none of them shows any readily recognizable relation to the observed geologic trends. We shall see, however, that strong magnetic trends show the same independence of visible structure and a comparable tendency toward a roughly east-west orientation.

Although there are local exceptions, such as those just mentioned, nearly all of the country from the Rockies to the Pacific is underlain by an upper mantle with an abnormally low-velocity P_n. Though most of this country is tectonically active, it does not necessarily follow that there is any relation between low-velocity P_n and active tectonism, for the equally or more active Coast Ranges of southern California overlie an upper mantle of normal seismic properties. Furthermore, there are no obvious differences in tectonic activity or style between the several smaller belts of dramatically different P_n outlined on the Herrin and Taggart map.

The elastic properties of the mantle may correlate, not with tectonism directly, but, as Pakiser (oral communication, 1965) has suggested, with the state of differentiation of the mantle. The area of low-velocity P_n roughly corresponds with that of Late Tertiary and Pleistocene volcanism, in which pockets of magna or potential magma may remain which are not completely separated from the rest of the mantle, whereas in nonvolcanic areas the mantle is perhaps more homogeneous and more basic because it has more thoroughly differentiated. The thick and high-velocity crust of the Great Plains province is underlain by a mantle of exceptionally high P_n. Is the crust there thick because the underlying mantle has been completely differentiated? And has the differentiation been so complete that more than usually mafic fractions have been transferred to the crust, thereby accounting for its higher rigidity?

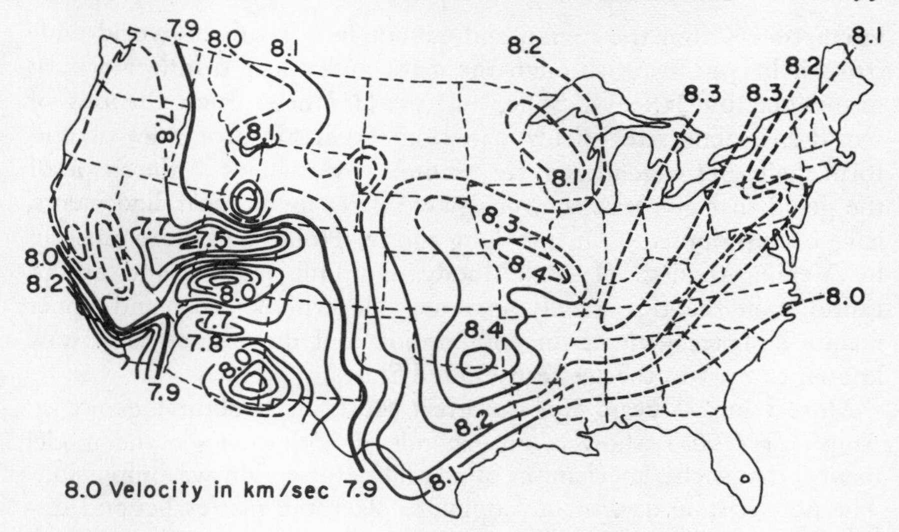

Fig. 3. Velocities of P_n in the United States. From Herrin and Taggart (1962).

Tectonic Significance of the Seismic Data

Whether or not the speculations as to the causes of the differing properties of crust and mantle in the several regions have any validity, some important conclusions may be drawn from the seismic data. The thickness of the crust varies from place to place by more than 150 percent, and the seismic velocities of both crust and upper mantle vary by nearly 10 percent. The Moho is offset by faults of as much as 5 km throw, and in places slopes as steeply as 6°; its irregularities are more than ten times greater than the mountainous irregularities of the topography in the western United States. Furthermore, although heat-flow data are scant, I think it can be safely inferred from the distribution of Cenozoic volcanism that there are great local variations in temperature, and hence plasticity, of both crust and mantle.

In view of these facts, it is no longer possible to seriously consider the once popular idea that much of the earth's crust may be regarded as a uniform elastic shell through which tectonic stresses can be transmitted for great, even semicircumferential, distances, a theory still defended by so distinguished a geophysicist as Harold Jeffreys (1952). Tectonic activity of a region must be connected with crust-mantle interaction within the region and cannot be a result of world-wide stress fields operating through the crust and upper mantle alone, as maintained by Jeffreys, Stille, and many others. Such theories of world-encircling stress fields, based on the elastic properties of uniform shells and engendered, for example, by changes in the shape of the geoid that give rise to world-wide patterns of shear lineaments, have been proposed by many—long ago by Hobbs and more recently by Vening-Meinesz (1947), Moody and Hill (1958), and others. I find such theories wholly incompatible with a crust and upper mantle so heterogeneous in composition and dimensions as is now known to exist in the western United States.

More than 60 years ago the great Austrian structural geologist, Ampferer (1906), showed, on grounds of both geologic and model theory, that such a mechanism of crustal deformation was impossible. The new seismic data, in my opinion, place the matter beyond dispute. A crust so weak locally as to deform in almost precise conformity to the weight of a water load nowhere more than 1,000 feet thick seems most unlikely to respond in the same way as one 50 km thick with a much higher rigidity as measured by seismic response. Under horizontal stress, which the thick rigid crust could withstand, the thin, plastic segment would squeeze out in diapirs. The crust is locally so weak, and both it and the mantle so heterogeneous, that yielding must be highly irregular. Stress trajectories in the crust of uniform trend over half a continent are thus inconceivable. Deformation of any crustal segment must surely depend on the immediately underlying mantle and not on a worldwide stress system involving the shell above a postulated level of strain beneath a contracting or warping geoid.

Thus the vertical upbowing of the San Gabriel Range (Gilluly, 1949) is going on concurrently with the purely strike-slip displacement of the San Andreas fault alongside (Noble, 1926), and the left-lateral White Wolf fault (Dibblee, 1955) is active concurrently with the purely dip-slip thrust of the Buena Vista Hills (Wilt, 1958) a few miles away across the San Joaquin Valley, while dip-slip normal faulting is going on along the Sierra Nevada boundary fault perhaps 50 miles to the east. In many parts of the Great Basin, dip-slip normal faults trending due east are contemporaneous with, and curve into, others striking due north. Elsewhere, notable strike-slip and dip-slip components are active in some Basin and Range faults. Buwalda (1952) called attention to the simultaneous upbowing, in Pliocene and Pleistocene times of the N-W trending California Coast Ranges, of the E-W trending Transverse Ranges, and the N-S trending Cascades, as well as the simultaneous development in Washington and Oregon of E-W and N-W trending folds and in Montana of N trending normal faults. At the same time, the Colorado Plateaus have been upraised over a broad area with only minor warping, and the southern Rockies have been elevated along a much narrower zone. All these evidences of inhomogeneity of stress fields within the continent are consistent with the great variability of crust and mantle revealed by the seismic studies, but they are totally incompatible with the existence of continent-wide stress trajectories in the crust and upper mantle.

Magnetic Trends (Fig. 4)

The magnetic maps of Affleck (1963), Fuller (1964), and Zietz (1969) generally reveal trends independent of the structure of the visible rocks. Over the Precambrian core of the northern Front Range,

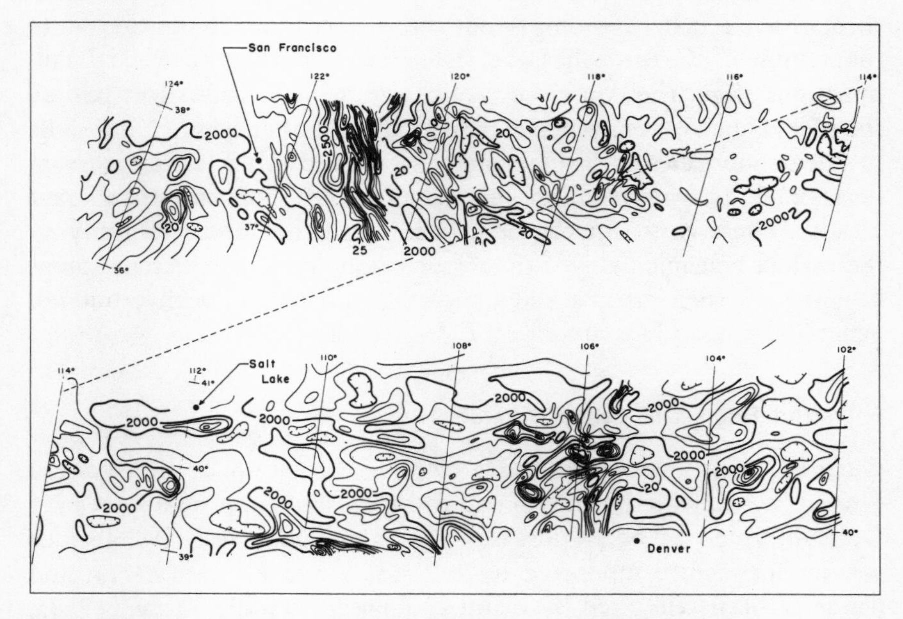

Fig. 4. Some magnetic trends in the western United States, as found by aeromagnetic surveys on a great circle course 1° broad between Washington, D.C., and San Francisco, Calif. Generalized from Zietz, *Bull. Geol. Soc. Am.* 80, 1703–1714, 1969.

the magnetic pattern does indeed agree with the trends of the gneissic structure, and the magnetic and tectonic patterns are also parallel near the Uinta Mountains and in the Sierra Nevada. Elsewhere, however, the maps of Fuller and of Zietz commonly show magnetic patterns of E-W trend that cross exposed tectonic elements without vertical or horizontal offset of geologic structures of widely ranging ages. For example, the Antler orogenic belt of early Mississippian age shows no deflection between latitudes 40°N and 41°N, where a pronounced E-W magnetic anomaly crosses it, but reappears to the north of the Snake River Plain in central Idaho, just where an extension of its gentle arc to the south would be expected. None of the Basin Ranges show either offsets or common termination at the magnetic anomaly.

These facts are puzzling, since many of the anomalies have steep lateral gradients that show they are due to relatively shallow magnetic features. Surely all are caused by rocks the temperatures of which are below the Curie point, and hence at depths less than 20 km. It is thus difficult to understand the complete independence of the magnetic trends from the trends of the visible tectonic features throughout a belt extending from the eastern side of the Sierras to the region of Great Salt Lake. Similarities in successive axial culminations along the Antler orogenic belt, which crosses this region, show that the Antler trends extend to depths of at least 12 km. The magnetic anomalies must therefore be generated in a relatively narrow depth zone, ranging from not less than 12 km to not more than about 20 km. There is no correlation of the magnetic trends with any of the various volcanic fields of the region, though many stocks and plugs of intrusive rock are marked by strong anomalies; locally, but not generally, these plugs are along E-W trends.

Strike-slip Faults and Tectonic Rotation

Becker (1933) first postulated a clockwise rotation of the western United States, which he attributed to right-lateral slip on a system of faults of which the San Andreas system is the most westerly. His idea was independently supported by Billingsley and Locke (1941), and has since been reiterated by many geologists, notably Carey (1958), Wise (1963), Watkins (1965), and Shawe (1965). Strike slip is of course prominent in the San Andreas, Death Valley–Furnace Creek, and Las Vegas–Walker Lane systems, and strike-slip components are conspicuous on many faults of the Basin and Range province near these structures (Callaghan and Gianella, 1935; Slemmons, 1957; Shawe, 1965). Regionally, however, strike slip is clearly unimportant in the Basin and Range fault process, as is evidenced by the characteristically abrupt curves and angles in the fault trends, by the disappearance of many faults along strike slip within exposed areas of mountain blocks (Gilluly, 1928; Gilluly and Gates, 1965), and by such features as the fault-surrounded, down-dropped blocks in southern Oregon described by Fuller and Waters (1929, 1931). Notably few faults of the Basin and Range type are associated with the San Andreas system except south of the Transverse Ranges; strike slip must have been slight, though detectable, along the boundary fault of the Sierra Nevada, for this fault has en echelon courses north and south of the monocline which interrupts it near Bishop (Bateman, 1951). The scarp of the Black Mountains bordering Death Valley south of Furnace Creek, where little or no strike slip can have taken place (Hunt, 1964) is fully as prominent, and the fault throw as great as they are to the north along the Funeral Range, which is fronted by the great Death Valley–Furnace Creek strike-slip fault. In short, some Basin and Range faults unquestionably have considerable strike-slip components, but just as unquestionably most do not; strike slip is not an essential element in Basin and Range faulting.

All the proponents of general clockwise rotation of the western edge of the continent by movements analogous to those along the San Andreas Rift have had trouble finding an eastern boundary of the area supposedly so deformed. No lineaments comparable to the Las Vegas–Walker Lane zone have been recognized east of that structure within the Great Basin. Certainly there is no evidence of any strike-slip movement along the Grand Wash and Hurricane faults.

The Olympic-Wallowa lineament of Raisz (1945) has been suggested as possibly marking a northeasterly limit of the clockwise rotation (Wise, 1963), but Raisz himself considered a left-lateral displacement more probable. As no geologic boundaries have been shown to show strike-slip offset anywhere along the lineament, it seems more likely to be a minor fortuitous feature rather than a major tectonic boundary. It is true that crustal properties to the northeast and southwest of the line differ in a general way, and Cantwell and co-workers (1965a and b) concluded from resistivity measurements that the Wallowa-Olympic lineament might be the boundary between the two crustal types. Skehan (1965) also considered the lineament as marking a boundary between differing crustal sections. Resistivity lines as long as 200 miles, however, are not very precise in locating boundaries; that between the differing crustal sections may not even be close to the Wallowa-Olympic lineament. On the projected position of the line near Puget Sound, the faults recognized by Danes and co-workers (1965) trend at a considerable angle to the supposed lineament. Dehlinger and associates (1965) found differences in travel time between the two sides of the Cascades, but no differences bounded by the Olympic-Wallowa line. None of the geologic boundaries mapped in the Cascades, nor any structure of the Columbia Plateau has been offset along this line. That the Hope and Osburn faults of northern Idaho, both with many miles of strike slip, are roughly parallel to the lineament is not pertinent. Both these faults have been inactive since the Columbia River flood basalts were emplaced; pre-Miocene movement would in any case be irrelevant to a continental rotation to which Basin and Range faulting is supposed to contribute. Taubeneck (1966), on the basis of absence of distortion of Miocene dikes and Mesozoic structures where they are crossed by the lineament, concludes that lateral offset has not yet been found in Miocene and younger rocks anywhere along the lineament. There can be little disagreement with his conclusion.

A regional study of the Basin and Range faults themselves fails to support any such large-scale rotation (Fig. 5). Thompson (1959) has shown that the Dixie Valley and Fairview Peak faults produced an extension of the crust of about 5 feet. He calculated that if the total structural relief of the valley had been produced by comparable displacements, the total distention across it would have amounted to about 1½ miles in roughly 15 million years—about 1 foot per century, which is well within that of historical fault displacements.

By using the tectonic map of the United States and measuring the azimuth of each 10 miles long normal fault segment, I have compiled an average trend for the faults within each square degree of the Basin and Range province (Fig. 6). On the assumption of a 60° dip and a 3 mile throw as the average for the Basin and Range faults (a very conservative throw in my opinion), each mile of fault length has added 1.5 square miles of area to the Basin and Range province. The tectonic map shows more than 20,900 linear miles of normal fault segments as much as 10 miles long. Accordingly, at least 30,000 square miles of area have been added to the province through distention. The

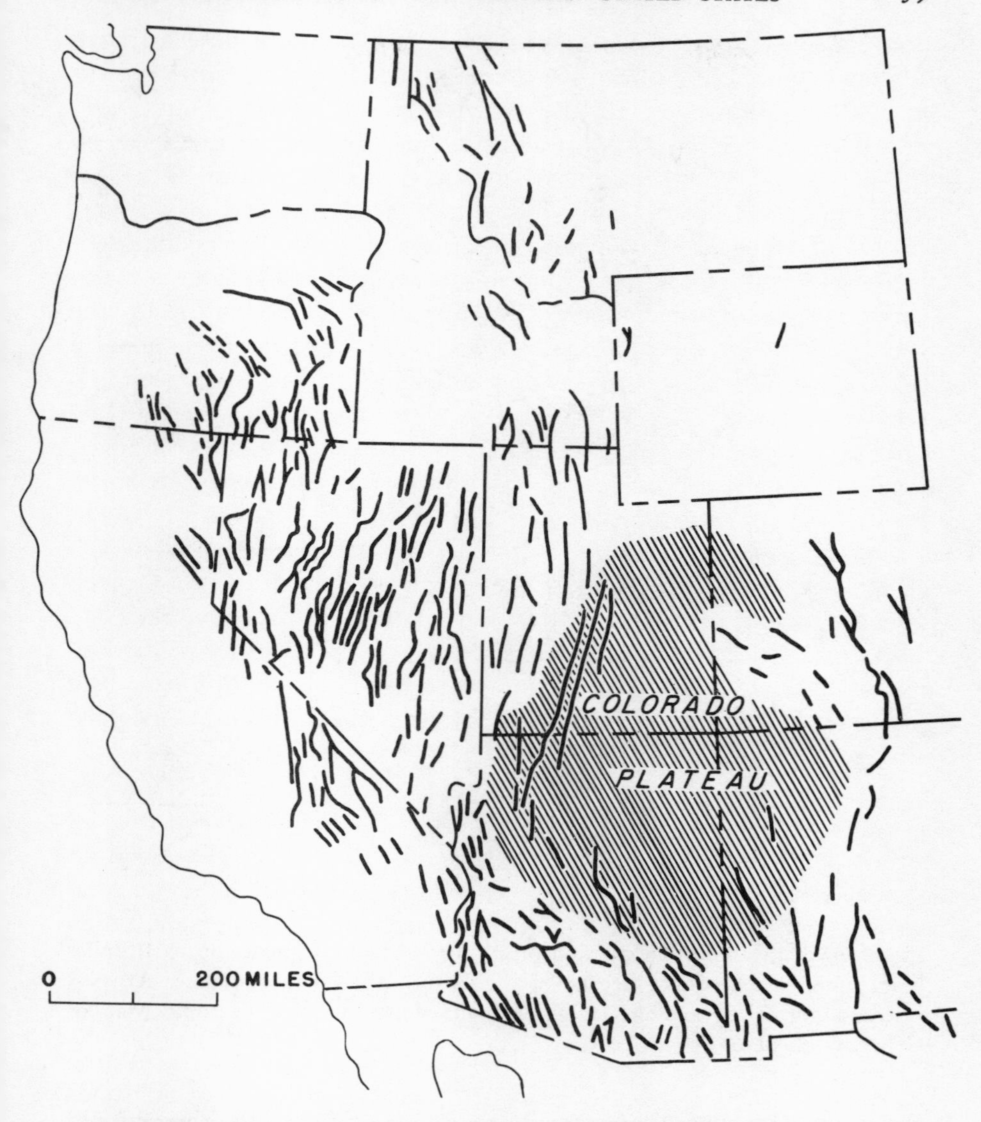

Fig. 5. Basin and Range faults of the western United States. After Cohee, et al. (1961). U.S. Geol. Survey, *Tectonic Map*.

present area is about 400,000 square miles, so the province has been enlarged by at least 7.5 percent. Actually, there must be many miles of normal faults concealed beneath the alluvium of the intermontane basins. Some of these have been demonstrated by drilling and others by gravity measurements. Although such concealed faults must ordi-

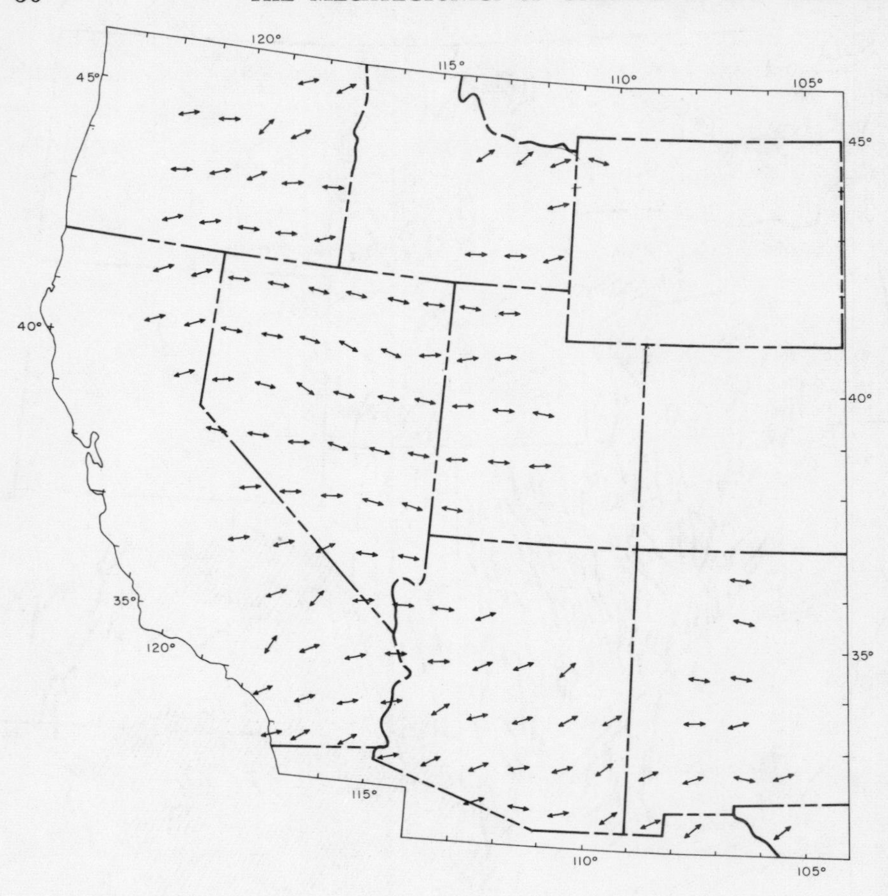

Fig. 6. Average trend of extension of all normal fault segments up to 10 miles long, within each square degree of the Basin and Range province. Arrows, oriented normal to average trend of faults shown on the Tectonic Map of Cohee, et al. (1961).

narily have smaller displacements than the visible faults, they would add considerably to the estimate of distention. When these and the many fault segments shorter than those measured are included, I think it likely that 10 percent of the province has been added since the beginning of the Miocene by spreading of the crust. This is an area the size of Kentucky. The spread would have been at the rate of 2,500 square miles per million years.

Large as this amount of spreading is, it is wholly inadequate to thin the crust from an assumed former average of 40 km to its present 30 km average thickness dimension. Other processes must have been at work. Since the mean crustal velocity of compressional waves in the province is lower than average, the unknown process cannot have been simply the deep erosion of the upper crust, for that should have increased the average density of the crust. In any event, such deep erosion is denied by the geology of the area.

When the distention is averaged across the width of the province, the nearly uniform latitudinal distention is striking (Fig. 7). Within each degree of latitude from 45°N to 32°N, the dilation ranges be-

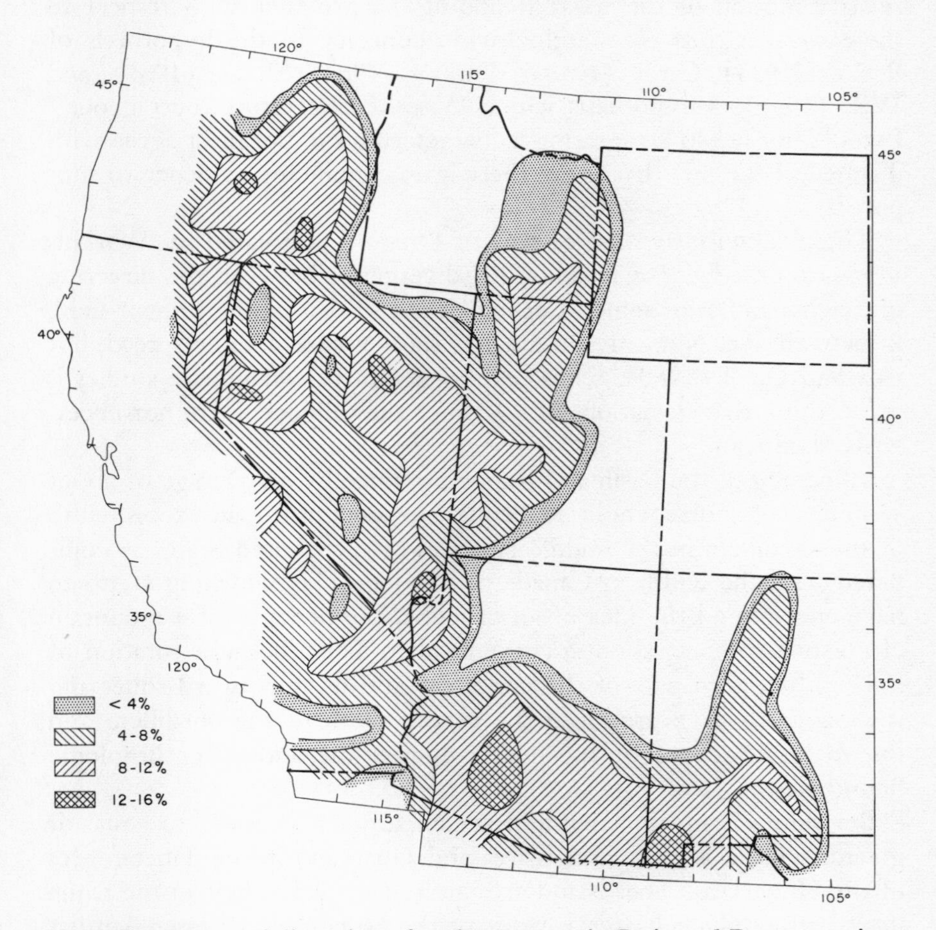

Fig. 7. Percentage of distention of each 1° square in Basin and Range province by normal faults, on the assumption that each fault has a throw of 3 miles and a dip of 60°.

tween 4 and 16 percent, counting only the measured fault segments. On the average, each isopleth is about equally distributed in latitude. I find it significant that the direction of spreading is not such as to favor the concept of clockwise rotation and shear as a cause of the reverse *S* curve in the outcrop of Mesozoic plutons and the quartz-diorite line of Moore (1959). Through most of the province, the dilation was nearly E-W. But in Oregon, at the extreme north of the province, the spreading is along SW-NE trends, not E-W or NW-SE, as might be expected were the rotation clockwise. The same is true in the southwestern part of the province in western Arizona. The relative motion of the western side of the province with respect to the eastern is thus counterclockwise, contrary to the hypothesis of Becker (1934), Carey (1958), Watkins (1964), Shawe (1965), and Wise (1963)—a hypothesis which Wise himself terms "outrageous." I would not use so denigrating a characterization, but must agree with Taubeneck (1966) that as yet there is no convincing evidence to support it.

The clockwise rotation of part of Oregon, postulated by Watkins (1964) on the basis of roughly 20° divergence between the direction of magnetization of some Miocene lavas and that of the present field, is inconclusive. Not only do his data show a considerable spread, but Cox and Doell (1960), whose experience in paleomagnetic studies is great, think that deviations of 20° since the Oligocene are not necessarily significant.

Widening of the Basin and Range province by an average of about 45 miles is of course wholly inadequate to account for the excess width of the cordillera in the middle latitudes of the United States as compared with the width in Canada or Mexico. This widening seems to have encouraged the idea of an orocline, the bending of a mountain chain so as to bring about a curve in plan. The clockwise rotation of the southwestern part of the continent and its northward squeezing is thought to have produced the excess width of the cordillera and the reverse *S* curve of the quartz-diorite line and other lithologic boundaries.

But similar widening, without accompanying *S* trends, to equal or greater widths, is present in Korea and Japan and in the Tibetan arcs of the Himalayas. That island arcs are curved is implicit in the name itself. Although such arcs as those of the Greater and Lesser Antilles seem to demand extreme late bending and disruption of formerly continuous structures, there is no reason to think that every orogenic belt on earth was originally straight and, if now curved, was bent later. In fact, straight mountain belts such as much of the Chilean Andes are exceptional, not normal, features. At any rate, Moore and coworkers (1961) have given good evidence, in the paleogeography of Washington and Oregon, that the curves in the quartz-diorite line and in the belt of plutons were in existence in Early Tertiary time; it is not a post-Miocene development.

Relation between Oceanic and Continental Structures

The topography of the continental borderland off southern California is much like that of the Basin and Range province, and it seems likely that this small offshore area has a crust as inhomogeneous as that of the western part of the continent. Raitt (1956) found that farther offshore most of the crust between Hawaii and the mainland is rather uniformly about 4.8 km thick, though in the mid-Pacific there were areas where the crust is up to 13 km thick. The innumerable linear magnetic anomalies, the presence of which permits measurement of displacement on the great fracture zones, show that the oceanic crust is not completely homogeneous. Nevertheless, the strikingly uniform width of the anomaly belts for hundreds of miles indicates a broad uniformity. The complexity of its structure cannot even remotely approach that of the continental crust, for the great fracture zones of the East Pacific extend in almost unswerving lines for many hundred miles. Even such relatively straight lineaments as the Rocky Mountain Trench and the Denali fault are sinuous compared with any of the Pacific Ocean fracture zones; there is nothing like them anywhere in the continental crust. Clearly, the oceanic crust has almost uniform mechanical properties over gigantic regions, far greater than any continental structural province on the face of the earth.

The relations of the great fracture zones of the East Pacific to the structures of the continent seem ambiguous. It has been claimed, for example, that the Clarion fracture zone is continuous with the trans-Mexican volcanic belt. Certainly the Clarion zone trends toward the western end of the volcanic belt and is itself characterized by many volcanoes, but the belt on land diverges considerably to the south of a prolongation of the Clarion zone. Is the fracture zone deflected at the edge of the continent? Does the fracture zone merely trigger an independent set of crustal forces on the continent, or is the intersection purely fortuitous?

A merely accidental connection seems contradicted by the fact that the Murray fracture zone is in line with the Transverse Ranges of California (Menard, 1955), but that its right-lateral slip of 154 km (Vacquier and co-workers, 1961) is in the opposite sense to the left-lateral movement on the Garlock fault (Smith, 1960) and to the left-lateral swerve of the San Andreas fault near its junction with the Garlock and Big Pine faults (Hill and Dibblee, 1953). Nor is the San Andreas offset at the south side of the Transverse Ranges. Surely there can be no possible connection of the Murray zone in the visible crust with the doubtful Texas lineament. In fact, east of the San Bernardino Range there are no signs of E-W tectonic features anywhere near its trend. Yet the remarkable alignment of the Murray zone with the only east-trending ranges of California is highly suggestive of some connection between the two features.

No E-W structures or interruptions of northwest-trending tectonic elements have been recognized on the strike of the Pioneer fracture, which should intersect the shore in the latitude of Ukiah. Its left-lateral displacement has not been traced to the shore and there is no sign of structures on this trend within the continent.

The relations of the Mendocino fracture zone to continental structures are also ambiguous. Part of the difficulty is caused by the uncertain position of the offshore continuation, if any, of the San Andreas Rift. Since the report of Lawson and co-workers in 1908, a number of investigators have believed that the rift, whose definite trace passes out to sea near Point Arena, curves parallel to the shore and finally curves out to sea again near Cape Mendocino. Shepard (1957) considered it likely that the rift curves into and follows the Gorda escarpment, which makes an angle of about 70° with the general trend of the fault on the continent. It thus would join with the Mendocino fracture zone. But this interpretation is contrary to the fact that the San Andreas Rift is a right-lateral structure, whereas the Mendocino zone has left-lateral displacement of 1,200 km. Yet the Mendocino zone seems to offset the continental slope 100 km in a right-lateral direction (Menard, 1960). Hurley (1960) points out that clusters of epicenters lie not only along the Gorda escarpment, but also along a direct line with the San Andreas fault, which he regards as the probable continuation of the rift, and along the base of the continental slope. Either of these courses conforms more closely with the known trend of the San Andreas system than does the course along the Gorda escarpment. But a lateral movement has not been traced along any of these trends. Irwin (1960), too, favors a direct continuation of the San Andreas out to sea along its known trend on land.

Whatever the correct interpretation of the continuation of the San Andreas fault on the sea floor, clearly neither the Mendocino fracture zone nor the San Andreas Rift system offsets the other. Ogle (1952, 1953) described a breccia zone of uncertain but easterly trend at False Cape, just north of Cape Mendocino. This may be a continuation of the Mendocino fracture zone, but, if so, it fails to cut or offset the structures a few miles to the east along the Mad and Eel Rivers. The fact that the continuation of this trend would pass between the southern Cascades and the northern Sierra Nevada was pointed out by Menard (1955). But the trend of the ranges is not offset. Gilliland (1962) has discussed other features far inland along this projected trend. The Uintas lie not far from this line, and farther east there are several sedimentary basins whose greatest depths lie near the 40th parallel and which might be thought to record a deep-seated zone of persistently renewed weakness virtually crossing the continent along the Mendocino trend. The trend would have originated in Precambrian time and have been rejuvenated sporadically ever since. The magnetic maps of Affleck (1963), Zietz (1969), and Fuller (1964) also show strong E-W anomalies along this general zone. However, it should also be mentioned that E-W magnetic anomalies of equal strength have been recognized in a broad belt far to the south of the projected trend of the Mendocino fracture zone.

On the other hand, the surface rocks of the Klamath Mountains strike directly into comparable rocks of the northern Sierra Nevada foothills. None of the Mesozoic or younger structures crossed by the Mendocino trend exhibit offset, and it would be gratuitous to assume any in the Paleozoic, for the trends and lithology are similar in these rocks as well. None of the Basin and Range faults, chiefly of Miocene and younger age, in eastern California, Nevada, and Utah are broken at the Mendocino line; all continue uninterruptedly across it. So, too, with the Antler orogenic belt of Late Devonian or Early Mississippian age. The basement rocks of the Medicine Bow and northern Front Ranges show no evidence of transcurrent faulting along this line.

Although the Cheyenne and Denver Basins are separated by a moderate high of about 2,000 feet of structural relief at the appropriate latitude, the north-trending course of their common western flank is unbroken; nor are there any facies changes in the Rockies that can be related to the Mendocino trend. In short, the surface geology shows no sign of interruption by transcurrent faulting along this trend and it seems likely that the continental plate and ocean floor are completely decoupled, merely interfering with each other for a few miles near their common interface. If fracture zones of the oceanic segment influence the continental structures it must be indirectly. Possibly the fractures go beneath the continental crust, wholly within the mantle, with only sporadic local interaction between crust and mantle along them. Certainly there is nothing comparable to another San Andreas Rift to record the continuations of the fracture zones across the continent.

One aspect of interrelations between oceanic and continental structures is brought out by consideration of the San Andreas–Gulf of California area. Hamilton (1961) has suggested that the San Andreas and associated faults have undergone an aggregate strike slip of about 350 miles since Late Cretaceous time, thereby opening up the Gulf of California. Two elements of the oceanic structures seem to oppose this suggestion: (1) The Clarion fracture zone and its belt of volcanoes do not seem to show an offset where the extended San Andreas zone should intersect it. (2) The Transverse Ranges, the only dominantly E-W structures of Paleozoic and older rocks in a large region, are nearly aligned and on the projected trend of the Murray fracture zone. It seems highly improbable that a displacement of 350 miles (or, for that matter, using Crowell's (1962) more modest estimate of about 130 miles) fortuitously leaves the two E-W ranges so nearly in alignment, not only with each other but with the projected Murray zone. Although mapping of much of western Mexico has been chiefly by reconnaissance, no E-W range comparable to the San Gabriel has been identified east of the Gulf, either 130 or 350 miles to the southeast of San Bernardino. Nor is the western continuation of the San Bernardinos found to the northwest. The geology of the Gulf of California, though known only from reconnaissance data (Anderson, 1950), seems not to favor such a great movement. The displacement of about 30 miles, recorded by Noble (1954), is much more likely to be near the total movement on the San Andreas fault in the area of the

Transverse Ranges area, being compatible with their common trend parallel to the Murray fracture zone. Allison (1964) has recently called attention to this point, though without mentioning the Murray fracture zone.

In summary, we now know from the gravity and seismic studies that the continental crust varies dramatically in thickness and elastic properties from place to place. The crust is especially thin and of low rigidity in areas of Late Cenozoic volcanic activity. The upper mantle, too, varies greatly in elastic properties and has especially low rigidity in the same areas. Much of the remarkably complete isostatic compensation is brought about below the crust. The sensitive yielding of the Bonneville Basin to a modest water load and its prompt relaxation time are such that the part of the mantle involved must be fully an order of magnitude less viscous than that involved in the isostatic rebound of Fennoscandia.*

There are few comparable tests of relaxation time elsewhere, but it may be safely inferred that generally mantle viscosity varies greatly from province to province. Province-wide variations of elastic properties and crustal thickness are certain. The Moho itself is faulted, with displacements of nearly 20 percent of its depth, and locally it slopes as much as 6°. All this evidence of inhomogeneity in continental crust and mantle, and local variations in elastic properties of both, in my opinion lead to the unavoidable conclusion that uniformly oriented stress trajectories cannot possibly be transmitted over extensive continental regions through crust and upper mantle, let alone over whole hemispheres, as proposed by such eminent geophysicists as Jeffreys (1952) and Vening-Meinesz (1947).

Theories of continent-wide stress patterns such as those of Hobbs (1911), Moody and Hill (1958), and others, are, I think, untenable. The yielding of the continental crust depends on the deformation of the immediately underlying mantle and not upon stress accumulations from distant sources. The geologic reasoning and model analyses of

* Professor Leon Knopoff has since informed me that the differences in viscosity might be explained by the fact that the area involved in the Lake Bonneville adjustment was very much smaller than that of Fennoscandia and therefore only a much thinner zone of the mantle was involved in flowage. Had the Bonneville loading been extended over a much larger area, so that deeper zones of the mantle were involved, viscosity approaching that shown beneath Fennoscandia might also have been identified.

Ampferer (1906) made this conclusion highly probable more than 60 years ago; we can now consider the matter settled by these geophysical measurements.

Whether or not the earth is contracting or expanding, and I know of no evidence that it is doing either, it is not changes in the earth's radius that are causing crustal deformation. The systematic regional disposition of deformation must be due to convective or advective motion within the earth's mantle. The crust deforms passively under the drag of subcrustal motion in the mantle, for it is too inhomogeneous and weak to transmit stress uniformly over wide regions. Mountains are made by currents in the mantle and not by compression of a world-wide shell above some hypothetical "level of no strain" in a contracting earth.

The thickness and elasticity of the oceanic crust must be far more uniform than those of the continental one; the remarkably straight and long fracture zones in the oceanic crust are strong evidence for uniform stress patterns over great regions. The properties of the two crusts are so very different, and the relative independence of their strain patterns so pronounced, there must be nearly complete decoupling of the two. Oceanic patterns seem locally, as in the Transverse Ranges of California, to be influencing the deformation pattern of the continent, but the influence does not extend far inland. If the E-W magnetic trends along the projection of the Mendocino zone are a reflection of the continuation of that structure at depth beneath the continent, it is noteworthy that this is an orientation almost at right angles to the magnetic pattern shown in the oceanic crust, where the magnetic trends are at high angles to the fracture zones, not parallel to them.

The expansion of the Great Basin area through normal faulting does not seem to conform to what would be expected had a great clockwise rotation occurred in the western States. Such a rotation, at least since the beginning of the Miocene, seems clearly denied by the agreement in pattern of the Early Tertiary shorelines and the quartz-diorite line, as pointed out by Moore and co-workers (1961), and by the continuity of Miocene dikes, cited by Taubeneck (1966). Provable movement on continental faults does not seem remotely comparable to that on the great oceanic fracture zones. The alignment of the Transverse Ranges (the only considerable structures of this trend) and of Paleozoic and older rocks for hundreds of miles along the San

Andreas fault with the Murray fracture zone seems to show that the fracture zone, despite its right lateral displacement, is somehow related to the Ranges. If so, the movement on the San Andreas zone is more likely to be the 30 miles demonstrated by Noble (1954) than the considerably greater movements postulated by many others.

There are many persuasive arguments for and against continental drift and continental rotation. Perhaps the strongest opposing evidence are the marked differences in the mechanical properties of the crust and mantle in the two domains. The structures of the western United States do not seem to be incompatible with drift or rotation on a continental scale, but the probable intracontinental rotation on such shear zones as the San Andreas, Las Vegas, and Death Valley–Furnace Creek is really trivial on a regional scale. I can see no evidence of major rotation in the geology of the western States.

References

Affleck, J.: Magnetic anomaly trend and spacing patterns. *Geophysics* 28:379–395, 1963.

Allison, E. C.: Geology of areas bordering the Gulf of California. *Am. Assoc. Petrol. Geologists*, Memoir 3, pp. 3–29, 1964.

Ampferer, O.: Über das Bewegungsbild von Faltengebirgen. *Jb. geol. Reichsanst. (Bundesanst.) (Vienna)* 56:539–622, 1906.

Anderson, C. A.: Geology of islands and neighboring land areas. *Geol. Soc. Am.*, Memoir 43, pp. 1–53, 1950.

Bateman, P. C.: Personal communication, 1951.

Becker, H.: Die Beziehungen zwischen Felsengebirge und Grossem Becken im westlichen Nordamerika. *Z. Deut. Geol. Ges.* 86:115–120, 1934.

Billingsley, P., and A. Locke: Structure of ore districts in the continental framework. *Trans. AIME* 144:9–64, 1941.

Buwalda, J. P.: Diverse but simultaneous orogeny. *Bull. Geol. Soc. Am.* 63:1322–1323, 1952.

Callaghan, E., and V. P. Gianella: The earthquake of January 30, 1934, at Excelsior Mountains, Nevada. *Bull. Seism. Soc. Am.* 25:161–168, 1935.

Cantwell, T., and A. Orange: Deep resistivity results from the Pacific Northwest. *Trans. Am. Geophys. Un.* 46:71, 1965.

Cantwell, T., P. Nelson, J. Webb, and A. S. Orange: Deep resistivity measurements in the Pacific Northwest. *J. Geophys. Res.* 70:1931–1937, 1965.

Carey, S. W.: A Tectonic Approach to Continental Drift. In *Continental Drift: A Symposium . . . , March, 1956*. Hobart, Geology Dept., University of Tasmania, 1958, pp. 177–355.

Cohee, G. V., et al.: *Tectonic Map of the United States*. Washington, D.C., U.S. Geol. Survey and Am. Assoc. Petroleum Geologists, 1961.

Cook, K. L.: The problem of the mantle-crust mix: Lateral inhomogeneity in the uppermost part of the earth's mantle. *Adv. Geophys.* 9:295–360, 1962.

Cox, A., and R. Doell: Review of paleomagnetism. *Bull. Geol. Soc. Am.* 71:645–768, 1960.

Crittenden, M. D., Jr.: New Data on the Isostatic Deformation of Lake Bonneville. U.S. Geol. Surv., Profess. Paper No. 454–E. Washington, D.C., Government Printing Office, 1963a.

Crittenden, M. D., Jr.: Effective viscosity of the earth derived from isostatic loading of Pleistocene Lake Bonneville. *J. Geophys. Res.* 68:5517–5530, 1963b.

Crowell, J. C.: Displacement along the San Andreas fault, California. *Geol. Soc. Am.*, Special Papers No. 71, 1962.

Danes, Z. F., et al.: Geophysical investigation of the southern Puget Sound area, Washington. *J. Geophys. Res.* 70:5573–5580, 1965.

Dehlinger, P., E. F. Chiburis, and M. M. Colver: Local travel time curves and their geological implications for the Pacific Northwest States. *Bull. Seism. Soc. Am.* 55:587–607, 1965.

Dibblee, T. W., Jr.: Geology of the southeastern margin of the San Joaquin Valley, California. *Calif. Dep. Nat. Resources, Div. Mines Bull.* 171, pp. 23–34, 1955.

Fuller, M. D.: Expression of E-W fractures in magnetic surveys in parts of the U.S.A. *Geophysics* 29:602–622, 1964.

Fuller, R. E.: The Geomorphology and Volcanic Sequence of Steens Mountains in Southeastern Oregon. *Washington Univ. Pub. Geol.*, vol. 3, no. 1, 1931.

Fuller, R. E., and A. C. Waters: The nature and origin of the horst and graben structure of southern Oregon. *J. Geol.* 37:204–238, 1929.

Gilbert, G. K.: Lake Bonneville. U.S. Geol. Surv., Mon. No. 1. Washington, D.C., Government Printing Office, 1890.

Gilliland, W. N.: Possible continental continuation of the Mendocino fracture zone. *Science* 137:685–686, 1962.

Gilluly, J.: Basin Range faulting along the Oquirrh Range, Utah. *Bull. Geol. Soc. Am.* 39:1103–1130, 1928.

Gilluly, J.: The distribution of mountain-making in geologic time. *Bull. Geol. Soc. Am.* 60:561–590, 1949.

Gilluly, J., and O. Gates: Tectonic and igneous geology of the northern Shoshone Range, Nevada, U.S. Geol. Surv., Profess. Paper 465. Washington, D.C., Government Printing Office, 1965.

Hamilton, W.: Origin of the Gulf of California. *Bull. Geol. Soc. Am.* 72:1307–1318, 1961.

Herrin, E., and J. Taggart: Regional variations in P_n velocity and their effect on the location of epicenters. *Bull. Seism. Soc. Am.* 52:1037–1046, 1962.

Hill, M. L., and T. W. Dibblee, Jr.: San Andreas, Garlock, and Big Pine faults, California. *Bull. Geol. Soc. Am.* 64:443–458, 1953.

Hobbs, W. H.: Repeating patterns in the relief and in the structure of the land. *Bull. Geol. Soc. Am.* 22:123–176, 1911.

Hunt, C. B.: Personal communication, 1964.

Hurley, R. J.: New evidence on the northward continuation of the San Andreas fault. *Bull. Geol. Soc. Am.* 71:1894, 1960.

Irwin, W. P.: Geologic reconnaissance of the Northern Coast Ranges and Klamath Mountains, California, with a summary of the mineral resources. *Calif. Dep. Nat. Resources, Div. Mines Bull.* 179, 1960.

Jeffreys, H.: *The Earth*, 3d ed. Cambridge University Press, 1952.

Lawson, A. C., et al.: *The California Earthquake of April 18, 1906.* Washington, D.C., Carnegie Institution, vol. 1, pt. 1, 1908.

Menard, H. W.: Deformation of the northeastern Pacific Basin and the West coast of North America. *Bull. Geol. Soc. Am.* 66:1149–1198, 1955.

Menard, H. W.: The East Pacific Rise. *Science* 132:1737–1746, 1960.

Moody, J. D., and M. J. Hill: Wrench-fault tectonics. *Bull. Geol. Soc. Am.* 67:1207–1246, 1960.

Moore, J. G.: The quartz-diorite line in the western United States. *J. Geol.* 67:198–210, 1959.

Moore, J. G., A. Grantz, and M. C. Blake, Jr.: The quartz-diorite line in northwestern North America. In U.S. Geol. Surv., Profess. Paper 424-C, pp. 87–90. Washington, D.C., Government Printing Office, 1961.

Noble, L. F.: The San Andreas Rift and some other active faults in the desert region of southeastern California. *Carnegie Inst. Washington Yb.* 25:415–428, 1926.

Noble, L. F.: The San Andreas Fault Zone from Soledad Pass to Cajon Pass, California, in Geology of Southern California. In *Calif. Dept. Nat. Resources, Div. Mines Bull.* 170, Chapter IV, pp. 37–48, 1954.

Ogle, B. A.: Major shear zone at False Cape, Humboldt Co., California. *Bull. Geol. Soc. Am.* 63:1341, 1952.

Ogle, B. A.: Geology of Eel River Valley area, Humboldt County, California. *Calif. Dep. Nat. Resources, Div. Mines Bull.* 164, 1953.

Pakiser, L. C.: Personal communication, 1965.

Pakiser, L. C., and J. S. Steinhart: Explosion Seismology in the Western Hemisphere. In *Research in Geophysics*, H. Odishaw, ed. Cambridge, Mass., The M.I.T. Press, 1964, Vol. 2, pp. 123–147.

Raisz, E.: The Olympic-Wallowa lineament. *Am. J. Sci.* 243-A:479–485, 1945.

Raitt, R. W.: Seismic refraction studies of the Pacific Ocean Basin. *Bull. Geol. Soc. Am.* 67:1623–1640, 1956.

Roller, J. C., et al.: A Preliminary Summary of a Seismic-refraction Survey in the Vicinity of the Tonto Forest Observatory, Arizona. *U.S. Geol. Surv. Technical Letter Crustal Studies* 23, 1964.

Shawe, D. R.: Strike-slip control of Basin-Range structure indicated by historical faults in western Nevada. *Bull. Geol. Soc. Am.* 76:1361–1377, 1965.

Shepard, F. P.: Northward continuation of the San Andreas Fault. *Bull. Seism. Soc. Am.* 47:263–266, 1957.

Skehan, J. W.: The Olympic-Wallowa lineament: A major deep-seated tectonic feature of the Pacific Northwest. *Trans. Am. Geophys. Un.* 46:71, 1965.

Slemmons, D. B.: Geological effects of the Dixie Valley-Fairview Peak, Nevada, earthquakes of December 16, 1954. *Bull. Seism. Soc. Am.* 47:353–375, 1957.

Smith, G. I.: Estimate of total displacement on the Garlock fault, southeastern California. *Bull. Geol. Soc. Am.* 71:1979, 1960.

Taubeneck, W. H.: An evaluation of tectonic rotation in the Pacific Northwest. *J. Geophys. Res.* 71:2113–2120, 1966.

Thompson, G. A.: Gravity measurements between Hazen and Austin, Nevada: A study of basin-range structure. *J. Geophys. Res.* 64:217–229, 1959.

Thompson, G. A., and M. Talwani: Geology of the crust and mantle, western United States. *Science* 146:1539–1549, 1964.

Vacquier, V., A. D. Raff, and R. E. Warren: Horizontal displacements in the floor of the northeastern Pacific Ocean. *Bull. Geol. Soc. Am.* 72: 1251–1258, 1961.

Vening-Meinesz, F. A.: Shear patterns of the earth's crust. *Trans. Am. Geophys. Un.* 28:1–61, 1947.

Watkins, N. D.: Paleomagnetism of the Columbia Plateau. *J. Geophys. Res.* 70:1379–1406, 1964.

Wilt, J. W.: Measured movement along the surface trace of an active thrust fault in the Buena Vista Hills, Kern County, California. *Bull. Seism. Soc. Am.* 48:169–176, 1958.

Wise, D. U.: An outrageous hypothesis for the tectonic pattern of the North American cordillera. *Bull. Geol. Soc. Am.* 74:357–362, 1963.

Woollard, G. P., and H. R. Joesting: Bouguer Gravity anomaly map of the United States (exclusive of Alaska and Hawaii). Washington, D.C., U.S. Geol. Surv., 1964.

Zietz, I., and co-workers: Aeromagnetic investigation of crustal structure for a strip across the Western United States. *Bull. Geol. Soc. Am.* 80:1703–1714, 1969.

5

Tectonics and Geophysics of Eastern North America *

PHILIP B. KING

Geologist, U.S. Geological Survey,
Menlo Park, California

The theme of this conference is "the evidence for or against existence, possible causes, and mechanics of fracturing, drifting, and rotating of continents and ocean basins." Let us, therefore, begin at once with this theme, instead of approaching it indirectly.

A reconstruction of the continents on opposite sides of the Atlantic Ocean, prepared by Sir Edward Bullard and his associates (1965) is shown in part in Figure 1. It is a geometric fitting of the opposing shapes of the outer edges of the continental blocks, using numerical methods, and without regard for the inner constitution of the blocks. At first glance, we would be tempted to borrow an adjective so dearly beloved by the physicists, and call this "an elegant reconstruction." We will return to the reconstruction at the end of the article, after reviewing various lines of geological and geophysical evidence, when we will conclude that, whereas some parts of the reconstruction are truly elegant, other parts are not so elegant, and some are downright incongruous.

At this point, I must make a disclaimer; I am neither a whole-hearted "drifter," nor a whole-hearted "antidrifter." Continental drift seems to provide answers for many of the puzzles of world geology and geophysics, but it leaves other puzzles unsolved, or even creates new puzzles. And I would be rash indeed to propose continental drift on the evidence of geology alone (which is my field of competence), when an equal or greater part of the evidence for or against drift is in the field of geophysics (on which I am not prepared to speak).

* Publication authorized by the Director, U.S. Geological Survey.

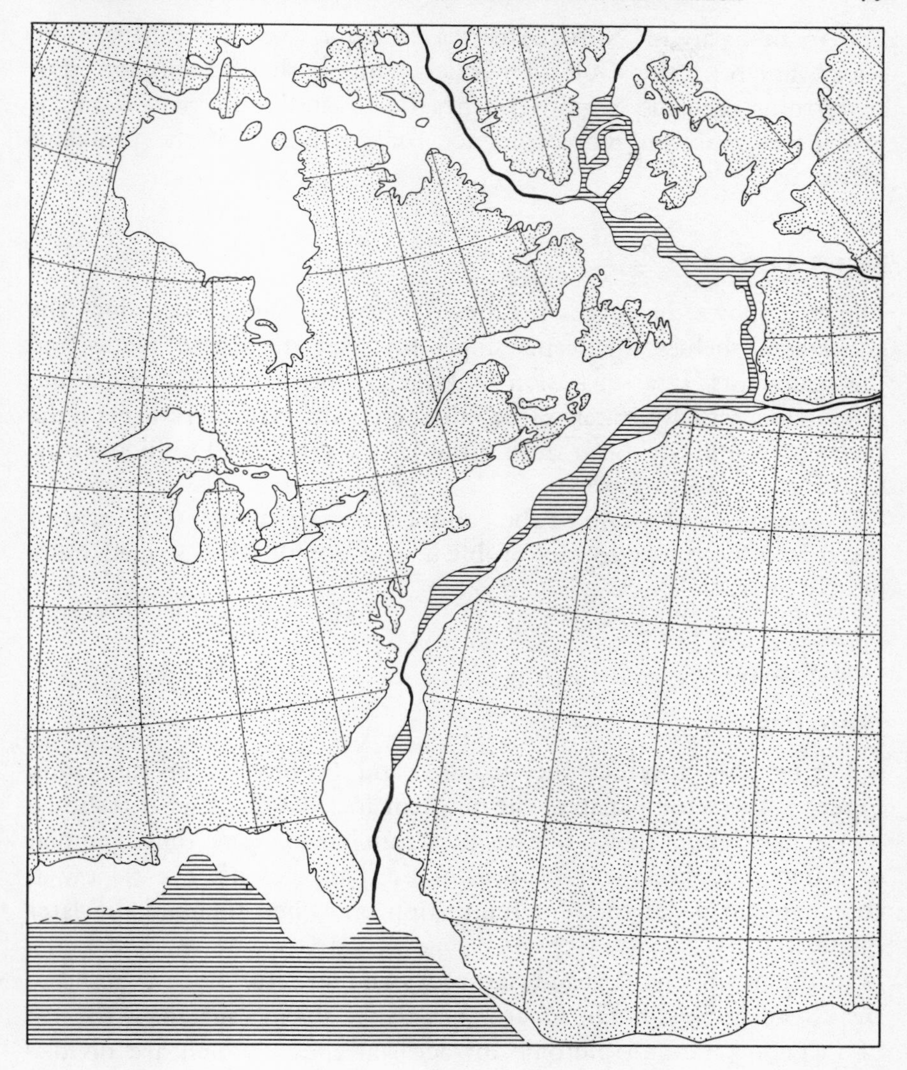

Fig. 1. The fit of the continents across the North Atlantic Ocean, as reconstructed by Bullard and co-workers (1965).

The subject of this conference is "What's New on Earth?", and my discussion deals with "What's new in eastern North America?" It will summarize at least the larger aspects of "what's new" in the geology, tectonics, and geophysics of that region—not all of which will bear directly on problems of drift, fracturing, or rotation. In order to keep the inquiry within workable limits, I have restricted the discussion to that part of North America which is east of the Mississippi River, north of the Gulf of Mexico, and south of Labrador, thus regretfully ignoring regions farther north and south which might provide as much or more evidence for a solution of the problems before us.

Subsea Features

Figure 2 shows the subsea features of the area under discussion: the continental shelves, the deeper slopes and rises, the abyssal plains, and the seamounts. If a separation of North America from Europe and Africa has actually taken place, all these subsea features except the shelves were newly born during the last few hundred million years, i.e., after the drifting began.

The nature and origin of the continental margin within the area of the map is pertinent to our problem, and it will be examined more closely later.

Land Features

TECTONICS

Let us now consider the geology and geophysics of the lands. Figure 3 is a tectonic map of the problem area. Much of what is shown on the map is not "new," but has been familiar for some time. Familiar or unfamiliar, the salient features of the tectonics deserve review at this point, in order to lay a foundation on which to build the later discussion.

In the north, the Precambrian rocks and structures are exposed in the Canadian Shield. The greater part of the shield's surface is formed of metamorphic and plutonic infracrustal rocks, which are divided on the map into a central area and a Grenville belt on the southeast. Smaller areas (marked by stipple) are formed of less disturbed sedimentary or volcanic supracrustal rocks, which are postorogenic to the deformation of the infracrustal rocks of their localities (Stockwell and co-workers, 1965). Southward, these Precambrian rocks pass beneath a platform cover of younger strata; here, the configuration of their upper surface is shown by meter contour lines (Flawn and co-workers, 1967).

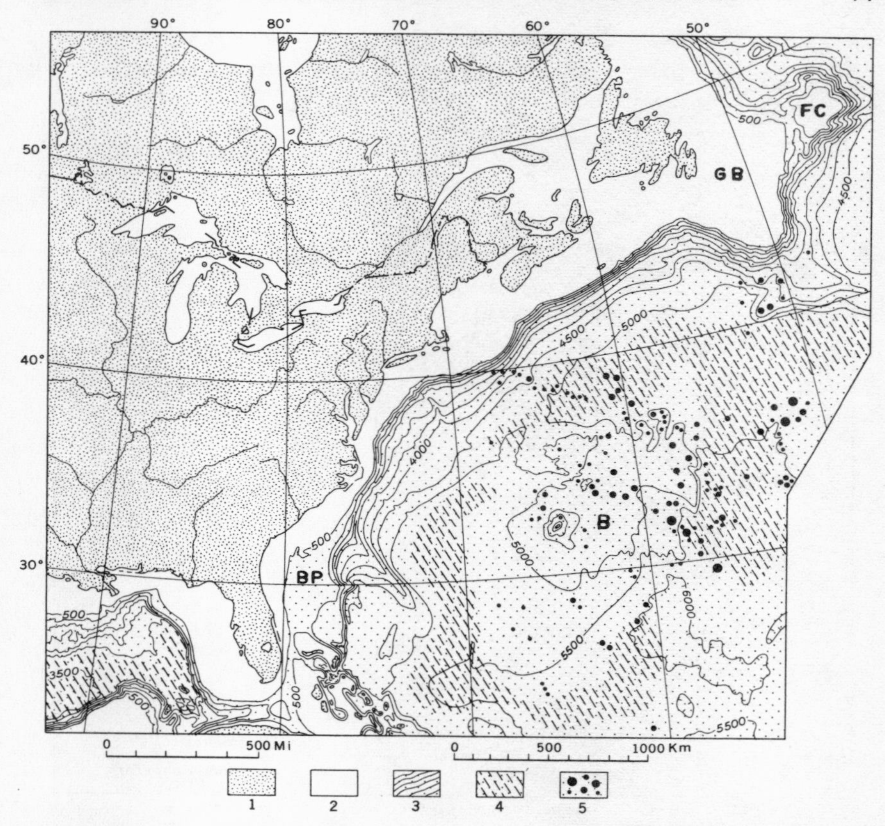

Fig. 2. Map of the region discussed; part of eastern North America and an adjoining part of the North Atlantic Ocean. Subsea features compiled from Uchupi (1965), Pratt (1968), and other sources. *B*, Bermuda; *BP*, Blake Plateau; *FC*, Flemish Cap; *GB*, Grand Banks. 1, land areas; 2, seas shallower than 500 meters (continental shelves); 3, seas deeper than 500 meters (ocean basins), bathymetry being indicated by contours at 500-meter intervals; 4, abyssal plains; 5, seamounts (mostly volcanic cones), diameter of circle indicating basal diameter of seamount.

Along the southeastern side of North America is the Appalachian foldbelt, deformed during the Taconian, Acadian, and Alleghenian orogenies of Paleozoic time. Figure 3 does not show the extent of the areas affected by these different orogenies, as was done by Neale and associates (1961, Fig. 2); instead, a narrow miogeosynclinal belt on the northwest, composed of folded and thrust-faulted sedimentary rocks is differentiated from a broad eugeosynclinal belt on the south-

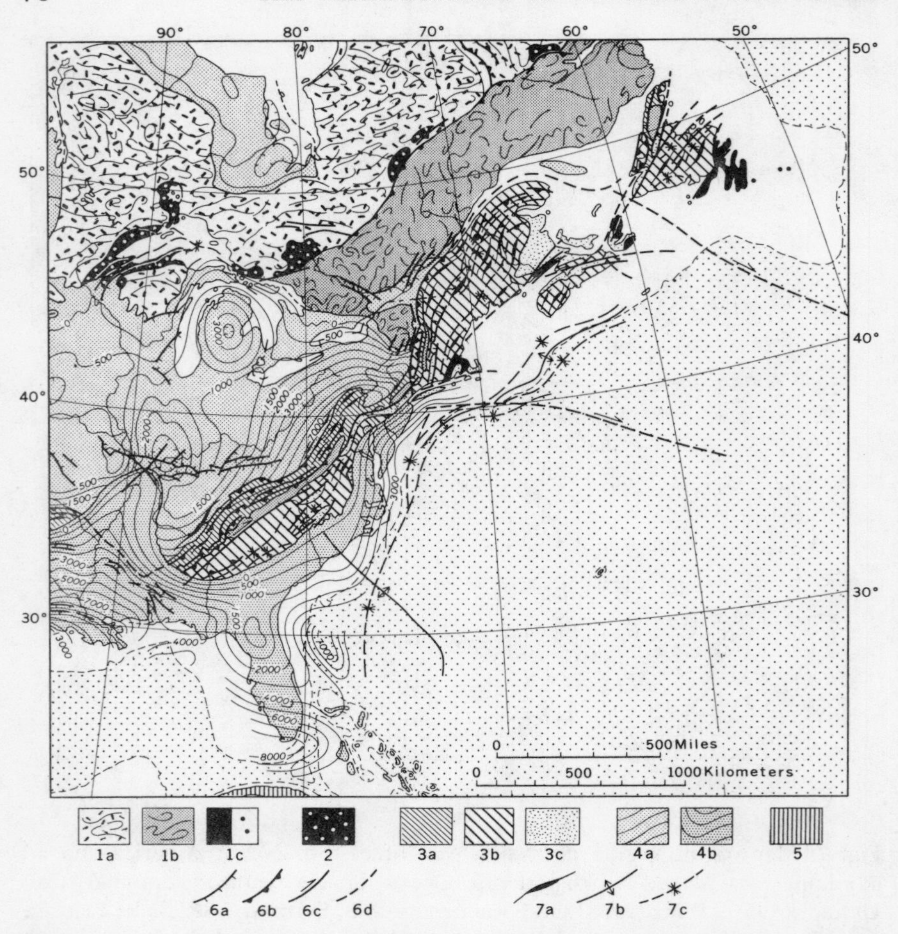

Fig. 3. Map of part of eastern North America, showing tectonics of land area. Generalized from Tectonic Map of North America (King, 1969). 1, Precambrian infracrustal rocks (metamorphic and plutonic), showing trend lines: *a*, Superior and adjacent provinces; *b*, Grenville province and related areas; *c*, Atlantic coastal area, with circles indicating known subsea occurrences. 2, Precambrian supracrustal rocks (sedimentary and volcanic), which are post-orogenic to Precambrian deformations of their localities. 3, rocks of Appalachian foldbelt: *a*, miogeosynclinal rocks; *b*, eugeosynclinal rocks; *c*, post-orogenic deposits (Carboniferous and Triassic). 4, platform areas, with contours on top of basement in depths below sea level at 500-meter intervals above 2,000 meters, and at 1,000-meter intervals below: *a*, on Precambrian basement; *b*, on Paleozoic basement. 5, rocks of Antillean foldbelt (in Cuba). 6, faults: *a*, normal; *b*, thrust; *c*, transcurrent; *d*, subsea. 7, folds; *a*, closely compressed anticline; *b*, anticlinorium; *c*, syncline or synclinorium.

east, composed of deformed and largely metamorphosed sedimentary and volcanic rocks, in which small to large masses of granitic and other plutonic rocks are embedded.

In addition to these two main categories, several less extensive units are shown in the Appalachian foldbelt:

(1) Along the core of the foldbelt between the miogeosynclinal and eugeosynclinal areas, Precambrian infracrustal rocks emerge in long strips, from Newfoundland to North Carolina, forming the Long Range, the Green Mountains, the Blue Ridge, and similar chains; they are uplifted parts of the basement on which the Paleozoic geosynclinal rocks were laid.

(2) Farther southeast near the coast, in both Canada and New England, are other Precambrian rocks, shown by a different pattern because of special features. Some of them (especially in Newfoundland) are unaltered geosynclinal sedimentary and volcanic rocks, but others are metamorphic and plutonic infracrustal rocks. Such Precambrian rocks must extend considerable distances southeastward from the land, beneath the continental shelf. They have been identified in the Grand Banks as far as 160 km offshore from Newfoundland, where they form the surface of shallow parts of the sea floor, that have been swept clear of later sediments (Lilly, 1966).

(3) Separately shown (stippled areas) are less deformed strata which are postorogenic to the climactic Appalachian deformations of their localities. Some of these, of Carboniferous age, form a wide area in New Brunswick and smaller areas elsewhere in Canada and New England. More extensive is the Upper Triassic Newark Group, patches of which occur from Nova Scotia to South Carolina, and even farther in subsurface (McKee and co-workers, 1959). The Carboniferous rocks are locally involved in moderate to steep compressional folding (Fyson, 1967), and the Triassic rocks are tilted, warped, and block-faulted (Sanders, 1963); nevertheless, none of their structures are of the magnitude of those in the rocks beneath.

Northeastward, the Appalachian foldbelt runs out to sea along the coast of Newfoundland, without diminution in strength of deformation, and surely must continue to the edge of the continental shelf, 100 km or so beyond. Southwestward, it plunges beneath the younger strata of the Gulf Coastal Plain in Alabama, again without diminution in strength of deformation; the Ouachita foldbelt which emerges from the Gulf Coastal Plain west of the Mississippi River probably lies on its continuation (Flawn and co-workers, 1961). In the Atlantic and Gulf Coastal Plains, the rocks of the Appalachian and Ouachita foldbelts are masked by a platform cover of Mesozoic and Tertiary strata. The upper surface of the two foldbelts slopes gently toward the Atlantic Ocean and Gulf of Mexico (as shown by meter contours); along parts of the Gulf Coast the surface lies at depths of more than 13,000 meters below sea level.

At the very bottom of the map, in Cuba, is a fragment of still another foldbelt, that of the Antilles, which is more extensive south of the map area.

Second-order tectonic features—the folds, faults, and trend lines—are shown on the map by symbols. Most of these are integral parts of the foldbelts in which they lie, but some have no such obvious relation.

ISOTOPIC AGES

The time relations of the tectonic features, mentioned in passing during the preceding discussion, become clearer on Figure 4, which shows the isotopic age provinces in the foldbelts.

Many such maps have been made, covering all or parts of North America, e.g., Gastil (1960); Engel (1963); Stockwell (1964); Goldich and co-workers (1966b). Figure 4 embodies all the published data known to me, with some interpretations of my own. Locations of isotopic age determinations are indicated by spot points, although in places so many have been made that they cannot all be marked on the map; the different ages are grouped into seven categories, ranging from 2,300 to 2,500 million years to 200 to 350 million years. These determinations are, in turn, generalized into provinces, the outcrop areas of which are shown by different patterns. Where the provinces are concealed by platform cover, their subsurface extent is suggested by age determinations that have been made on the basement rocks.

The validity of these provinces is attested by the fact that in each of them the determined ages tend to cluster within distinctive ranges. While by far the greater number of ages in each province fall within these clusters, a few ages lie outside. These are "exceptions that prove the rule," and result from relic of earlier events that were not quite overwhelmed by the dominant event of the province, or from minor superposed subsequent events.

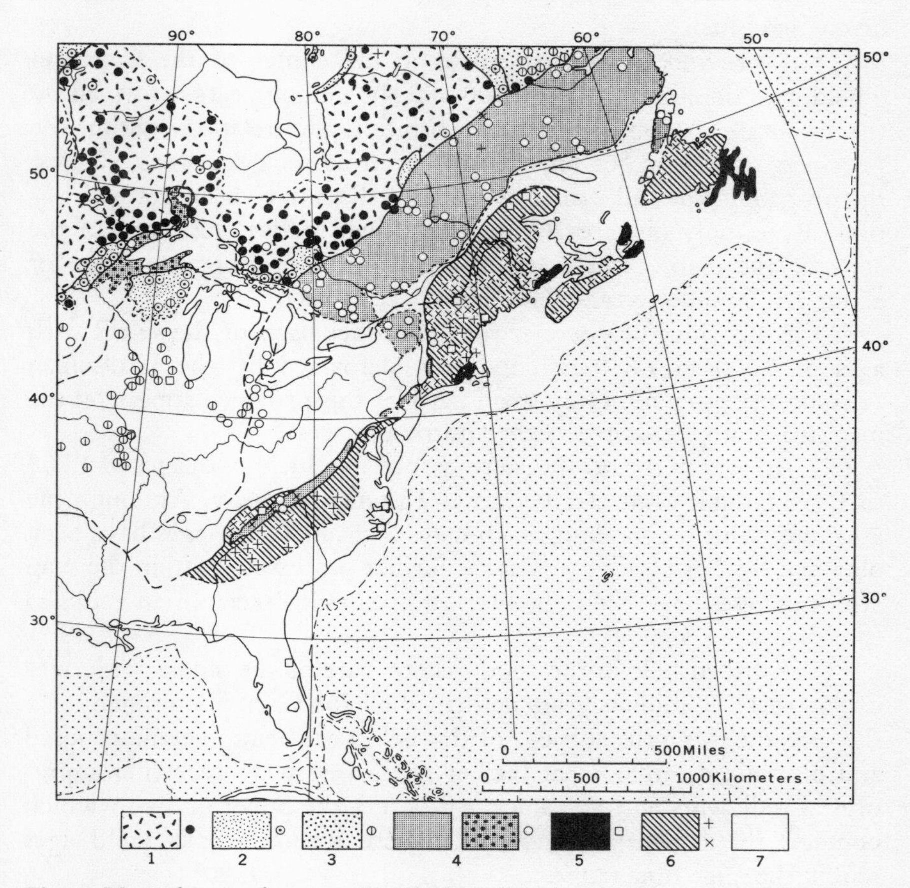

Fig. 4. Map of part of eastern North America, showing isotopic age provinces in the foldbelts. Compiled from many sources, in particular the Age Determinations and Geological Studies papers of the Geological Survey of Canada (1960 through 1966), tabulations by the Committee on Nuclear Science of the National Academy of Science–National Research Council (1965), and Lidiak et al. (1966). Ages in millions of years. 1, 2,300 to 2,500; 2, 1,600 to 1,850; 3, 1,200 to 1,450; 4, 800 to 1,100 (infracrustal rocks shaded, supracrustal ones stippled); 5, 500 to 800; 6, Paleozoic age provinces undifferentiated, with symbols indicating ages of 350 to 500 and 200 to 350; 7, areas where foldbelts are concealed by platform cover, with spot points in these areas indicating where age determinations have been made on basement rocks penetrated in drill holes.

On the basis of the isotopic age determinations, the following can be distinguished:

(1) A province of ancient rocks in the center of the Canadian Shield with dominant ages of 2,300 to 2,500 million years deformed by the Kenoran (Algoman) orogeny. Stratigraphic relations within the province, such as major unconformities in its rock sequence, suggest the probability of still earlier events, but these have been verified isotopically in only one area (Minnesota River Valley, just west of the edge of the map), where several determinations suggest an age of about 3,550 million years (Goldich et al., 1966a).

(2) A province of younger rocks in other parts of the shield with ages of 1,600 to 1,850 million years deformed by the Hudsonian (Penokean) orogeny. Only small parts of this province appear on the map; it is more extensive farther north and west.

(3) An area in the northeastern part of the shield with ages of 1,200 to 1,450 million years ascribed to an Elsonian orogeny. Although the area in the shield containing such dates is small, similar ages have been obtained from rocks in many areas farther southwest; within the map area such ages have been obtained from buried Precambrian rocks in the Mississippi Valley.

(4) A broad belt in the southeastern part of the shield with ages of 800 to 1,100 million years deformed by the Grenville orogeny. At the time the infracrustal rocks of this belt were being metamorphosed and plutonized, other areas farther northwest were receiving supracrustal sediments and volcanics, e.g., the Lake Superior Basin which contains the Keweenawan Series, where igneous rocks yield ages within the same time range.

Metamorphic and plutonic rocks with ages like those in the Grenville belt extend southeastward and southwestward. Within the map area, they are exposed in the core of the Appalachian foldbelt, from Newfoundland to North Carolina. They are also known in subsurface west of the Appalachians in Ohio, West Virginia, and Tennessee. Southwest of the map area, metamorphic and plutonic rocks have yielded similar ages through Texas (Muehlberger and co-workers, 1966), and into southern Mexico (Fries, 1962).

Clearly, a wide and lengthy belt of eastern North America underwent a significant orogenic event between 800 to 1,100 million years ago that was not shared by the remainder of the continent.

(5) Between the category of Grenville dates and the categories of

Paleozoic dates, I have created a category of dates between 500 and 800 million years which bridges late Precambrian and early Paleozoic time. Within this time range, there are only a small number of age determinations in eastern North America, but they are widely distributed. For the most part, the dates are not significant, representing such features as stray dikes, or genuine earlier dates that have been downgraded by later events (Long and co-workers, 1959). In general, the conclusion of Bickford and Wetherill (1965) is valid that "at least for this continent the interval between the end of the Grenville orogeny (900–1,000 m.y. ago), and the earliest Appalachian events about 500 m.y. ago was one of remarkable igneous and metamorphic quiescence."

Nevertheless, I am becoming persuaded that a few of these dates, southeast of the Grenville belt and nearer the Atlantic Coast, express a genuine event, and I have accordingly marked the Precambrian rocks in which they occur as a separate isotopic age province. The event, or possible orogeny, in this province has appropriately been termed the Avalonian (Lilly, 1966, p. 572; Rodgers, 1967, pp. 409–410).

The reality of the Avalonian event is most convincingly displayed in the Avalon Peninsula of southeastern Newfoundland, where the Holyrood Granite and the adjacent Harbour Main Volcanics have yielded ages of 537 to 580 m.y. (McCartney and co-workers, 1966, p. 947). The Holyrood Granite itself is overstepped by fossiliferous Lower Cambrian, but in surrounding areas a very thick late Precambrian sequence intervenes (Williams, 1964, p. 947). Similar ages have been obtained by the Geological Survey of Canada from basement rocks in Cape Breton Island and southern New Brunswick (C. H. Stockwell, written communication, 1966), although some of the rocks from which they were obtained have been traditionally classed as "Archean."

The Avalonian event seems also to be represented in the crystalline rocks of Rhode Island and nearby Massachusetts, where the Dedham Granodiorite, which underlies Lower Cambrian, has yielded an age of 560 m.y. (Isachsen, 1964, pp. 815–816). Indications of the event can be found farther south, in buried crystalline rocks close to the Atlantic Coast. Granite gneiss penetrated by several wells near Albemarle and Pamlico Sounds, North Carolina, has been dated as 585 to 590 m.y. (Socony Mobil, Nos. 1 and 3 State of North Carolina; Denison and others, 1967). The gneiss may be a basement of the adjoining terrane, as other wells encountered schist, rhyolite, etc., which have yielded younger dates. More speculatively, the little-known buried crystalline rocks of south-central Florida might include an Avalonian basement; a microdiorite, possibly a late intrusive, has been dated at 480 m.y. (Sub Oil Co., No. 1 Powell Land Co.; Muehlberger and co-workers, 1966, p. 5419).

Both the Grenville and Avalonian provinces, their metamorphic and plutonic rocks, and the events which created them, have interesting implications in any synthesis of the growth of the North American continent, or of its former trans-oceanic connections. The Avalonian event has much more extensive counterparts in other continents. It corresponds closely to the Pan-African thermo-tectonic episode of Kennedy (1964), which produced orogenic deformation of geosynclinal sediments in some parts of Africa, and a reactivation of the earlier basement in many others; in southern Africa it includes a Damaran episode at 450 to 570 m.y., and a Katangan episode at 580 to 680 m.y. (Clifford, 1967, pp. 47–55). A similar episode can be recognized in the crystalline rocks of the Brazilian Shield near the Atlantic Coast of South America, according to recent determinations of P. M. Hurley and others.

(6) On the map (Fig. 4), I have given only cursory treatment to the complex array of Paleozoic dates in the Appalachian belt, grouping them merely into one set with a range of 350 to 500 million years (probably reflecting the Taconian and Acadian orogenies), and another set with a range of 200 to 300 million years (reflecting the Alleghenian and other late Paleozoic orogenies); moreover, both sets of dates have been placed in a single province. On the scale of the map, it has not been feasible to make a more detailed separation; besides, the 300-million year age span of the whole is short compared with the age ranges in the Precambrian foldbelts which have just been discussed.

EARTHQUAKE EPICENTERS

Skipping 200 million years or so of geologic time, let us examine the modern tectonic state of the region, which we can demonstrate by the pattern of the earthquake epicenters that have been recorded during the last 300 years (Fig. 5).

In such a representation, various human factors, such as the clustering of epicenters around Boston and New York, which for a long time have been centers of population, of learning, and of perceptive observation will need to be discounted. Similarly, the sparsity of epicenters in the Canadian Shield and in the Atlantic Ocean, where there are few or no people to make observations, might be regarded with suspicion. But even if we allow for these sources of error, generaliza-

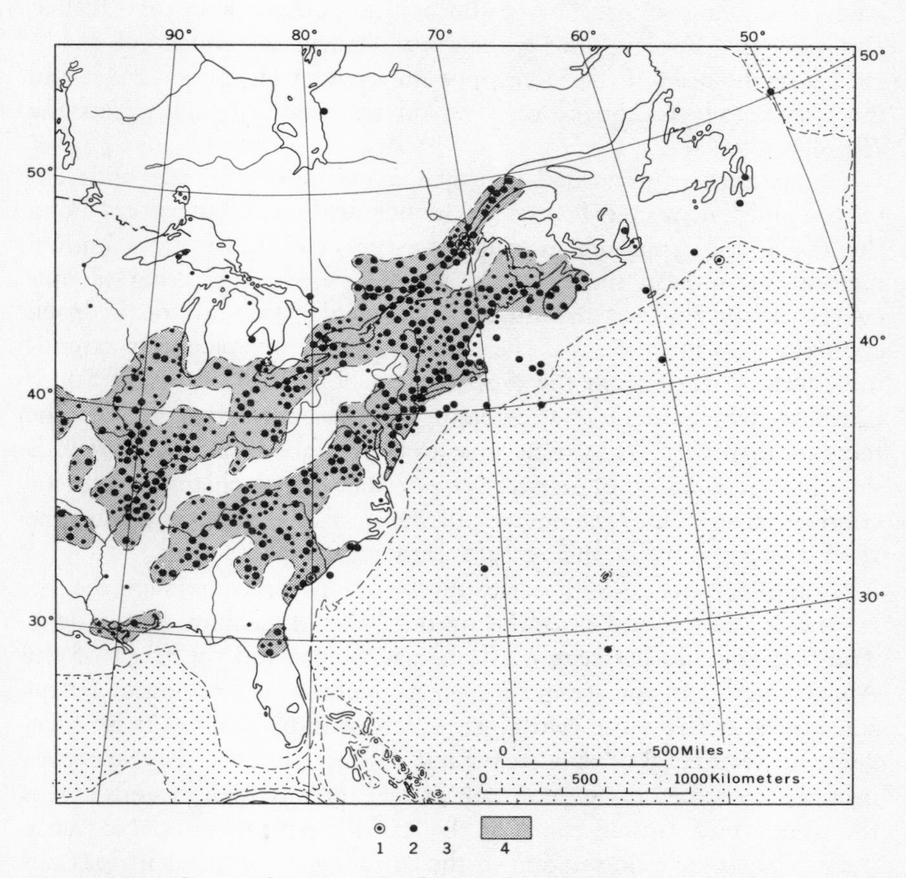

Fig. 5. Map of part of eastern North America, showing locations of recorded earthquake epicenters from 1600 A.D. to the present. Compiled from Woollard (1958), Smith (1962), and Eppley (1965). 1, major earthquake; 2, moderate to strong earthquake; 3, minor earthquake; 4, area of moderate to high earthquake intensity.

tions can be made about relative concentrations of earthquake frequency (Fig. 5, shaded areas).

Certainly, most of this region is not notably unstable, at least as compared with the western part of North America. The largest number of the earthquakes whose epicenters are plotted were rather minor; they rattled dishes on the shelves, or toppled a few gravestones. Nevertheless, the region does contain the epicenters of some of the strongest shocks recorded in North America: the Trois-Rivières earthquake of 1663 in the St. Lawrence Valley of Quebec (Smith, 1962; Eppley, 1965); the New Madrid earthquake of 1811–1812 near the head of the Mississippi embayment (Eppley, 1965); and the Charleston earthquake of 1886 on the coast of South Carolina (Eppley, 1965).

The pattern of recorded earthquake epicenters reveals both expected and unexpected features. A concentration of epicenters along the axis of the Appalachians might be expected, and follows known tectonic features in the bedrock. Even here, the clustering of epicenters along the axis has no apparent explanation in the bedrock geology. Also mysterious is the transverse spur of epicenters extending southeastward from the Appalachian belt to the coast of South Carolina, which includes the major Charleston earthquake. On the tectonic map (Fig. 3), a transverse arch with about the same trend is shown to extend southeastward from the Appalachians across the continental shelf into the ocean, but the spur of epicenters lies on the southwestern flank of this feature, rather than on its crest.

Even more mysterious is the clustering of epicenters in the St. Lawrence Valley and near the head of the Mississippi embayment. The first area lies between the Canadian Shield and the front of the Appalachian belt, the second is near a much faulted region of Paleozoic rocks. Perhaps the two clusters can be explained by these local geologic features, but it is surprising to discover that they are connected by a somewhat weaker belt of epicenters which extends across the nearly undisturbed rocks of the middle eastern states. Not only does this belt have no relation to the surface geology, but it does not seem to be related to any other geophysical features (Figs. 6, 7, 8). The belt of epicenters that extends from the St. Lawrence Valley to the Mississippi embayment is broadly parallel to the belt of epicenters which follows the axis of the Appalachians, but between them is a belt with a much lower incidence of earthquake frequency.

GRAVITY ANOMALIES

We will now consider the available geophysical data for the region. Figure 6 shows Bouguer gravity anomalies, contoured at 50-milligal intervals. The most prominent feature on the map—positive gravity anomalies in the oceanic area, and mainly negative gravity anomalies in the continental area, is to be anticipated. Unfortunately, the small scale of the map requires a large contour interval, which eliminates much of the detail of gravity variations which has been worked out in the continental area (for further data, see the gravity

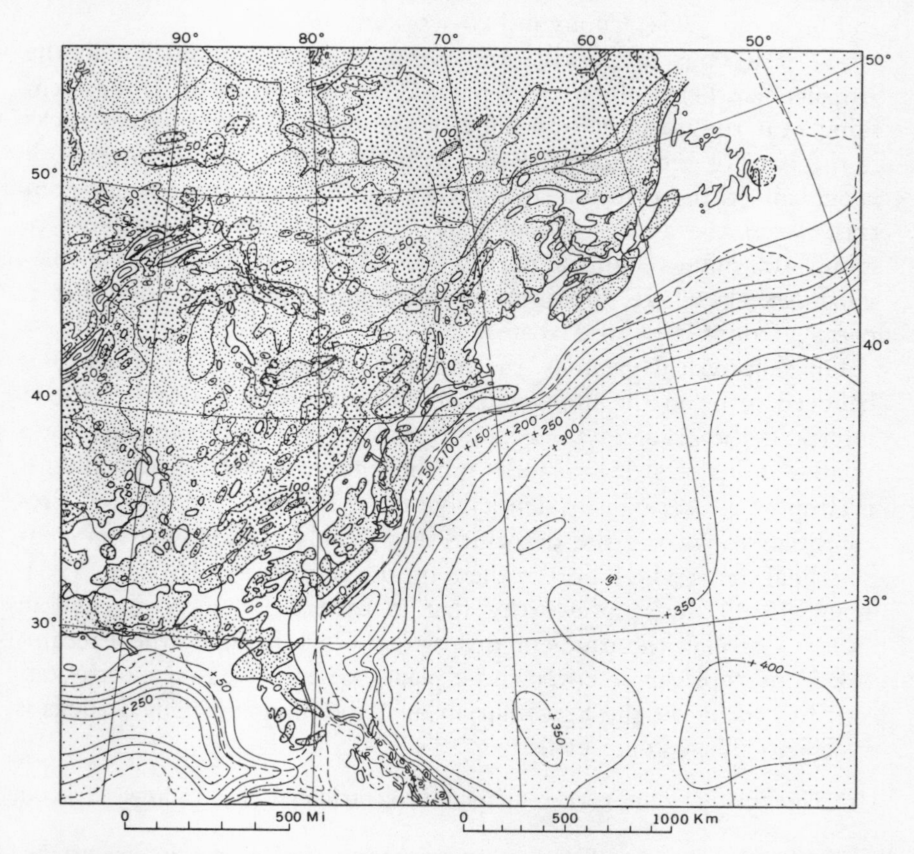

Fig. 6. Map of part of eastern North America, showing Bouguer gravity anomalies. Compiled from Canada Dominion Observatories (1957), Woollard and Monges Caldera (1956), Woollard and Joesting (1964), and Worzel (1965). Contour interval, 50 milligals; areas with negative gravity anomalies, shaded in increasingly deeper tints at 50 milligal intervals.

maps of the United States and Canada: Woollard and Joesting, 1964; Canada Dominion Observatories, 1957). In eastern North America, surface relief is moderate, and gravity anomalies largely reflect underlying tectonic and crustal features, thus differing from western North America, where the anomalies nearly mirror the high mountain relief. The contours on the map, despite their large interval, bring out several major gravity features:

(1) The belts of strong positive and negative gravity anomalies near and south of Lake Superior, which have an obvious relation to the rocks and structures of the underlying basement (Thiel, 1956; Zietz and Griscom, 1964; Coons and co-workers, 1967).

(2) The belts of positive and negative gravity anomalies in the Appalachian belt, which are separated by an abrupt linear gradient lying near the northwestern edge of the uplifted Precambrian rocks in the core of the foldbelt. Even with the large contour interval, it is evident on the map that in North Carolina there is a gravity contrast across the gradient from plus 50 milligals on the southeast to minus 100 milligals on the northwest. The gravity contrast is somewhat less elsewhere, but more detailed contour maps show that it persists from Gaspé to Alabama.

On geophysical grounds, the top of the source of the anomaly is believed to be in the upper part of the crust, and Griscom (1963) interprets the gradient as "a major break in the crust, representing a relative uplift of at least 4 miles on the southeast side." The geologist is tempted to ascribe it to density contrasts between the uplifted Precambrian rocks and the deformed and deeply downfolded miogeosynclinal rocks which border them on the northwest (Isachsen, 1964; King, 1964). It is noteworthy that in eastern Pennsylvania and in North Carolina the gradient drops to the southeast of the Precambrian outcrops–in precisely those places where extensive northwestward thrusting of the Precambrian over the miogeosynclinal rocks is indicated by geologic evidence.*

* The following dissenting opinion is contributed by Andrew Griscom (1967):

The density contrast between the miogeosynclinal Paleozoic sedimentary rocks and the exposed Precambrian rocks is surely insufficient to explain the gravity anomaly. Bromery finds little or no density contrast in the Reading Prong area of Pennsylvania between the Cambro-Ordovician carbonate and shale sequence and the Precambrian rocks (little or no means less than 0.1

(3) A small replica of the negative gravity anomaly of the Appalachians in the Ouachita foldbelt west of the Mississippi River (at west edge, Fig. 6), with an abrupt gravity gradient south of it. Curiously, the belt of negative gravity anomalies virtually disappears in the intervening area, where the foldbelts are concealed beneath Coastal Plain strata, although the presence of the foldbelts beneath these strata is indicated by drill data.

(4) A linear negative gravity anomaly in the eastern part of the Canadian Shield, along the northwestern border of the Grenville province. I do not know whether this feature is comparable to the negative gravity belts in the Appalachians and Ouachita areas or has originated from other causes.

MAGNETIC ANOMALIES

As one of the geophysical exhibits, it would be desirable to include a map showing magnetic anomalies comparable to the gravity map. However, although maps showing magnetic anomalies are available for many parts of the problem area, for example, E. R. King, 1959; Zietz et al., 1966; Gough, 1967, a map of the entire region or its larger parts has not been published. Nevertheless, significant magnetic features of the coastal and offshore area have been sketched on the basis of reconnaissance surveys (King, E. R., and co-workers, 1961; Drake and others, 1964; Heirtzler and Hayes, 1967; Taylor and co-workers, 1968), and are copied on Figure 7.

The figure shows that in the coastal and offshore areas anomaly belts extend northeastward for long distances, not only parallel to each other, but also parallel to the edge of the continental shelf, the coastline, and the Appalachian foldbelt farther inland. In the north, where many seismic surveys have been made in the offshore area, the anomaly belts are known to coincide with ridges and troughs in the basement rocks and their sedimentary cover (Drake and co-workers, 1963). The northeast-trending anomaly belts are conspicuously

---

gm/cc). For comparison, it takes approximately 25,000 feet of sedimentary rocks with a density contrast of 0.25 gm/cc relative to crystalline rocks in order to generate a gravity low of about 80 milligals. If you halve the density contrast to 0.12 gm/cc, you must double the rock thickness in order to generate the same gravity anomaly. This is the reasoning behind my rejection of your geologic temptations, although I cannot be completely certain until the maximum thicknesses and densities of the sedimentary rocks are better known.

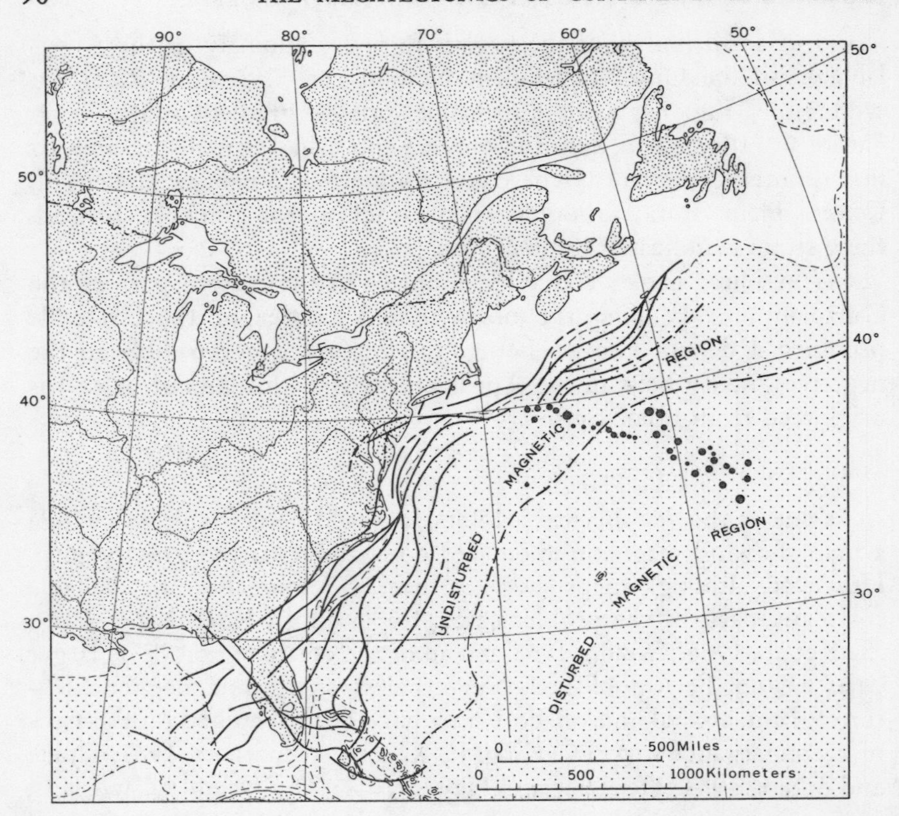

Fig. 7. Map of part of eastern North America, showing magnetic anomalies. Solid lines indicate high-intensity magnetic trends (after Drake and others, 1963, Fig. 4; Gough, 1967, Fig. 2). Solid circles are seamounts, many of which show magnetic anomalies. Dashed line is boundary between undisturbed and disturbed magnetic regions (after E. R. King and others, 1961; Heirtzler and Hayes, 1967).

interrupted in two places by transverse belts: one near the 40th parallel, the other in southwestern Florida.

The transverse anomaly belt near the 40th parallel is traceable 500 km offshore, past the edge of the continental shelf. A chain of seamounts, themselves showing magnetic anomalies (the Kelvin or New England seamounts), prolong the same trend farther to more than 1,500 km offshore (Northrop and co-workers, 1962). Individual northeast-trending anomaly belts at sea seem to be correlative across the transverse belt, but both they and the rock structures which they follow are deflected right-laterally 150 km.

The transverse anomaly belt in southwestern Florida crosses the northeast-trending anomaly belts which characterize most of the peninsula (E. R. King, 1959). The latter occur in a part of the peninsula where the pre-Mesozoic basement lies at relatively shallow depths, and the transverse belt nearly coincides with a zone of abrupt downwarping and thickening of the Mesozoic cover, southwest of which the basement has not been reached by drilling. Southwest of the transverse belt in the Gulf of Mexico, northeast-trending anomaly belts reappear (Fig. 7) (Heirtzler and co-workers, 1966; Gough, 1967). A northeast-trending band of highly irregular anomalies, suggestive of a volcanic province in the basement, is recognizable on both sides of the transverse belt, and shows no apparent lateral offset. Published magnetic surveys are not available to indicate whether the transverse belt continues northwestward from Florida into the Gulf Coast States.

Seaward from the belt of linear magnetic anomalies is a belt about 500 km broad that is nearly featureless magnetically (undisturbed magnetic region of Fig. 7), whose outer edge is near the base of the continental rise. From here eastward, the ocean floor has many magnetic anomalies (disturbed magnetic region of Fig. 7), and so continues beyond the edge of the figure, and thence across the Mid-Atlantic Ridge to another undisturbed magnetic region of about the same width along the European and African coasts (Heirtzler and Hayes, 1967). The magnetically disturbed region contains linear belts of positive and negative anomalies parallel to the Mid-Atlantic Ridge. Vine and Matthews (1963) interpret these and similar linear anomaly belts elsewhere as located over bodies magnetized in the direction of the earth's present field, and in the direction of a reversed magnetic field, and propose that they were formed during later geologic time by spreading of the sea floor away from the mid-ocean ridge. Heirtzler and Hayes (1967, p. 187) suggest, further, that the sea floor in the undisturbed magnetic regions on the two sides of the North Atlantic Ocean dates from late Paleozoic time, when no magnetic reversals occurred for about 50 m.y.

CRUSTAL THICKNESS

A final geophysical item is the thickness of the crust, for which available data are shown in Figure 8. Determinations of the thickness of the crust, in kilometers, which have been obtained from deep crus-

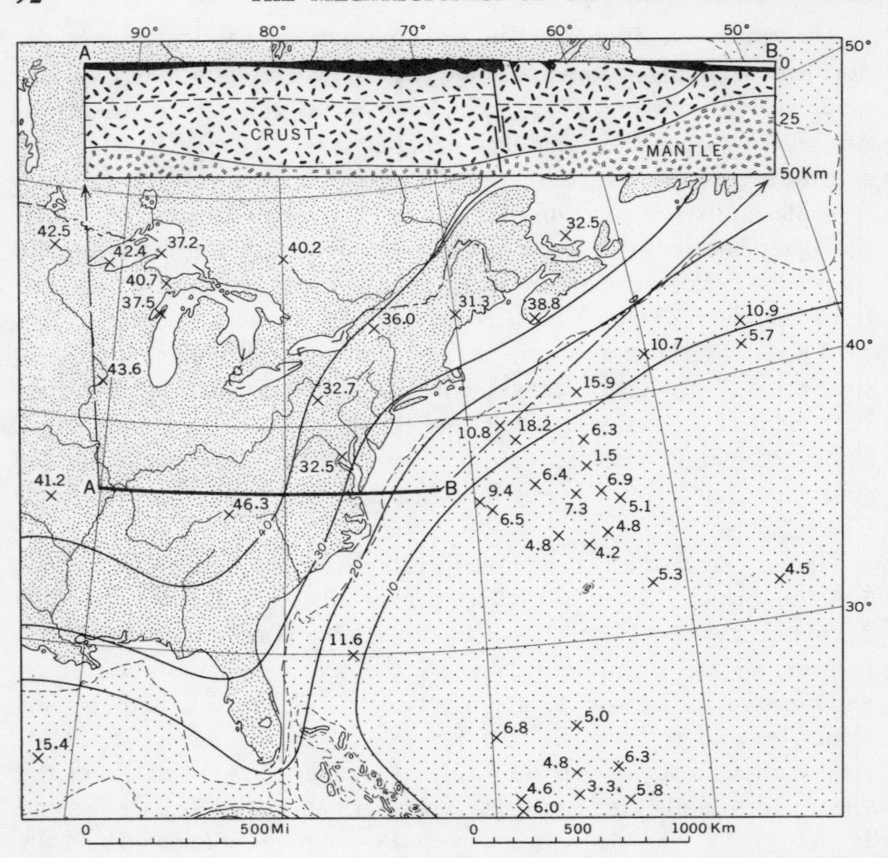

Fig. 8. Map of part of eastern North America, showing crustal thickness. Thicknesses, as determined by deep crustal seismic profiles, are in kilometers, and are summarized by contours at 10-km intervals. Crustal thicknesses along the 37th parallel are shown in cross section *AB* at the top; vertical scale, ×5. Thickness data from Steinhart and Meyer (1961) and McConnell and coworkers (1966); cross section from Hamilton and Pakiser (1965), redrawn with different vertical scale.

tal seismic refraction profiles, are spotted widely on the map. To bring these determinations into focus, I have sketched approximate isopach contours on the map at 10-km intervals. These data emphasize that the crust thins from more than 40 km in the continental interior to much less than 10 km in the oceanic area.

Further refinement of the contours is not feasible from the available data. It is true that more elaborate maps purporting to show crustal thickness have been prepared, based on gravity data and other evidence (for example, Steinhart and Meyer, 1961, Figs. 10.11–10.13), but these are largely hypothetical and have not been verified by instrumental determinations (Pakiser and Zietz, 1965). Most of these maps show a marked thickening of the crust to more than 40 km beneath the Appalachian foldbelt, but this remains to be confirmed. If any thickening occurs, it is more likely to be moderate, as illustrated in the cross section along the 37th parallel shown at top of Figure 8 (Hamilton and Pakiser, 1965; Pakiser and Zietz, 1965).

These moderate variations in crustal thickness beneath the Appalachian foldbelt and in other parts of eastern North America contrast greatly with the crustal thicknesses determined by seismic refraction profiles in the Cordilleran foldbelt of western North America, where there are abrupt variations from less than 20 km to more than 50 km from one tectonic unit and the next. Orogeny in the Cordilleran foldbelt reached its climax in the last half of Mesozoic time, with activity continuing to the present; the strong relief of the base of the crust beneath the foldbelt is the product of these orogenies. In the Appalachian foldbelt, orogenies occurred much earlier, and this foldbelt has been nearly stable during the last 200 million years.

During and shortly after the orogenic climax in the Appalachian foldbelt, the crust beneath it may well have had a thickness as variable as that beneath the Cordilleran foldbelt; if so, it has since been smoothed by one or more subterranean processes. According to Pakiser and Zietz (1965): "The crust and upper mantle [of this region] may be thought of as mature. Much mafic material has already been added to the crust by intrusion and volcanism, and much of the silicic upper crust has already been removed by erosion and transported to the oceans."

LINEAR FEATURES

One of the topics of this conference is "possible fracturing and rotation of continents and ocean basins." The lengthy transverse and longitudinal faults mentioned earlier are pertinent to this topic.

Drawing great linear and more or less hypothetical faults across a map is a favorite pastime of many geologists. In the Appalachian region, one of the first practitioners of the sport was Hobbs (1904, especially Plate 45) nearly 70 years ago, but it is being carried on today by his successors. I myself am as addicted to the sport as any-

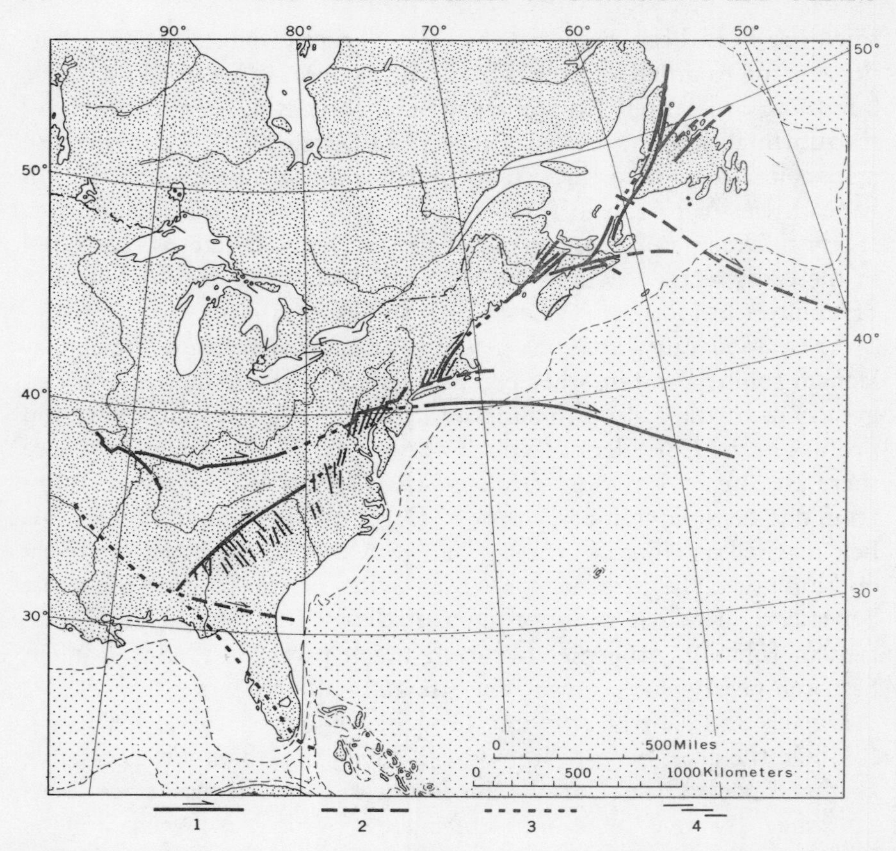

Fig. 9. Map of part of eastern North America, showing known or inferred linear features of great length. Many of these are faults, but their known manifestations are of various ages and senses of displacement. Also shown are dikes of Triassic age in the crystalline rocks of the Appalachian foldbelt. Linear features compiled from many sources; dikes after King (1961). Linear features or faults: 1, known; 2, inferred; 3, suspected; arrows show reported or suspected directions of strike-slip displacement; 4, dikes of Triassic age.

one, and in Figure 9 I have represented the lines which seem most plausible to me. Lines have been drawn in other places and directions by other geologists. I reject many of these lines; these other geologists should feel as free to reject mine. Most of the lines in the figure also appear on the tectonic map (Fig. 3), but a few others have been added from subsurface data.

One of the most troublesome problems of line drawing is the age relations of the features. Often, the lines "put too many eggs in one basket," connecting features whose ages might be anywhere from Precambrian to Cenozoic, depending on locality. The dedicated drawer of lines dismisses the problem by saying that the lines represent deep, fundamental crustal fractures, which might be manifested in the upper crust or at the surface by faults of different ages or different kinds, or even by no break at all. But this dismissal supplies fuel to the skeptic, who can object that lines so evanescent must exist as deep, through-going fractures solely in the minds of their proponents.

Equally troublesome is the problem of the sense of displacement of the features, provided they are actually faults. It is easy to assume that all or many such lengthy faults are transcurrent, but the amount, and even the existence, of strike-slip displacement on any fault is difficult to prove, as shown by the lively debate which still attends so large and obvious a feature as the San Andreas fault of California. On many of the faults in Figure 9 only dip-slip displacement can be proved, and some geologists are convinced that this is the only displacement that has occurred. At least, we can say that some indication for strike-slip displacement exists along many of the lines shown on the map.

The map shows two longitudinal zones: the Cabot fault zone extending from Newfoundland to New England, and the Brevard fault extending from North Carolina to Alabama. It also shows three transverse zones: one south of Newfoundland and the Grand Banks, another near the 40th parallel, and a third in Florida and adjacent states.

The Cabot fault was interpreted by Wilson (1962) as a continuous feature extending from Newfoundland, through Nova Scotia and New Brunswick, into southeastern New England, formed during Late Devonian and Carboniferous time, and probably with left-lateral displacement. It is true that an unusual concentration of high-angle faults occur in the areas mentioned, but the faults form a zone which in places is as much as 100 km broad. Northeast-trending parallel and branching faults are present within the zone, crossed by others in different directions. A connection between the faults in Newfoundland and those on the mainland seems well established from geophysical surveys (Cameron, 1966), but geophysical surveys in the Gulf of Maine have failed to reveal indications of faulting (Tagg and Uchupi, 1966). Although most of the faults are probably of late Paleozoic age, many of those near the Bay of Fundy displace Triassic rocks. The faults have been interpreted as high-angle thrusts with little strike-slip displacement (Cameron, 1966), but evidence for left-lateral displacement has been presented for some faults in southeastern New Brunswick (Webb, 1963); elsewhere there are indications of small to large right-lateral displacements. The Cabot fault zone thus includes faults of diverse trends and ages, whose gross sense of displacement remains to be established; it seems unlikely that the zone includes any single, through-going fracture.

The Brevard fault, unlike the Cabot fault zone, is a single, continuous fracture throughout its known extent (Reed and co-workers, 1961; King, 1964). Juxtaposition of contrasting formations across the fault cataclastic structures in the fault zone, and other evidence indicate that it is a transcurrent fault, probably with as much as 135 miles of right-lateral displacement (Reed and Bryant, 1964). Apparently it formed in late Paleozoic or early Mesozoic time, after the climactic deformation of the crystalline rocks which it transverses. The fault has not been mapped northeast of North Carolina, and if it continues into Virginia its surface indications are not as prominent as they are farther southwest.

A transverse fault south of Newfoundland and the Grand Banks is sketched in Figure 9 to explain the marked eastward offset of the edge of the continental shelf in the Grand Banks, and a similar apparent offset of pre-Carboniferous structures in Newfoundland with respect to the mainland, both of which suggest a large right-lateral strike-slip displacement (King, 1951; Drake and Woodward, 1963). As both the Carboniferous formations and the Cabot fault zone are seemingly continuous between Newfoundland and the mainland, any major transverse displacement could only have been earlier. Seismic refraction surveys south of Newfoundland indicate a fault in about the position sketched, probably downthrown to the south in post-Paleozoic time (Press and Beckmann, 1954). However, the proposal that there was a displacement on the fault at the time of the Grand Banks earthquake of 1929 is erroneous; breaking of cables after this earthquake is now known to have been caused by a turbidity current (Heezen and Ewing, 1952).

The existence at sea of a transverse fault zone near the Fortieth Parallel is suggested by the line of Kelvin Seamounts on the ocean floor, and by magnetic and seismic discontinuities on the continental shelf and continental slope (Drake and co-workers, 1963, pp. 5270–5271; Drake and Woodward, 1963, pp. 53–54) (Fig. 7). Detailed magnetic surveys demonstrate a right-lateral deflection of geological and geophysical features across the zone by about 150 km, possibly before Eocene time (Taylor and co-workers, 1968, p. 774). A similar situation exists on the prolongation of the zone onto the land to the west, where there is a right-lateral deflection of structural belts of the northern Appalachians with respect to those of the southern Appalachians (Fig. 3), but no continuous surface faulting.

However, as pointed out by Woodward (in Drake and Woodward, 1963, pp. 56–60), discontinuous surface faults occur (Fig. 7). The Triassic border fault trends east-west in southeastern Pennsylvania, and the Paint Creek-Irvine and Rough Creek faults of Kentucky trend in the same direction, but are a little farther to the south. Evidence for strike-slip displacement in the surface rocks along the Rough Creek faults has been cited (Clark and Royds, 1948, pp. 1741–1749). Even greater subsurface displacement is possible on the faults in Kentucky; the basement is believed to stand higher on the south side than the north side (Flawn and co-workers, 1967), and a belt of positive gravity anomalies is offset right-laterally across them by about 100 km (Woollard and Joesting, 1964) (not well shown on Fig. 6, because of the large contour interval); the gravity anomalies may express a subcrop belt of Precambrian lava or iron formation.

Some geologists and geophysicists have been impressed with the fact that both the Mendocino fracture zone off the Pacific Coast and the transverse zone off the Atlantic Coast are near the Fortieth Parallel and have searched for indications of a connection between them in the continental area (Gilliland, 1962; Fuller, 1964, pp. 613–615), but the results impress me as inconclusive.

In summary, a transverse zone seemingly exists near the Fortieth Parallel in the eastern United States and the adjacent Atlantic Ocean. It is expressed in one place by aligned seamounts, in others by deflection of geological and geophysical features, and in a few places by surface or subsurface faults, these features being of various ages. Whether this diverse array expresses a truly fundamental structure or is merely coincidence, I will leave to the predilections of the reader. Certainly presently available evidence gives scant comfort to extending the zone farther westward across the continent. If the oceanic fractures and transverse zones do extend into the continents, there must have been decoupling at the base of the continental crust, so that they are expressed only intermittently or imperfectly in the rocks near the surface.

The transverse belt in southern Georgia is drawn along the north edge of the buried Paleozoic rocks of the Suwannee Basin, to account for the seeming truncation of trends in the Piedmont crystalline rocks at the edge of the basin (Flawn and co-workers, 1961). The transverse belt in southwestern Florida is drawn along the transverse magnetic anomaly mentioned earlier (Fig. 7), but the tectonic significance of this feature is uncertain. It has been suggested that the transverse anomaly follows the southeastern prolongation of the front of the Ouachita foldbelt (E. R. King, 1959). However, surveys in the Gulf of Mexico southwest of the transverse anomaly show the same northeastward-trending magnetic anomalies as those northeast of it, and apparently these are not offset (Gough, 1967). Possibly, the two transverse zones join and continue northwestward to offset right laterally the fronts of the Ouachita and Appalachian foldbelts; this is a long-discredited proposal (King, 1951) which now seems to deserve further appraisal.

The summarized longitudinal and transverse linear features do not seem to be closely related to the foldbelts in which they lie. Whether they played some role in the disruption or rotation of the continental and oceanic crust is as yet uncertain. Eventual answers will depend largely on the amount of strike-slip displacement which can be proved to have taken place along them.

Figure 9 also shows the trends of mafic dikes in the crystalline rocks of the Appalachian region, which have a remarkably systematic pattern, changing from N-W in the south to N-E in the north. They express, in some manner, the stress patterns existing near the edge of the continent during a moment in time—late in the Triassic and toward the close of the Appalachian deformation.

CONTINENTAL MARGIN

Finally, let us consider the continental margin of eastern North America, expressed by the continental shelf and continental slope.

When I was considering the possibility of a continental separation, I speculated about two odd features along the continental margin: the Flemish Cap east of Newfoundland and the Grand Banks (*FC*, Fig. 2) and the Blake Plateau east of Florida (*BP*, Fig. 2), which have the superficial appearance of being fragments of the continental crust that have either slipped off, or have dropped down, toward the ocean basin, perhaps during fragmentation of the continents. However, these suggestions are not confirmed by oceanographic surveys.

Flemish Cap is a semidetached flat-topped bank, rising to within 100 miles of the surface at the edge of the Atlantic Ocean Basin, and separated from the Grand Banks to the west by a channel more than 1,000 meters deep (Heezen and co-workers, 1959). Recent magnetic surveys indicate that it has a strongly positive magnetic anomaly, probably caused by mafic igneous rocks, some of which nearly reach the sea floor (Hood and Godby, 1965). It is unlikely that the Flemish Cap is formed of continental crust, and it may be a truncated volcanic center.

The Blake Plateau is a smooth-surfaced tract about 300 km broad, lying at a depth of 600 to 1,000 meters. Between it and the land to the west is a narrow continental shelf which rises above it in a scarp; the plateau itself breaks off eastward in another scarp facing the Atlantic Ocean Basin (Ewing and co-workers, 1966). The plateau has been extensively investigated by seismic surveys, sparker profiles, and drill holes (Hersey et al., 1959; JOIDES, 1965; Ewing and co-workers, 1966; Sheridan et al., 1966). These investigations confirm the inference that the plateau is formed of continental crust, but fail to reveal any evidence of downfaulting with respect to the continental shelf. Instead, the contrast in depth between plateau and shelf seems to be entirely a depositional feature. The Mesozoic and Paleocene strata are continuous and unbroken across them from one to the other, but the younger Tertiary strata are much thicker on the shelf (Fig. 10). Apparently the plateau and the shelf were one feature until the end of Paleocene time; thereafter, sedimentary upbuilding continued on the shelf, but was greatly reduced or nearly ceased on the plateau.

A more normal part of the continental margin of the Atlantic Ocean Basin occurs in the long segment between the Blake Plateau and the Grand Banks. Here, the continental shelf is a seaward extension of the Atlantic Coastal Plain, and shares many of its features. The crystalline rocks of the Appalachian foldbelt pass southeastward beneath a thickening wedge of Mesozoic and Tertiary strata, which extends seaward across the coastal plain and shelf to the continental slope.

The structure of the shallower layers on the continental shelf and

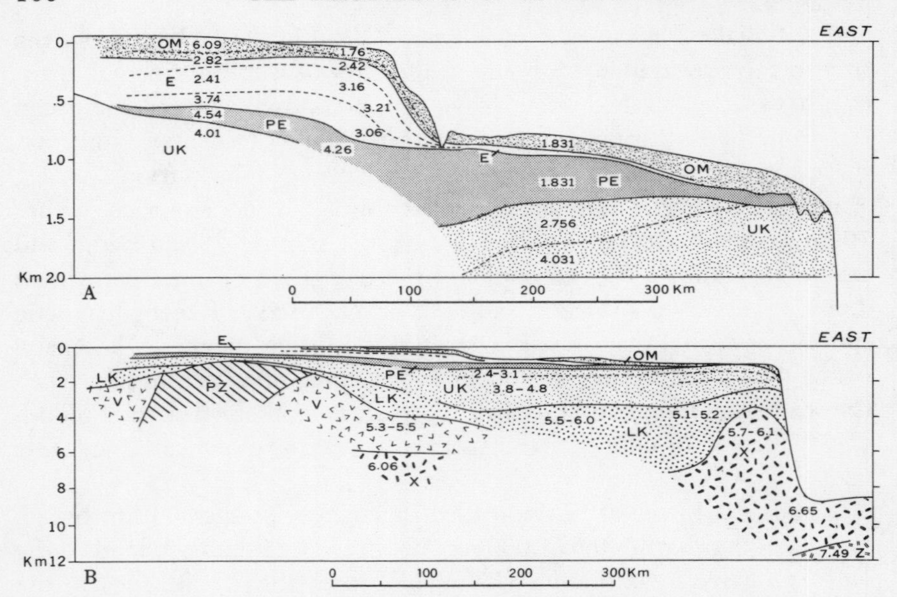

Fig. 10. Structure sections across continental shelf and Blake Plateau, along a line extending eastward from St. Augustine, Florida; based on drill records, where available, and on seismic profiles elsewhere. *A*, section showing details of shallow layers; *B*, section showing whole sedimentary sequence down to basement. After Ewing and others (1966) and Sheridan et al. (1966). *Z*, oceanic crust; *X*, crystalline basement; *V*, volcanic rocks; *PZ*, Paleozoic sedimentary rocks; *LK*, Lower Cretaceous; *UK*, Upper Cretaceous; *PE*, Paleocene; *E*, Eocene; *OM*, Oligocene and Miocene. Numbers indicate velocities of layers in kilometers per second.

slope is illustrated by continuous seismic-reflection ("sparker") profiles, of which representative examples are shown on Figure 11*A*. The profiles demonstrate the complexity of the origins of the shelf and slope, but in general confirm the long-held concept that they are depositional surfaces, caused by upbuilding and by prograding of sediments toward the ocean basins (Uchupi and Emery, 1967). Only in a few places is the slope an erosional surface truncating the strata, in the manner proposed by Heezen and co-workers (1959).

The structure of the deeper layers of the shelf and slope is shown by the many seismic refraction profiles that have been made by the Lamont Geological Observatory and other organizations. Above the basement there is a thick lower body of semiconsolidated strata, prob-

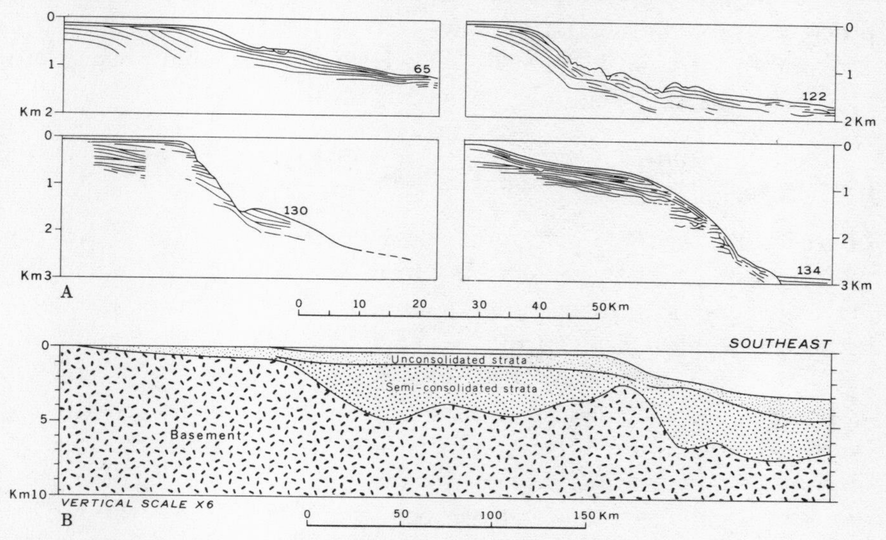

Fig. 11. Sections across continental shelf and slope east of the Middle Atlantic States. *A*, sections showing structures of shallow layers, based on continuous seismic-reflection ("sparker") profiles. Strata shown are Cenozoic and Upper Cretaceous, but have not been identified in detail; profiles 65 and 122 show prograding of strata on the continental slope, profiles 130 and 134 show partial truncation of strata. Profile 65, south of Long Island, N.Y.; profile 122, east of Cape Henry, Va.; profile 130, east of Cape Hatteras, N.C.; profile 134, south of Cape Lookout, N.C. After Uchupi and Emery (1967). *B*, section southeastward from New York City, showing whole sedimentary sequence down to the basement, based on seismic profiles. After Drake and co-workers (1959).

ably largely of Mesozoic age, overlain by a thinner body of poorly consolidated or unconsolidated strata of Tertiary age. The basement surface beneath them commonly forms a ridge near the edge of the continental shelf, separating troughs on each flank which are filled with as much as 7,000 meters of sediments (Drake and co-workers, 1959) (shown in plan by axis symbols in Figure 3, and in profile in Figure 11*B*).

Drake and co-workers (1959) compared these sediment-filled troughs with those of the earlier Appalachian geosyncline, showing that they are much alike in both width and depth, and perhaps even in the nature of their sediments. They suggested that the offshore troughs may have resulted from a continuation of geosynclinal processes, now shifted farther away from the continental interior, and not formed until after the Appalachian geosyncline had been filled with sediments and closed by deformation.

Comments on Geologic and Geophysical Data

In recapitulating these diverse geological and geophysical data, it is worth asking: To what extent has North America behaved throughout its history as an independent continent, and to what extent does its history seem to demand the existence of lands to the east, which have now vanished or drifted away?

There is some evidence, at least, for an independent growth of North America toward the Atlantic Ocean through time—the theory of continental accretion. Commonly cited as an indication of accretion is the building of the Grenville belt against the older Precambrian rocks of the Canadian Shield and continental interior, followed by the building of the Appalachian belt against the Grenville belt. To this history we can add the proposed belt of post-Grenville Avalonian basement, and the post-Appalachian geosynclinal troughs along the continental margin, both of which suggest a continuation of continent-forming processes nearer the Atlantic Ocean Basin than the processes which formed the earlier belts.

The only feature in eastern North America not truly compatible with an independent continental history is the northeastward termination of the Appalachian foldbelt. The Appalachian belt extends with full force through Newfoundland, and there is every indication that it continues to the edge of the continental shelf not far away; here it must be broken off abruptly, for no comparable structures are known in the deep ocean basin beyond. It could not possibly bend aside for there is not enough room on the continental shelf to bend it, and no place to bend it to.

Trans-Atlantic Comparisons

Finally, let us test the geologic possibilities of drift or separation of the continents by comparing the tectonics on opposite sides of the North Atlantic Ocean. These possibilities are illustrated by Figure 12, which is a generalized tectonic map of the opposing continents, plotted on the reconstruction made by Bullard and co-workers (1965) (*see* Fig. 1).

A former continuity between the rocks of the Appalachian foldbelt in Newfoundland and those of the Paleozoic foldbelts in the British Isles is very plausible. There is also merit in the comparison between the high-angle longitudinal faults in Newfoundland and those in the northern British Isles (Wilson, 1962), although a precise correlation between the Cabot fault on one side and the Great Glen fault on the other seems less likely. On the small scale of the figure, it has not been possible to subdivide the Paleozoic foldbelts of the two regions according to age, which would require distinction of areas involved in the Taconian, Acadian, and Alleghenian orogenies in North America, and the Caledonian and Variscan orogenies in Europe; a subdivision on this basis might increase or decrease the plausibility of the reconstruction.

Comparisons are also possible between the Appalachian foldbelt of North America and the Paleozoic foldbelt along the northwest coast of Africa (Sougy, 1962). The African foldbelt was primarily deformed after Devonian time, probably during one of the Variscan orogenies. It is the mirror image of the Appalachian foldbelt, with an internal zone of crystalline rocks and an external zone that was thrust eastward toward the Sahara Shield in the interior of the continent. The internal zone of the Appalachian foldbelt contains indications of a late Precambrian-early Paleozoic Avalonian event, and the internal zone of the African foldbelt contains indications of the nearly contemporaneous Pan-African event (Kennedy, 1964, figure). If the Appalachian and African foldbelts were once joined, their later separation was longitudinal rather than transverse to the structural grain, hence fewer direct comparisons can be made than with the foldbelts of the British Isles. Comparisons between the two foldbelts are further obscured because the internal parts of each are concealed along the coasts by wide belts of Mesozoic and younger strata.

Other trans-Atlantic matching is less plausible. The Grenville belt, which is broad and massive throughout the length of eastern North America virtually disappears in Europe (Fig. 12; *see also* Miller, 1965, Fig. 1; Fitch, 1965, Fig. 4). In northeastern North America, the Appalachian foldbelt is thrust northwestward over a Grenville basement with isotopic ages of 800 to 1,100 million years; in northwestern Scotland, the Caledonian foldbelt is thrust northwestward over a

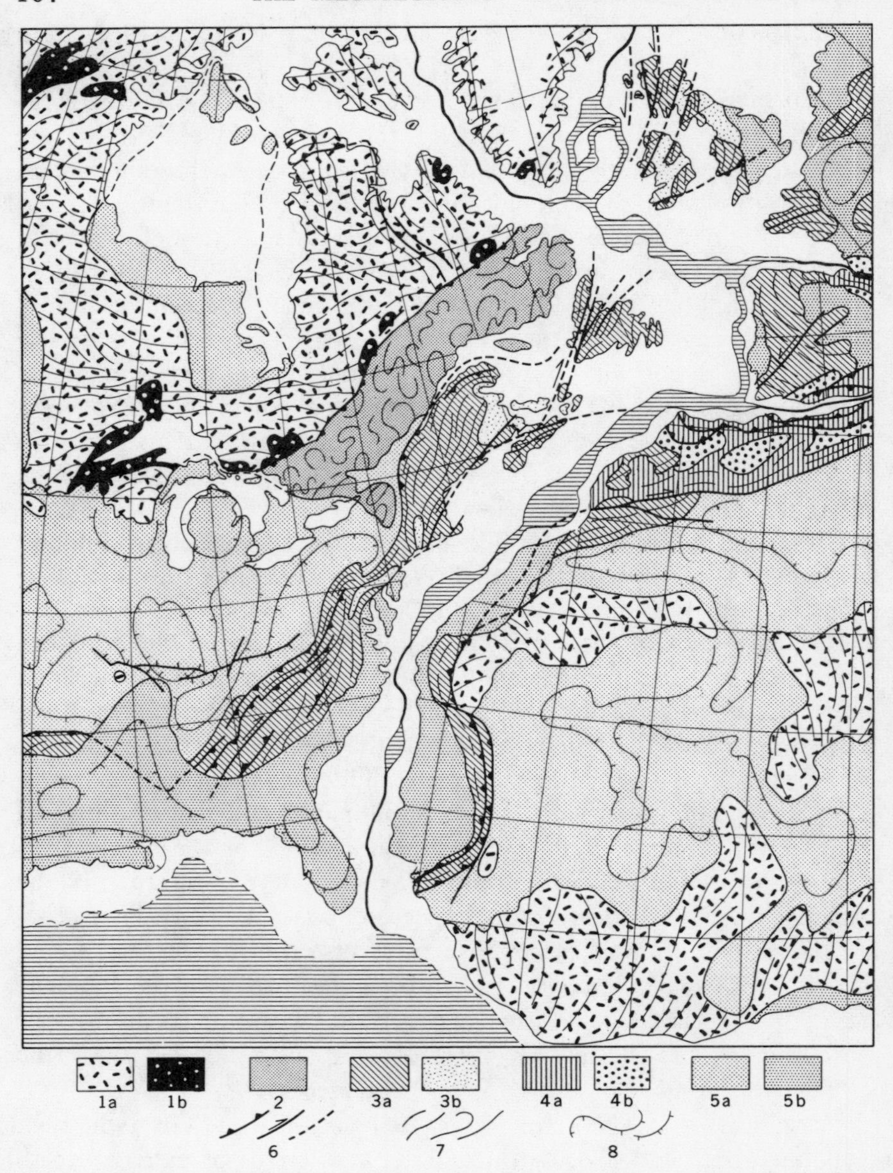

Fig. 12. Generalized tectonic map of the continents on opposite sides of the Atlantic Ocean, plotted on the reconstruction by Bullard and co-workers (1965) (compare Fig. 1). Compiled from Tectonic Map of North America (King, 1969), Tectonic Map of Europe (Commission for the Geological Map of the World, 1962), Sougy (1962), and other sources. 1a, early Precambrian foldbelts; 1b, their postorogenic deposits. 2, late Precambrian (Grenville)

Lewisian basement with isotopic ages between 1,500 and 2,500 million years, i.e., ages which only occur much farther N-W in North America. This might be rationalized by assuming that in Europe the Grenville belt is entirely overwhelmed by the Caledonian belt (Harland, 1965, p. 71 and fig. 7), just as the Appalachian belt partly overwhelms the Grenville belt in North America, and it is true that some relic Grenville dates have been found in the Caledonides of Scotland and Norway. Nevertheless, it seems odd that the actual cross-over of the two foldbelts should be hidden from view on the continental shelves, nearly at the line of separation between the two continental masses.

The reconstructed position of Spain (Figs. 1 and 12) creates problems. It is reasonable to rotate Spain with respect to the rest of Europe in order to close up the triangular gore of oceanic crust in the Bay of Biscay, but it is unreasonable to shift Spain several hundred miles eastward with respect to Morocco and Algeria, since the Alpine structures are continuous from one to the other across the Strait of Gibraltar. If a shift occurred, perhaps it was farther south, along transcurrent faults between the Atlas foldbelt and the Sahara Shield (Rod, 1962).

But if Spain and Morocco were once juxtaposed against the Maritime Provinces and New England, it is surprising that the latter contain no traces of the Mesozoic and Cenozoic Alpine structures which are so prominent in the former. Perhaps the Alpine foldbelt curves back on itself from Spain to Morocco and does not continue farther west or perhaps the Alpine structures were formed after the continents had been separated.

The Bullard reconstruction creates an even greater incongruity southwest of the area of Figures 1 and 12 where, in order to fit North America and South America to the continents of the Eastern Hemisphere, it has been necessary to eliminate all of Mexico and Central America south of the 20th Parallel, despite the fact that southern Mexico and northern Central America have a crystalline basement of Precambrian and Paleozoic ages.

---

foldbelt. 3a, Paleozoic foldbelts; 3b, their postorogenic deposits. 4a, Mesozoic and Cenozoic foldbelts; 4b, their postorogenic deposits. 5, platform areas: a, with Precambrian basement; b, with Paleozoic basement. 6, faults: a, thrust faults; b, high-angle faults (arrows show strike-slip displacement); c, concealed faults. 7, trend lines in foldbelts. 8, form lines in platform areas.

An alternative reconstruction of the North Atlantic continents has recently been suggested by Drake and Nafe (1967, Fig. 15), which has more to recommend it geologically. The continental shelves off Newfoundland and the British Isles remain in juxtaposition, as do eastern Brazil and the Bight of Africa—both regions where the trans-oceanic resemblances are most compelling. In the intervening part of the North Atlantic, however, the fit is not made at the continental shelves but farther out, at the edges of the undisturbed magnetic regions (Fig. 7), following the suggestion of Heirtzler and Hayes (1967) that these regions represent a late Paleozoic ocean floor.

By this reconstruction, the Appalachian and northwestern African foldbelts were never joined, but are contemporaneous structural systems on the opposing coasts of the primitive ocean. Further, Spain can now be returned to its proper position with respect to Morocco, and the crystalline basement of southern Mexico and northern Central America can be juxtaposed against the basement of the northern Andes in Colombia and Venezuela. During Mesozoic and Cenozoic time, while sea-floor spreading was creating the disturbed magnetic region down the middle of the North Atlantic Ocean, Mexico and northern Central America parted company from South America; the intervening lands of the Antilles and the Panama-Costa Rica isthmus were then newly born from the broad area of oceanic crust that was opening up (Donnelly, 1964)—instead of being swirled shreds of primaeval continental flotsam, as some "drifters" have proposed.

Despite the geological plausibility of this alternative reconstruction, its authors caution us that it fails to resolve several geophysical problems, so that its vindication remains as a challenge for the future. I must therefore return to my original judgment that, while continental drift (or continental separation) explains many of the puzzles of the geology and geophysics of the world, it leaves other puzzles still unsolved. Finally, any discussion concerning the larger problems of the world should be qualified by the cautions of an eminent geophysicist.

Unwary readers should take warning that ordinary language undergoes modification to a high-pressure form when applied to the interior of the earth. Thus, the high-pressure form of dubious is "certain," of perhaps is "undoubtedly," of vague suggestion is "positive proof," and of trivial objection is "unanswerable argument" (Birch, 1952).

References

Bickford, M. E., and G. W. Wetherill: Compilation of Precambrian geochronological data for North America. *Nat. Acad. Sci.–Nat. Res. Council, Nucl. Sci. Ser. Report* No. 41:21–27, 1965.

Birch, A. F.: Elasticity and constitution of the earth's interior. *J. Geophys. Res.* 57:227–286, 1952.

Bullard, E., J. E. Everett, and A. G. Smith: The Fit of the Continents Around the Atlantic. In *A Symposium on Continental Drift. Phil. Trans. Royal Soc. London. A*, 258:41–51, 1965.

Cameron, H. L.: The Cabot Fault Zone. In *Continental Drift*, G. D. Garland, ed. *Royal Soc. Canada Spec. Pub.* 9, pp. 129–140, 1966.

Canada Dominion Observatories: *Gravity Anomaly Map of Canada* (to end of 1956). Ottawa: Canada Dominion Observatories, 1957. Scale, 1:6,336,000.

Clark, S. K., and J. S. Royds: Structural trends and fault systems in the Eastern Interior Basin [U.S.]. *Bull. Am. Ass. Petrol. Geologists* 32:1728–1749, 1948.

Clifford, T. N., 1967, The Damaran episode in the Upper Proterozoic-lower Paleozoic structural history of southern Africa: Geol. Soc. America Spec. Paper 92, 77 pp.

Commission for the Geological Map of the World, Subcommission for the Tectonic Map of the World. *Carte tectonique internationale de l'Europe*. Internat. Geol. Cong., Moscow, 1962, scale 1:2,500,000.

Coons, R. L., Woollard, G. P., and Hershey, Garland, 1967, Structural significance and analysis of mid-continent gravity high: Am. Assoc. Petroleum Geologists Bull., v. 51, pp. 2381–2399.

Denison, R. E., Raveling, H. P., and Rouse, J. T., 1967, Age and descriptions of subsurface basement rocks, Pamlico and Albemarle Sound areas, North Carolina: Am. Assoc. Petroleum Geologists Bull., v. 51, pp. 268–272.

Donnelly, T. W., 1964, Evolution of eastern Greater Antillean island Arc: Am. Assoc. Petroleum Geologists Bull., v. 48, pp. 680–696.

Drake, C. L., M. Ewing, and G. H. Sutton: Continental Margins and Geosynclines–the East Coast of North America North of Cape Hatteras. In *Physics and Chemistry of the Earth*. New York, Pergamon Press, 1959, vol. 3, pp. 110–198.

Drake, C. L., J. Heirtzler, and J. Hirshman: Magnetic anomalies off eastern North America. *J. Geophys. Res.* 68:5259–5275, 1963.

Drake, C. L., and Nafe, J. E., 1967, Geophysics of the North Atlantic region (preprint), *in* Symposium on continental drift, emphasizing the history of the South Atlantic area, Montevideo, Uruguay, October 16–19, 1967: United Nations Scientific and Cultural Organization, 55 pp.

Drake, C. L., and H. P. Woodward: Appalachian curvature, wrench faulting, and offshore structures. *Trans. N.Y. Acad. Sci.*, s. 2, 26:48–63, 1963.

Engel, A. E. J.: Geologic evolution of North America: *Science* 140:143–152, 1963.

Eppley, R. A.: *Earthquake History of the United States*—Pt. 1, *Stronger Earthquakes of the United States* (exclusive of California and western Nevada). U.S. Coast and Geodetic Survey, S.P. 41–1, 1965.

Ewing, J., M. Ewing, and R. Leyton: Seismic-profiler survey of the Blake Plateau. *Bull. Am. Ass. Petrol. Geologists* 50:1948–1971, 1966.

Fitch, F. J.: The structural unity of the reconstructed North Atlantic continent. *Phil. Trans. Roy. Soc. London* s. A, 258:191–193, 1965.

Flawn, P. T., August Goldstein, Jr., P. B. King, and C. E. Weaver: *The Ouachita System.* Texas Univ. Pub. 6120, 1961.

Flawn, P. T., and co-workers: *Basement Map of North America, Between Latitudes 24° and 60° N.* Washington, D.C. Am. Ass. Petrol. Geologists and U.S. Geol. Survey, 1967. Scale, 1:5,000,000.

Fries, C., Jr., ed.: *Estudios Geochronologicos de Rocas Mexicanas.* Mexico (City), Univ. Nac. Inst. Geol. Bol. No. 64, 1962.

Fuller, M. D.: Expression of E W fractures in magnetic surveys in parts of the U.S.A. *Geophysics* 29:602–622, 1964.

Fyson, W. K.: Gravity sliding and cross folding in Carboniferous rocks, Nova Scotia. *Am. J. Sci.* 265:1–11, 1967.

Gastil, R. G.: The distribution of mineral dates in time and space: *Am. J. Sci.* 258:1–35, 1960.

Gilliland, W. N.: Possible continental continuation of the Mendocino fracture zone. *Science* 137:685–686, 1962.

Goldich, S. S., et al.: Geochronology of the midcontinent region, United States, 2. Northern area: *J. Geophys. Res.* 71, no. 22:5389–5408, 1966a.

Goldich, S. S., and co-workers: Geochronology of the midcontinent region, United States, 1. Scope, methods, and principles: *J. Geophys. Res.* 71:5375–5388, 1966b.

Gough, D. I., 1967, Magnetic anomalies and crustal structure in eastern Gulf of Mexico: Am. Assoc. Petroleum Geologists Bull., v. 51, pp. 200–211.

Griscom, A.: Tectonic significance of the Bouguer gravity field in the Appalachian system. *Geol. Soc. Am. Special Paper*, No. 73:163–164, 1963.

Hamilton, W., and L. C. Pakiser: *Geologic and Crustal Cross-Section of the United States Along the 37th Parallel, a contribution to the upper mantle project.* U.S. Geol. Survey Misc. Geol. Inv. Map 1–448, 1965. Scale, 1:2,500,000.

Harland, W. B.: Discussion [continental reconstructions], *Phil. Trans. Roy. Soc. London* s. A, 258:59–75, 1965.

Heezen, B. C., and W. M. Ewing: Turbidity currents and submarine slumps and the 1929 Grand Banks [Newfoundland] earthquake. *Am. J. Sci.* 250:849–873, 1952.

Heezen, B. C., M. Tharp, and W. M. Ewing: The North Atlantic—Text to Accompany the Physiographic Diagram of the North Atlantic. *Geol. Soc. Am. Special Paper*, No. 65, 1959.

Heirtzler, J. R., Burckle, L. H., and Peter, G., 1966, Magnetic anomalies in the Gulf of Mexico: Jour. Geophys. Research, v. 71, pp. 519–526.

Heirtzler, J. R., and Hayes, D. E., 1967, Magnetic boundaries in the North Atlantic Ocean: Science, v. 157, pp. 185–187.

Hersey, J. B., and co-workers: Geophysical investigation of the continental margin between Cape Henry, Virginia, and Jacksonville, Florida. *Bull. Geol. Soc. Am.* 70:437–465, 1959.

Hobbs, W. H.: Lineaments of the Atlantic border region. *Bull. Geol. Soc. Am.* 15:483–506, 1904.

Hood, P., and E. A. Godby: Magnetic profile across the Grand Banks and Flemish Cap off Newfoundland. *Can. J. Earth Sci.* 2:85–92, 1965.

Isachsen, Y. W.: Extent and configuration of the Precambrian in the northeastern United States. *Trans. N.Y. Acad Sci.* s. 2, 26:812–829, 1964.

Joint Oceanographic Institutions' Deep Earth Sampling Program (JOIDES): Ocean drilling on the continental margin. *Science* 150:709–716, 1965.

Kennedy, W. Q., 1964, The structural differentiation of Africa in Pan-African (500 m.y.) tectonic episode: Leeds Univ. Research Inst. African Geol., 8th Ann. Rept., 1962–1963, pp. 48–49.

King, E. R.: Regional magnetic map of Florida. *Bull. Am. Ass. Petrol. Geologists* 43:2844–2854, 1959.

King, E. R., Zietz, Isadore, and Dempsey, W. J., 1961, The significance of a group of aeromagnetic profiles off the eastern coast of North America: U.S. Geol. Survey Prof. Paper 424-D, pp. D299–D303.

King, P. B.: *The Tectonics of Middle North America; Middle North America East of the Cordilleran System.* Princeton, N.J., Princeton University Press, 1951.

King, P. B.: *Systematic Pattern of Triassic dikes in Appalachian Region.* Washington, D.C., U.S. Geol. Surv. Profess. Paper No. 424-B:B93–B95, 1961.

King, P. B.: Further Thoughts on Tectonic Framework of Southeastern United States. In *Tectonics of the Southern Appalachians*, W. D. Lowry, ed. Virginia Polytechnic Inst., Dept. Geol. Sci. Mem. No. 1:5–31, 1964.

King, P. B., comp.: *Tectonic Map of North America*. Washington, D.C., U.S. Geol. Survey (1969). Scale 1:5,000,000.

Lidiak, E. G., and co-workers: Geochronology of the midcontinent region, United States. 4. Eastern area. *J. Geophys. Res.* 71:5427–5438.

Lilly, H. D.: Late Precambrian and Appalachian tectonics in the light of submarine exploration on the Great Bank of Newfoundland and in the Gulf of St. Lawrence; preliminary views. *Am. J. Sci.* 264:569–574, 1966.

Long, L. E., J. L. Kulp, and F. D. Eckelmann: Chronology of major metamorphic events in the southeastern United States [Appalachians]. *Am. J. Sci.* 257:585–603, 1959.

McCartney, W. D., and co-workers: Rb/Sr age and geological setting of the Holyrood granite, southeast Newfoundland. *Can. J. Earth Sci.* 3:947–957, 1966.

McConnell, R. K., Jr., R. N. Gupta, and J. T. Wilson: Compilation of deep crustal seismic refracton profiles. *Rev. Geophysics* 4:41–100, 1966.

McKee, E. D., et al.: *Paleotectonic Maps of the Triassic System*. Washington, D.C., U.S. Geol. Survey, Misc. Geol. Inv., Map 1–300, 1959.

Miller, J. A.: Geochronology and continental drift—the North Atlantic. *Phil. Trans. Roy. Soc. London*, s. A, 258:180–191, 1965.

Muehlberger, W. R., and co-workers: Geochronology of the midcontinent region, United States. 3. Southern area. *J. Geophys. Res.* 71:5409–5426, 1966.

National Research Council, Committee on Nuclear Science. *Geochronology of North America*. Nat. Acad. Sci.–Nat. Res. Council, Nuc. Sci. Ser. Report No. 41, 1965.

Neale, E. R. W., and co-workers: A preliminary tectonic map of the Canadian Appalachian region based on age of folding. *Can. Mining Met. Bull.* 54:687–694, 1961.

Northrop, J., R. A. Frosch, and R. Frassetto: Bermuda-New England seamount arc. *Bull. Geol. Soc. Am.* 73:587–593, 1962.

Pakiser, L. C., and I. Zietz: Transcontinental crustal and upper-mantle structure: *Rev. Geophysics* 3:505–520, 1965.

Pratt, R. M.: Physiography and Sediments of the Deep-Sea Basin. *In* The Atlantic continental shelf and slope of the United States. Washington, D.C., U.S. Geol. Surv. Profess. Paper No. 529-B, 1968.

Press, F., and W. C. Beckmann: Geophysical investigations in the emerged and submerged Atlantic Coastal Plain, Pt. 8, Grand Banks and adjacent shelves. *Bull. Geol. Soc. Am.* 65:299–313, 1954.

Reed, J. C., Jr., and B. Bryant: Evidence for strike-slip faulting along the Brevard Zone in North Carolina. *Bull. Geol. Soc. Am.* 75:1177–1196, 1964.

Reed, J. C., Jr., and co-workers: *The Brevard Fault in North and South Carolina.* Washington, D.C., U.S. Geol. Surv. Profess. Paper No. 424-C: C67–C70.

Rod, E.: Geological note on fault pattern, northwest corner of Saharan shield. *Bull. Am. Ass. Petrol. Geologists* 46:529–552, 1962.

Rodgers, John, 1967, Chronology of tectonic movements in the Appalachian region of North America: Am. Jour. Sci., v. 265, pp. 408–427.

Sanders, J. E.: Late Triassic tectonic history of northeastern United States. *Am. J. Sci.* 261:501–524, 1964.

Sheridan, R. E., and co-workers: Seismic-refraction study of continental margin east of Florida. *Bull. Am. Ass. Petrol. Geologists* 50:1972–1991, 1966.

Smith, W. E. T.: *Earthquakes of Eastern Canada and Adjacent Areas, 1534–1927.* Dominion Observatory Ottawa Pub., 26:271–301, 1962.

Sougy, J.: West African foldbelt. *Bull. Geol. Soc. Am.* 73:871–876, 1962.

Steinhart, J. S., and R. P. Meyer: *Explosion Studies of Continental structure—University of Wisconsin, 1956–1959.* Washington, D.C., Carnegie Institute Pub. 622, 1961.

Stockwell, C. H.: Fourth Report on Structural Provinces, Orogenies, and Time-Classification of Rocks of the Canadian Precambrian Shield. In *Age Determinations and Geological Studies,* Pt. 2, Geological studies. Canada Geol. Survey Paper 64–17, pp. 1–21, 26–29, 1964.

Stockwell, C. H., and co-workers: *Tectonic Map of the Canadian Shield.* Canada Geol. Survey, Tectonic Map of Canada Comm., Map 4–1965, 1965. Scale 1:500,000.

Stockwell, C. H.: Personal communication, 1966.

Tagg, A. R., and E. Uchupi: Distribution and Geologic Structure of Triassic Rocks in the Bay of Fundy and the Northeastern part of the Gulf of Maine. Washington, D.C., U.S. Geol. Surv. Profess. Paper No. 550-B: B95–B98, 1966.

Taylor, P. T., Isadore Zeitz, and L. S. Dennis: Geologic implications of aeromagnetic data for the eastern continental margin of the United States. *Geophysics* 35:755–780, 1968.

Thiel, E.: Correlation of gravity anomalies with the Keweenawan geology of Wisconsin and Minnesota. *Bull. Geol. Soc. Am.* 67:1079–1100, 1956.

Uchupi, Elazar: *Map showing relation of land and Submarine Topography, Nova Scotia to Florida.* Washington, D.C., U.S. Geol. Survey, Misc. Geol. Inv., Map 1–451, 1965. Scale 1:1,000,000.

Uchupi, Elazar, and Emery, K. O., 1967, Structure of continental margin off east coast of United States: Am. Assoc. Petroleum Geologists Bull., v. 51, pp. 223–234.

Vine, F. J., and Matthews, D. H., 1963, Magnetic anomalies over oceanic ridges: Nature, v. 199, pp. 947–949.

Webb, G. W.: Occurrence and exploration significance of strike-slip faults in southern New Brunswick, Canada. *Bull. Am. Ass. Petrol. Geologists* 47:1904–1927, 1963.

Williams, H.: The Appalachians in northeastern Newfoundland—a two-sided symmetrical system. *Am. J. Sci.* 262:1137–1158, 1964.

Wilson, J. T.: Cabot fault, an Appalachian equivalent of the San Andreas and Great Glen faults and some implications for continental displacement. *Nature* 195:135–138, 1962.

Woollard, G. P.: Areas of tectonic activity in the United States as indicated by earthquake epicenters. *Am. Geophys. Union Trans.* 39:1135–1150, 1958.

Woollard, G. P., and H. R. Joesting: *Bouguer Gravity Anomaly map of the United States* (exclusive of Alaska and Hawaii). Washington, D.C., Am. Geophys. Union and U.S. Geol. Survey, 1964. Scale, 1:2,500,000.

Woollard, G. P., and J. Monges Caldera: Gravedad, geologiá regional y estructura cortical en México. *Mexico (City), Univ. Nac. Inst. Geol. Anal.* 2:60–112, 1956.

Worzel, J. L.: *Pendulum Gravity Measurements at Sea, 1936–1959.* New York, John Wiley & Sons, Inc., 1965.

Zietz, I., and A. Griscom: *Geology and Aeromagnetic Expression of the Midcontinent Gravity High.* Geol. Soc. Am. Special Paper No. 76:184, 1961.

Zietz, I., and co-workers: Crustal study of a continental strip from the Atlantic Ocean to the Rocky Mountains. *Bull. Geol. Soc. Am.* 77:1427–1448, 1966.

6

Developments in Seismology and Georheology

LEON KNOPOFF

Institute of Geophysics and
Planetary Physics, University of
California, Los Angeles

Until recently, most of our information regarding the nature of the earth's interior has come from calculations of the distributions of densities and seismic velocities. The distributions of the latter have been obtained principally by travel-time methods, using both body waves and surface waves.

Energy is conserved in a gravitational field. In the models of the elastic wave fields from which travel-time interpretations are made it suffices to assume that energy is also conserved. Such physical properties as the attenuation factors of seismic waves and rheologic coefficients, including viscosities when appropriate, represent parameters characteristic of processes in which energy is dissipated. So far, little information has been available about the distribution of these properties in the earth's interior. Some data concerning the parameters describing dissipative processes in the earth's interior can be obtained by studying the variation of the amplitudes of motions from place to place at or near the earth's surface, especially after such energy-conserving effects as diffraction and scattering have been removed from the observations. One of the most striking recent developments has perhaps been the realization that, by studying the amplitudes of seismic-wave motions, it is possible to determine an additional set of physical parameters describing the earth's interior; these parameters are attenuation factors and are supplementary to those characteristic of fields in which energy is conserved: densities and elastic wave velocities.

Seismology, up to a few years ago, has been concerned with the nature of the vibratory motions of the earth's interior, whether produced by earthquakes or explosions. Recently the term has been broadened to include what has been called "zero-frequency seismology" (Press, 1965). It is now possible to discuss and to measure the motions of parts of the earth relative to one another, whether these are vibratory in character or not. The science of rheology is principally concerned with the study of flow processes in matter. It may be assumed that such processes in the earth are descriptive of motions of parts of the earth relative to one another on the longest possible time scale. Accordingly, the study of the motions of the earth's interior may be catalogued according to the time scale involved in such motions: the motions involved in ordinary seismologic investigations, such as short period body-wave seismology and longer period surface-wave seismology, fall on the high frequency end of the scale; deformations on a geologic time scale, involving motions with time constants of the order of millions of years, fall on the low frequency end of the scale.

Although the seismologist has broadened his interests beyond those of traditional travel-time seismology to look at somewhat more general problems involving the general motions of parts of the earth's interior relative to one another, recent activity by seismologists has not neglected the more traditional areas of investigation. Indeed, the development of new techniques of gathering and processing data and their application to problems of seismology has made considerable new information available concerning the nature of the earth's interior, especially with regard to the distributions of densities and elastic wave velocities. Below, I propose to discuss developments of the past five to ten years, first with regard to new results describing the distributions of velocities and densities in the earth's interior. Next, a description of some of the properties of the distribution of parameters of non-conservative fields such as attenuation factors and viscosities will be presented. Finally, some observations will be made describing recent activity in the area of interest involving motions of parts of the earth relative to one another on time scales of the order of years or even millions of years and still longer.

Earth Models

Models of the earth's interior, especially those constructed a few years ago, have been based on the concept of spherical symmetry on a global scale, or lateral homogeneity on a regional or local scale. It has long been known that such models are oversimplifications. The nonuniform distributions of earthquakes and volcanoes clearly indicate the existence of lateral heterogeneity, at least to depths where the deepest earthquakes are found, about 700 km. Oceanic and continental thicknesses differ significantly. Within the crust, interpretation of seismic profiles is ordinarily based on the assumption of lateral homogeneity, and even on an assumption of homogeneous discrete layers. Yet the crustal profiles which are obtained from the interpretations differ widely, even in adjacent localities. Until recently, there were few measurements of the lateral variation of physical properties below the crust. The velocity of seismic compressional waves seemed generally to be about 8.1 km per second under both continental and oceanic crusts. Measurements in recent years have made possible the assignment of numerical values to some of the deviations from the value formerly thought to be universal for the subcrustal velocity. The actual departures from the model of lateral homogeneity may not be great in percentage terms, but they hold the key to an understanding of many of the large-scale processes that have created the major features of the earth. For the solution of some problems, however, the use of a model of lateral homogeneity continues to be satisfactory. For the lower mantle, we have little quantitative or qualitative evidence bearing on the postulate of radial symmetry.

Many of the attempts to describe the physical and chemical nature of the earth's interior have been strongly influenced by the models proposed by Bullen (1953) to explain the travel-time curves of seismic events (Fig. 1). From travel-time data, and a small number of postulates, Bullen concluded that the earth's interior could be subdivided into regions which he labeled by the letters A through G. Except for region A, the models bear the strong imprint of the postulate of radial symmetry.

Region A is the thin crust of the earth which has different thicknesses under oceans and continents. Region B occupies the zone from the bottom of the crust to a depth of about 400 km into the mantle; this region was assumed to be essentially homogeneous. Beneath this lies region C, a transitional region not necessarily physically or chemically homogeneous. The lower mantle, region D, was postulated to be more or less homogeneous; considerable elements of chemical homogeneity were built into the construction of region D (Birch,

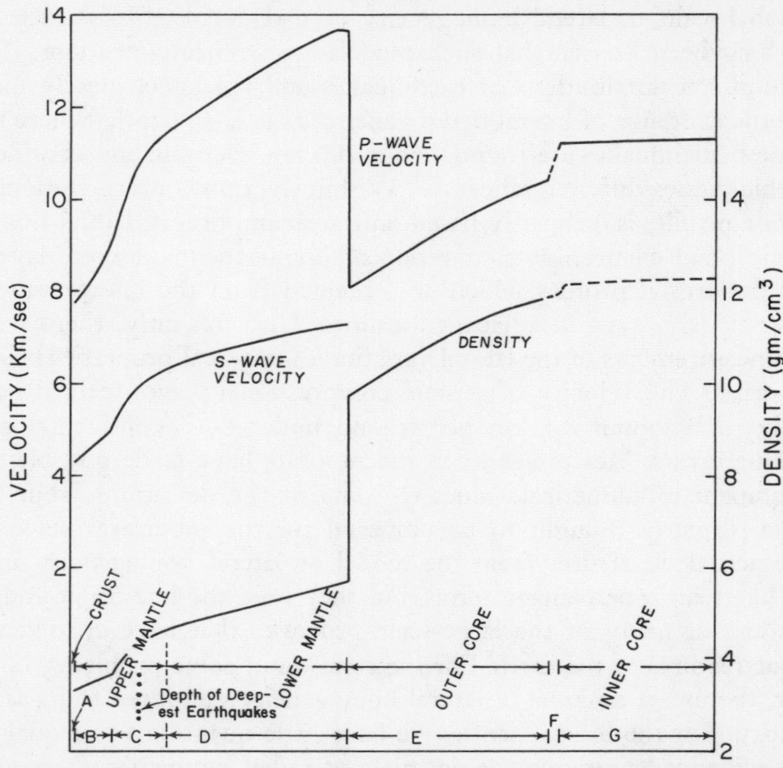

Fig. 1. Distribution of compression (*P*) and shear (*S*) wave velocities (*A*) and density distribution (*B*) in the mantle. After Bullen (1953).

1952). The lower mantle occupies the region from about 1,000 km below the surface to the boundary of the earth's core at a depth of 2,900 km, approximately half the radius of the earth. Region E, the outer core of the earth, is most likely fluid in nature since *S* waves have not been observed to traverse it. The outer core extends down to a point approximately 5,200 km below the surface where, after crossing an ill-defined, narrow zone called region F, the inner core, region G, is entered. The existence of the inner core was discovered by Lehmann (1936) from observations of the reflections from this region when illuminated with seismic energy from deep-focus earthquakes. In Bullen's model, little was known about detailed structure in any of the regions postulated to be chemically homogeneous, i.e., regions B, D, E, and G. The contact between the lower mantle and the outer core is a sharp one; a decrease in the velocity gradient is found at the bottom of the lower mantle. Not much could be said about the nature of the transition zones in the core and the mantle.

Bullen's model is an excellent point of departure for the discussion of new results in seismologic research. It should be pointed out that the models were constructed without the introduction of a low-velocity, shear-wave channel in the upper mantle at depths of 80 to 250 km. There is a high likelihood that this channel will be found worldwide.

Crust-Mantle Interface

The word Moho is a shortened form of the phrase, Mohorovičić discontinuity. For 30 to 50 years after the Moho was identified, the statement that the Moho was a worldwide seismic discontinuity was plausible; above it lay the crust, and below it the mantle. Compressional waves on the top of the mantle are called P_n; the velocity of P_n seemed everywhere to be close to 8.1 km per second. Compressional wave velocities in the crust were at least 15 percent lower. In principle, the Moho could be readily identified by a sharp arrival of refracted seismic waves with the appropriate velocity. The depth to Moho under the oceans was found to be generally about 10 to 12 km; under continents, 30 to 50 km. This seemed in accord with the principle of isostasy, which means essentially that the outermost rocks of the earth are in some kind of buoyant equilibrium.

In fact, many investigators assumed that the light crustal rocks were exactly in buoyant equilibrium in the denser mantle, thereby accounting for the greater elevation of the continental surface and the greater depth of the Moho, relative to oceanic values.

Under the conditions of crustal isostasy, the higher the elevation of the land, the deeper the crustal root. By measuring the Bouguer gravity anomaly at the surface and making some elementary reductions of the data, the depth to the bottom of the crust could be calculated. This faith in the postulate of crustal isostasy led to the construction of a number of maps of crustal thickness based on gravity data alone. Investigations in the past few years indicate that calculations of crustal thickness based on this simple procedure are not reliable. It follows that notions of simple crustal isostasy must be modified.

Of all the concepts surviving as remnants of the earlier simplified models, that of the Moho itself remained inviolate. In the classic definition, it represents the separation between crust and mantle; it is a more or less sharp discontinuity at relatively small depth compared with the earth's radius. The seismic-phase P_n velocity rides along the mantle on the underside of the Moho. One of the principal, recent developments in research on the upper mantle has been the appearance of new evidence for considerable radial and lateral heterogeneity. In particular, the velocity of P_n in the mantle immediately beneath the crust has been shown to be variable.

The P_n velocity has been determined by the techniques of crustal explosion seismology, including modern array methods, and, less precisely, from anomalies in the travel times of seismic waves from nuclear explosions. Subsequent measurements by crustal refraction methods have led to confirmation, in the main, of most of the features obtained in the interpretation of data collected by the latter method. It has been found that the crust and the top of the mantle in the western part of the United States, extending roughly from the Rockies to the west coast, have a completely different character than the crust and upper mantle under the eastern part of the United States. The P_n velocities are higher under the crust of the eastern United States than under the western part; under the former, the P_n velocities are as high as 8.3 or 8.4 km per second (Warren and coworkers, 1966). In the central part of the western United States, the velocities are as low as 7.7 km per second (Herrin and Taggart, 1966) (Fig. 2). Similar lateral variations in P_n velocities are found at sea: lower P_n velocities under island arcs and the midoceanic ridges, and higher P_n velocities under the deep ocean basins (Kosminskaya and Riznichenko, 1964).

The regional variations in the velocity of P_n appear to be strongly correlated to the nature of the topography. In the mountainous west, P_n is lower than in the east. Thus, the crustal blocks are not simply decoupled from the mantle below; instead, variations in the structure of the crust seem to be correlated with the properties of the top of the mantle beneath it. Hence, the crust can no longer be considered as independent of the mantle, floating in it in isostatic equilibrium, as an iceberg might do in a sea, the properties of which are independent of the type of iceberg above it.

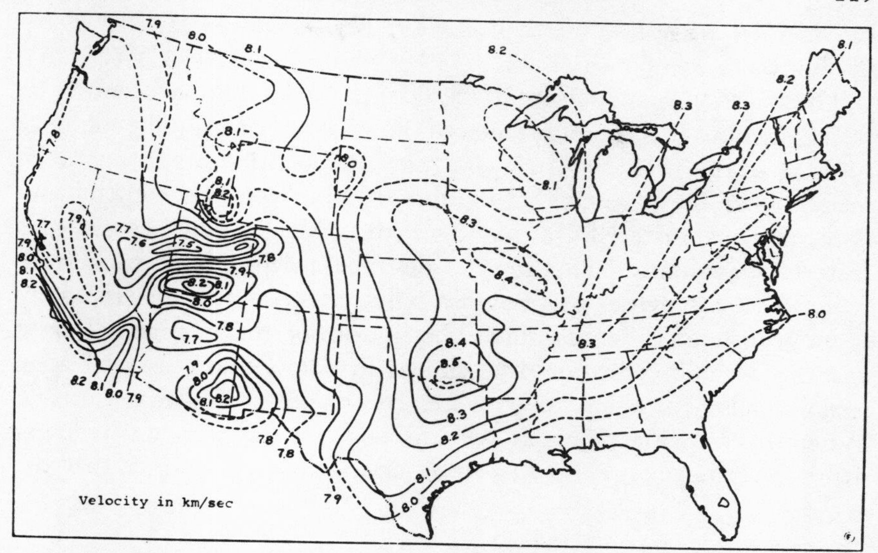

Fig. 2. Apparent P_n velocities for the United States. After Herrin and Taggart (1966).

Where Is the Moho?

As I have noted, compressional waves in the uppermost part of the mantle are called P_n; their velocity was supposedly near 8.1 km per second. Velocities as low as 7.7 km or as high as 8.4 km came to be accepted as P_n. Recent results force a careful reexamination of the definition of the Moho in seismic terms.

By using data obtained from nuclear explosions, Press (1960) found a distinct horizon overlying material with P wave velocity of 7.7 km per second at a depth of about 24 km below the surface between the Nevada Test Site and Pasadena. Below this, another discontinuity with an 8.1 km velocity layer is found at a depth of about 50 km. In eastern Nevada, from the Nevada Test Site north to the Snake River Plain of Idaho, a region of relatively high elevation, the depth to material with P_n velocity of 7.8 km per second is 26 km (Pakiser and Hill, 1963). No deeper discontinuity has been found beneath this zone, in contrast with the situation in southern Nevada and California.

In the Ivrea zone of western Italy, where there is a large positive gravity anomaly, material with a velocity of 7.4 km per second comes within 10 km of the surface at a sharp discontinuity and is underlain by material with a velocity of 8.1 km per second in another sharp discontinuity at a depth of about 55 km (Fuchs et al., 1963) (Fig. 3).

The question of which interface defines the crust may seem to be pedagogic, but it is of importance to the problem of crustal evolution. So long as the velocity of P_n was found to be 8.1 km per second, it seemed plausible to assert that the top of the mantle was a chemically homogeneous material. In many of the older geology texts, this material was simply called sima to describe a dense, basic rock from which the less dense, more acidic sialic crust could be derived by some process of differentiation. The bottom of the continental crust and the oceanic crust could be similar in composition, as well as in seismic velocity, since they were, presumably, similarly derived. When P_n velocities such as 7.8 km began to appear at shallower depths than normal, it became necessary to invoke the fact that the

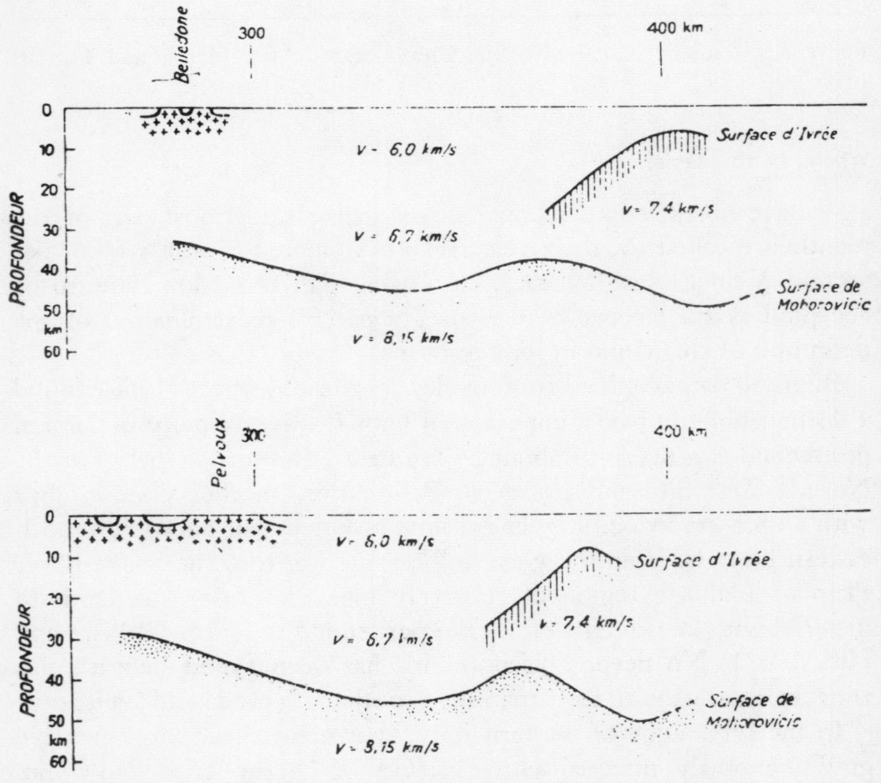

Fig. 3. Cross section of the outer parts of the earth in the Ivrea zone of western Italy. After Fuchs, et al. (1963).

pressure and temperature gradients were probably such that the lower P_n velocity represented the same mantle material but in different physical conditions than "normal." However, when two equally appropriate crustal boundaries are found overlying one another, the conjecture that there is only one type of material in the upper mantle, or that one process of formation of crust occurs, becomes suspect. Based on the southern California and Ivrea cross sections, at least two geochemical complexes of structural interrelations can coexist between crust and upper mantle. Thus, it is somewhat less certain that when a P_n is found with a velocity of 7.8 km per second, as in eastern Nevada, underlain by no deeper first order discontinuity within a few tens of kilometers, it is the same mantle material as "normal" material with P_n of 8.1 km in some other physical environment. It further follows that two regions with the same P_n, if widely separated geographically, do not necessarily have the same composition of the upper mantle nor the same process by which the crust was derived.

The notions of crustal isostasy are thus shaken. If, in the two cases described, the upper horizon is defined as the Moho, the crust has a relatively small thickness, less than the thicknesses in the eastern United States (Warren and co-workers, 1966), and cannot be in isostatic equilibrium at the Moho.

Why should we be concerned with the definition of the Moho? Because, if the definition of the bottom of the crust is not so certain as it might be, and if we are not so certain of the type of density distribution required for the crust from density-velocity relations (Woollard, 1959; Birch, 1961; Nafe and Drake, 1965; Knopoff, 1967) for given composition, simple Bouguer gravity data obtained at the surface cannot be used unambiguously to determine the properties of the outermost few tens of kilometers of the earth's radius.

The terms "crust," "Moho" and "P_n" are useful for some purposes, but, in view of the complexities just outlined the use of these terms in certain cross sections is no longer unambiguous. Extreme care must be exercised in discussion of such terms as "crustal thickness," "depth to Moho," and "P_n."

How Far Down Does the Crust Extend?

To what extent does the influence of surface features penetrate into the mantle? As we have seen, the influence of the surface features ("crustal" features, for want of a better word) penetrate somewhat more deeply than does the traditionally defined "crust." Evidence bearing on the depth of this penetration is forthcoming from surface-wave observations.

Seismic surface waves are dispersive; because of the heterochromatic nature of the source, both group and phase velocity dispersion can be measured. Surface waves of greater wavelength penetrate to greater depth, i.e., the character of the dispersion is influenced by the elastic properties of the earth between the surface and a depth of the order of the wavelength. The depth of penetration, where the particle displacements fall to a small fraction of the displacements near the surface, is roughly one-half to one-third the wavelength. Surface waves of a given wavelength, and passing through a thicker crustal cross section, will have a lower phase velocity than those passing through a thinner crust; the thicker crust has more low-velocity material in the range of the depth of surface-wave penetration (Ewing and co-workers, 1951; Ewing and Press, 1959).

Surface-wave velocities are closely correlated with the Bouguer gravity anomaly for Rayleigh waves with a 30 sec period (Ewing and Press, 1959; Knopoff and co-workers, 1966). A good correlation is to be found between the densities of silicate rocks and their elastic velocities (Woollard, 1959; Birch, 1961; Nafe and Drake, 1965; Knopoff, 1967). Hence, a positive or negative Bouguer anomaly at a point in the surface, which indicates a mass excess or deficiency in the column of matter beneath, implies a higher or lower seismic velocity, at least in the column containing the mass excess or deficiency.

The existence of a correlation between surface-wave phase velocities and the Bouguer anomaly indicates that the mass excess or deficiency exists, at least to the order of depths reached by the surface-wave particle motions.

Surface waves of different wavelengths penetrate to different depths in the mantle. Inquiries about the wavelengths of the phase velocities at which the correlation with the gravity anomalies disappears are thus pertinent. The answer to this question is not clear cut because surface waves of a given wavelength sample earth materials over a broad band of depths (Anderson, 1964a) and the "depth of penetration" plays the role of a "skin depth." Under the Alps and the western Mediterranean, there seems to be little correlation of Rayleigh wave phase velocities at 70 sec periods with Bouguer anomaly

(Knopoff and co-workers, 1966). Rayleigh waves of 30 sec periods sample the earth down to the top of the mantle, well above the low-velocity channel, while 70 sec waves explore down to the low-velocity channel itself. Thus, the mass deficiencies and excesses characteristic of variations in crustal thickness are also found in the mantle immediately beneath the crust, but these anomalies become relatively small at depths of the order of the low-velocity channel.

The depths to which the variations in crustal properties penetrate into the mantle can be demonstrated by investigating the low-velocity channel in the mantle itself for shear waves (*S* waves). Surface-wave techniques have now been used for studies to discover whether there are variations from region to region in the structure of the low-velocity channel itself. The low-velocity channel is found approximately between 50 to 220 km below the surface. The *S* wave velocity in the channel is somewhat less than in the mantle immediately above and below it. The thickness of the channel is not precisely the same from region to region nor is it found at uniform depths below the surface; the *S* wave velocity in the channel also varies. In short, the profile of the channel changes regionally.

For example, beneath the Alps and the western basin of the Mediterranean Sea, the velocity of *S* waves in the channel is about the same, nearly 4.15 km per second, although underlying considerably different structures at the surface (Berry and Knopoff, 1967). This velocity is significantly less than the velocities of 4.5 km found in the channel under shield areas (Brune and Dorman, 1963). Yet between the Alps and the sea to the south, roughly along the margin of the southern coast of France and along the western edge of Corsica and Sardinia, the velocity in the channel has approximately the shield value of 4.45 km (Fig. 4). Thus, variations of *S* wave velocity in the channel seem to occur over short distances laterally. The velocity in the channel is apparently not correlated with the topography of the earth's surface, and it therefore seems unlikely that the effects of lateral inhomogeneity of the crust and top of the mantle extend significantly into the low-velocity channel. Possibly, the forces which have produced the topography at the earth's surface originate at such shallow depths in the mantle that mountains and ocean basins are not "imaged" in the low-velocity channel.

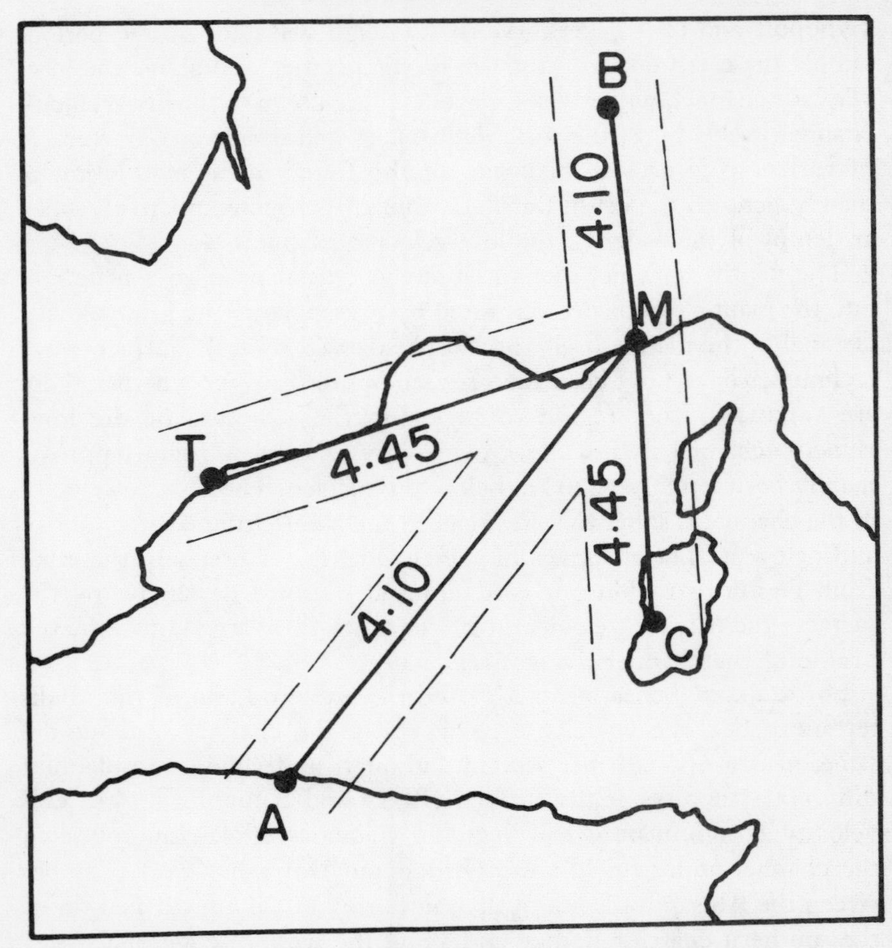

Fig. 4. *S* wave velocities in the low-velocity channel under the western Mediterranean and adjoining areas. M, Monaco; T, Tortosa; C, Cuglieri; A, Algeria; B, Besançon. After Berry and Knopoff (1967).

Evidence for lateral heterogeneity in the mantle has been cited by Oliver (1962). The dispersion of Love waves under oceans and continents was considerably different at periods less than roughly of 100 sec (Fig. 5). Surface waves of 100 sec periods penetrate to depths of 100 to 150 km; thus, the properties of the mantle underneath oceans and continents are different from one another, at least to depths of this order. Once again we seem to find that the lateral heterogeneity, associated in the outer regions of the earth with topography

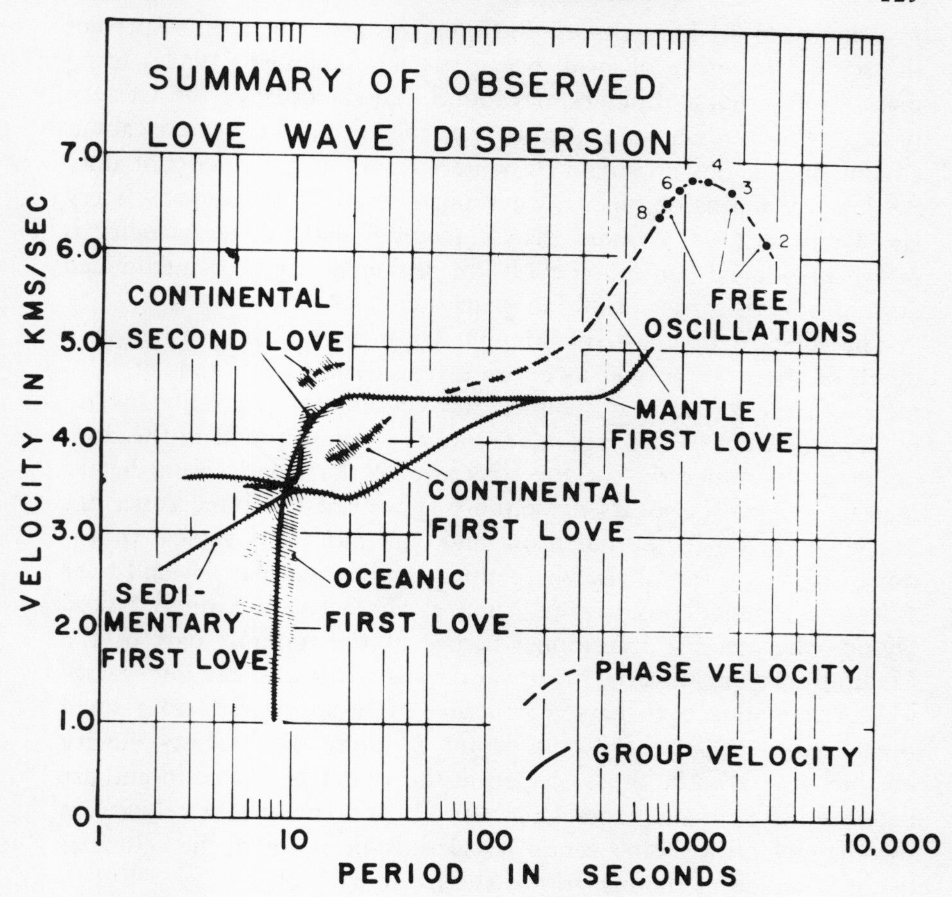

Fig. 5. Love wave (*Q* wave) dispersion under oceans and continents. After Oliver (1962).

and with variations in crustal properties, is not so pronounced at the depths corresponding to the low-velocity channel.

Structure of Small Ocean Basins

One of the principal results of marine geophysical exploration has been the determination that the structure of the crust in the deep oceans is relatively uniform. In the parts under deep water, remote from the continental shelves and from the midoceanic ridges, the crust is thin (10 to 1 km thick), and is underlain by material with a

P_n velocity of 8.1 km per second. Could this result be used to predict the seismic structure of small ocean basins? Fahlquist (1963), using marine refraction techniques, has found that the crust of the western basin of the Mediterranean Sea is relatively thin, as expected (about 11 km thick), but underlain by material with a P_n velocity of only 7.8 km. By means of surface-wave phase velocity techniques, Berry and Knopoff (1967) found that, if material with higher velocity is to be introduced beneath the 7.8 km region, it must be introduced at depths of the order of 30 km or greater.

The western basin of the Mediterranean shelves into water at a depth of about 3 km over its entire extent within a few tens of kilometers of the shore. The Bouguer gravity anomaly is positive. Many of the properties of this region are similar to those found in the deep ocean basins. Nevertheless, the structure of the mantle immediately beneath the crust appears to be more typical of the Ivrea zone, not many kilometers to the north on land, and indeed of typical transitional structures found elsewhere along continental margins and over extensive mountainous regions, such as the western United States. Despite the seeming transitional nature of the topmost part of the Mediterranean mantle, the mantle at depths of those of the low-velocity channel appears to have a structure more appropriate to a small ocean basin; the structures to depths of those of the low-velocity channel appear to be dome-shaped in the center of the basin and are depressed toward the edges (Fig. 6). The extremely low velocity in the channel in the basin's center has already been noted; the velocities in the channels at the margins of the basin are higher.

Additional Evidence for Heterogeneity in Earth's Upper Mantle

The *P* wave structures at depths of the order of the low-velocity channel are not known in the same detail as those for *S* wave distributions. Surface wave methods are not sensitive to variations in the compressional wave velocities in the mantle. By using body waves from nuclear explosions, Lehmann (1962; 1964a, b) discovered a first order discontinuity in the *P* wave velocity distribution at a depth of 225 km in the western United States. The velocity of *P* waves appears to jump to a value of 8.4 or 8.5 km per second. The presence of a sharp discontinuity was later substantiated by a compilation of observations of reflections from this horizon (Knopoff and co-workers, 1966). Leh-

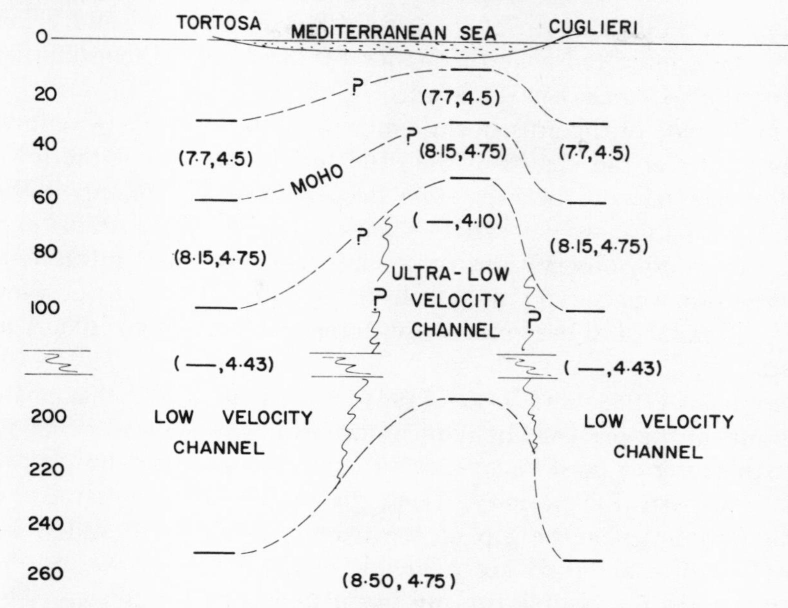

Fig. 6. Structure of upper mantle under the western Mediterranean and adjoining areas. After Berry and Knopoff (1967).

mann (1959, 1961) has found a similar horizon in Europe, and Carder and co-workers (1966) have also found it in the western Pacific from observations of nuclear explosions. A discontinuity with a similar velocity is found at a depth of 120 km under the eastern United States (Carder and co-workers, 1966; Healy, 1968).

These observations of a discontinuity in *P* wave velocity at a depth of the order of several hundred kilometers indicate, at least for those regions of the world where they have been observed, that a postulate formerly made with regard to the mantle may be untenable. Until recently, it had been a strong assumption that there were no first order discontinuities in physical properties of material between the bottom of the crust and the core, and one could therefore argue about the degree of chemical homogeneity of the mantle. Carder and co-workers (1966) have also indicated that they have observed a first order discontinuity in *P* wave velocity at a depth of about 640 km, from nuclear explosions under western North America and the western Pacific. Johnson (1967) and Archambeau and co-workers (1969) have pointed out the existence of two major steps in the *P*-wave depth distribution, at depths of 360 km and 600 km. Engdahl and Flinn (1969) have indicated that the upper of these two discontinuities appears to be sharper than the lower.

As in the case of the crust-mantle interface, observations of seismic discontinuities at the depths of the order of the bottom of the low-velocity channel, and perhaps even deeper, have bearing upon the mineralogic nature of the earth's mantle. Are the sharp transitions which permit the observations of seismic reflections and refractions phase transitions between polymorphs of the same mineral, or do they represent interactions between two different chemical constituents in contact?

Verhoogen (1965) and Tozer (1965) have pointed out that phase transitions in the presence of temperature gradients such as occur in the earth's interior must be spread out over a broad zone and cannot occur at a sharp discontinuity. Thus, the problems of the nature of the discontinuities at the top of the mantle, at 225 km, and those deeper (if these are real) are reducible to the determination of the zone of depths permissible for any postulated phase transitions. The discontinuities at the bottom of the "crust" appear to take place over seismically short intervals of depth, probably of the order of a few kilometers. The deeper discontinuities may also be consistent with occurrence of phase transformations, although Anderson (1967) finds that these must also be accompanied by changes in chemical composition. If chemical changes do occur, then convection of mantle material cannot take place across these horizons since convection is a homogenizing process. If these discontinuities are phase transformations the likely high viscosity of the denser phases are probably sufficient to inhibit any significant convective transport of mantle material across these horizons.

Is the Lower Mantle Homogeneous?

Because of the lack of available pertinent seismic data, region D of Bullen's model, the lower mantle, was constructed under conditions which bear the strong imprint of a postulate of homogeneity for this region. Recent seismic observations have cast doubt on the assumption that the lower mantle is homogeneous. Toksöz and Chinnery (1967) used the large aperture seismic array (LASA) in Montana to observe earthquakes distributed along the major earthquake belt to the west, along the Aleutians, Kamchatka, and the Kuriles. They show that there are at least three ripples in the velocity-depth curve within the lower mantle (Fig. 7). These ripples are inconsistent with any postulate of smoothly varying velocity where the velocities increase only because of the change in pressure and temperature, and suggest that the lower mantle is at least radially inhomogeneous. A conclusion cannot be drawn concerning lateral inhomogeneity at these depths on the basis of one set of observations. There is as yet no evidence for the order of these discontinuities; the observations do not exclude the possibility of low-velocity channels in the lower mantle.

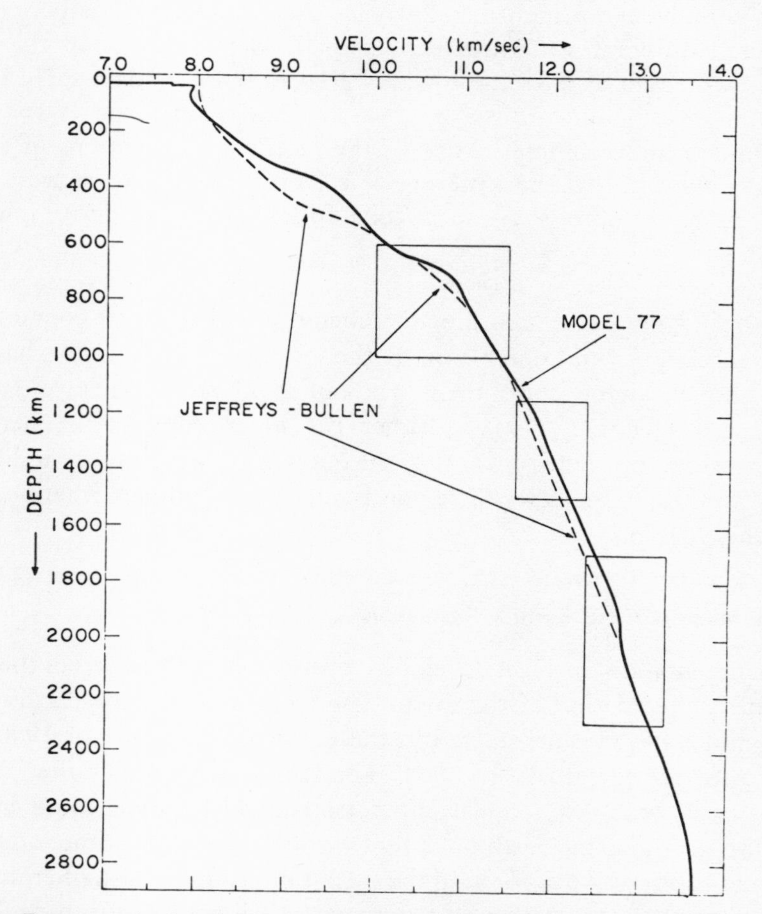

Fig. 7. *P* wave velocity-depth function for the mantle; note anomalous values in the three boxed areas. After Toksöz and Chinnery (1967).

Inner Core–Outer Core Boundary

The interface between the inner and outer core, Bullen's region F, was the subject of disagreement between Gutenberg, on the one hand, and Jeffreys and Bullen on the other, with regard to the manner in which the velocity-depth curve should be drawn through this region. The J-B curve had a minimum in it; Gutenberg's curve was drawn smoothly to connect the velocities at the bottom of the outer core and at the top of the inner core. However, Gutenberg discerned several arrivals for some of the waves grazing the surface of the outer core, each with distinct principal frequencies. Accordingly, Gutenberg constructed different structures for region F, depending upon the frequency content of the seismic waves. Recent sensitive data analysis by several investigators has shown that the travel-time curves for this region are quite complex. Most of the recent interpretations of these curves indicate that the structure of region F consists of at least two and perhaps more steps, probably followed by velocity minima (Fig. 8).

The existence of at least two steps in region F places some severe constraints upon the nature of the chemical and physical contact between the inner and outer core. If the inner core is the solid phase of the molten outer core, then any multifold step structure is not expected (Knopoff, 1959). Either the inner core has a chemical composition distinct from that of the outer core, which seems unlikely, or the stepwise structure of region F indicates a complex fractionation of this region.

Free Modes of the Earth's Oscillation

Instrumentation has been developed sensitive enough to detect the free modes of oscillation of the earth. The frequencies of the earth's free vibrations have been detected in the Chilean earthquake of 1960 and the Alaskan earthquake of 1964. The free modes of vibration are of two basic types: spheroidal and torsional. The various lines in the spectroscopy of the earth have finite widths due to the finite attenuation of elastic wave motions in the earth's interior. In the higher modes, the spectral lines overlap and the spectral distribution approaches a continuum. In this range, the free modes are indistinguishable from

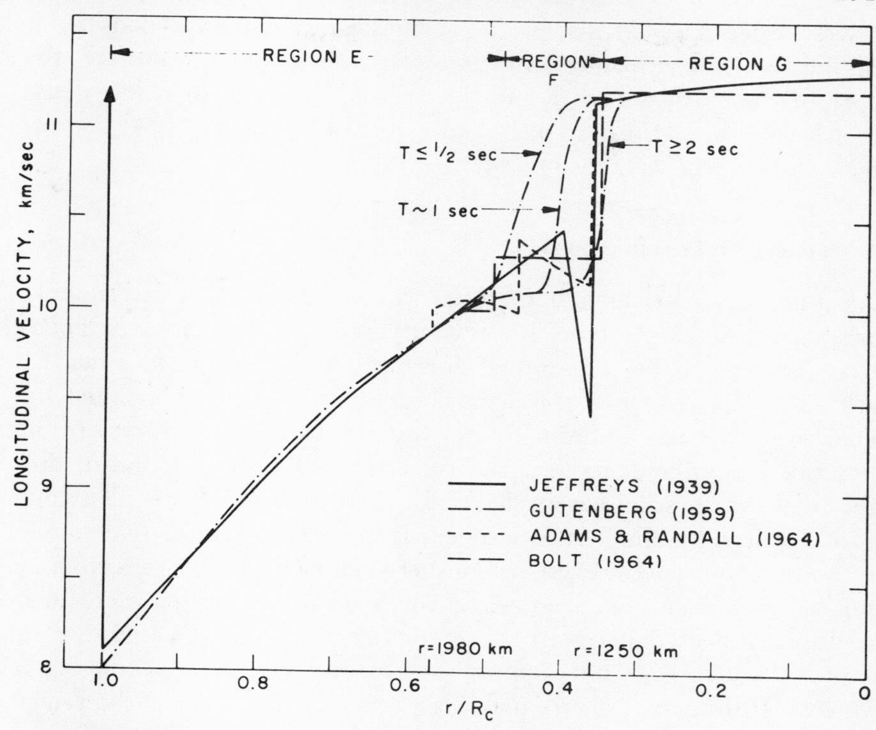

Fig. 8. Models of velocity structure for the core. After Hannon and Kovach (1966).

propagating surface waves; the torsional modes at high frequencies degenerate into Love waves, while the spheroidal modes degenerate into Rayleigh waves. Because of the rotation of the earth, the spectral lines are split, much in the same way as atomic line spectra are split in the presence of a magnetic field, due to the Zeeman effect.

Precise observations of the frequencies, line-splitting parameters, and line widths have led to refined methods for estimating the velocity distributions and densities in the deeper parts of the earth's interior. The lowest ordered free modes of oscillation of the earth represent the most potent way of sampling the lower mantle and the core (the latter in the case of the spheroidal modes only). The spectral line widths have provided information about the attenuation in the earth's interior, and have led especially to the result that the attenuation factors in the lower mantle are less than those in the upper one. The precise determination of the spectral frequencies, especially in the lowest ordered modes, has made previous estimates of the densities at the bottom of the mantle and the top of the core inappropriate; the core of the earth is denser and the mantle of the earth is less dense than suspected on the basis of Bullen's model (Anderson, 1964; Landisman and co-workers, 1965).

Attenuation of Seismic Waves

As indicated in my introductory remarks, the problem of determining seismic attenuation factors has recently received attention. Attenuation factors can be determined for both body and surface waves. Although, in principle, the attenuation factors can be obtained by comparing the amplitudes of these waves at two different stations, in practice it is difficult to eliminate the effects of scattering and of differential transmission across sharp discontinuities in the earth's structure. Perhaps the simplest experiment is to measure the relative amplitudes of body waves reverberating between two perfect reflectors at normal incidence. This can occur for *S* waves at normal incidence, multiply reflected between the earth's surface and the liquid core (since the core does not support shear, it is a perfect reflector for *S* waves). If deep-focus earthquakes are used and isotropy of the radiation pattern about the focus is assumed, the attenuation factors for the mantle above and below the shock can be compared. The lower mantle attenuates seismic waves much less than does the upper mantle (Anderson and Kovach, 1964). Thus, not only is there a significant difference in seismic-wave velocity between the upper and lower mantles, but also in the ability to attenuate elastic waves and convert the vibratory motions into heat. Most of the efficient conversion of *S*-wave energy into heat takes place in the upper mantle.

Although the actual elastic-wave velocities are normally used as parameters characteristic of a given region, dimensionless attenuation factors are more appropriate to describe these characteristics of different regions of the earth's interior. The specific attenuation factor is given the symbol $1/Q$; the quantity Q being the number of periods for a damped wave to fall to an amplitude of $e^{-\pi}$. Similarly, Q is equal to the ratio of the peak energy in a cycle of an oscillatory signal to the energy lost in that cycle.

One of the principal results of recent investigations has been the observation that, when elastic waves undergo attenuation in solids in the laboratory, *Q* seems to be independent of frequency (Knopoff, 1964a), a behavior which is not typical of liquids. This result suggests that the mechanism of the attenuation process is basically nonlinear, probably associated with stress reversals in the incident elastic wave in the presence of dislocations of the solid (Knopoff and MacDonald, 1960). When the result that *Q* is independent of frequency is applied to the interpretation of the attenuation of *Q* waves, the distribution of *Q* for *S* waves with depth can be estimated. The attenuation factors correlate quite well with those obtained from the observations of body waves. The *Q* values for the upper mantle appear to be about 110, while for the lower mantle they are much larger, perhaps as large as 1,500 (Fig. 9). An intermediate zone of high attenuation lies between depths of about 300 km and 600 km; the attenuation in this zone is greater than in the regions above or below (Knopoff, 1964a). Anderson and Archambeau (1964) place this zone somewhat higher in the mantle, probably at depths of the order of the low-velocity channel. The possibility exists that the observation that *Q* is independent of frequency applies only at low pressures, where dislocations in crystals are appropriate.

At present, little can be said about *Q* values for *P* waves in the mantle, or for *S* waves in the crust. The *Q* values for *S* waves are not particularly high, nor is it possible to establish whether the stratification which has been described occurs in discrete layers or is gradual.

Viscosity in Earth's Mantle

The problems of elastic-wave propagation can be studied once an equation of state has been constructed; for elastic deformations of a solid, the appropriate equation of state is Hooke's law, at least for those cases in which the strains of deformation are small. Similarly, in order to understand the nature of the flow of fluids in response to external forces, an equation of state is necessary. Perhaps the most familiar equation of state is that which states that a fluid under a constant shear stress will flow at a constant rate of strain. The constant of proportionality is called the viscosity, and a fluid which exhibits this type of deformation is said to be a Newtonian viscous fluid. Most probably, the forces in the earth tending to produce deformations are reasonably steady, at least over periods of thousands of years. Is the rheologic

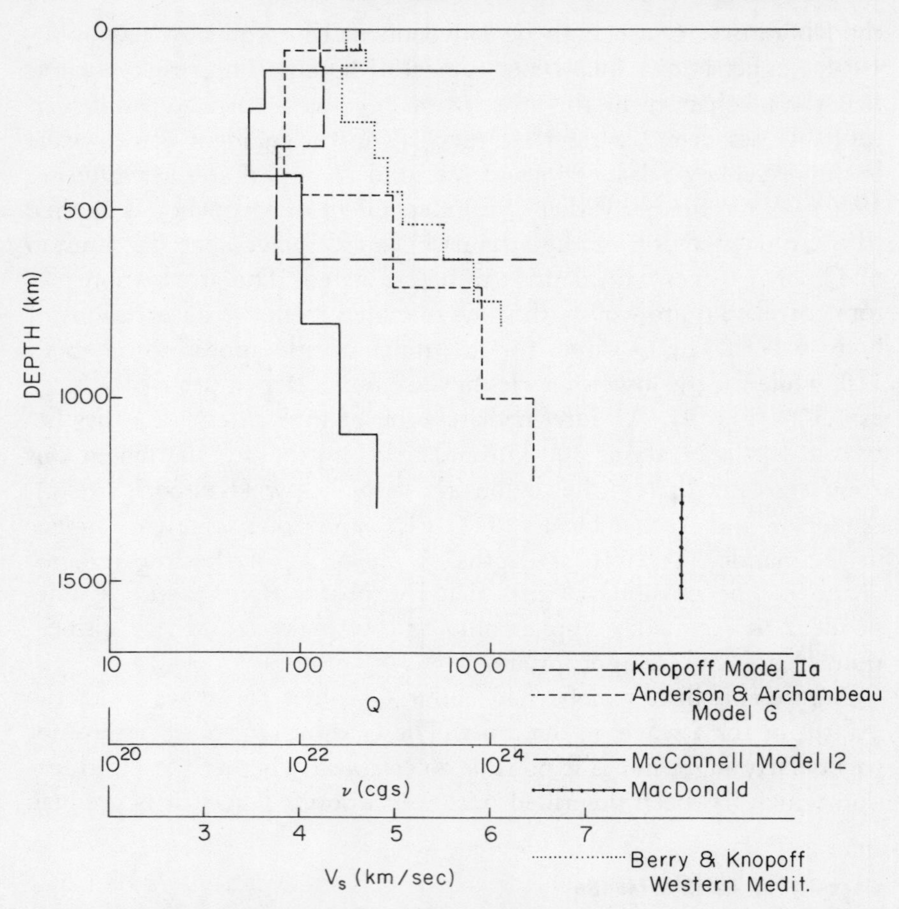

Fig. 9. Viscosity-depth and attenuation-depth functions for the mantle. After Knopoff (1964).

behavior of the earth, taken as a fluid, one which exhibits Newtonian viscosity? Some parts of the earth certainly do not behave as a Newtonian fluid; in those areas where earthquakes occur, the flow can hardly be considered steady, the motion in these regions being jerky rather than continuous and smooth.

Actually, the appropriate equation of state for the deformation of different parts of the earth appears to be a significant function of the magnitude of the applied deformational stress and the temperature. Different behavior is expected, for example, in the presence of low temperatures and high deforming stresses, high temperatures and low deforming stresses, or high temperatures and high stresses. If the deforming stresses are too small to cause movement of dislocations and the temperature is high enough, the only plausible mechanism for the deformation of a solid is that associated with the diffusion of atoms or vacancies through the crystal lattice or along grain boundaries (McKenzie, 1966). This type of deformation, called Herring-Nabarro creep, is indeed one in which the strain rate depends linearly upon the stress for a given constant temperature. In this type of deformation, therefore, the flow may be said to be that of a Newtonian viscous fluid.

At lower temperatures and at higher stresses, creep takes place by the movement of dislocations. In such case, a solid has a finite yield strength and brittle fracture is a possible mode of deformation. Thus, although some initial doubts may have been expressed concerning the applicability of the model of Newtonian viscosity to much of the mantle, apparently, except where earthquakes also occur close to the surface, the model is appropriate, at least for long-term deformations under small loads. From an estimate of the sizes of the deformational energies, it seems reasonably certain that the two basic experiments in which measurements have been made of the viscosity of the mantle as a parameter of response to small shear deformations—the rebound of postglacial Fenno-Scandia and post-Pleistocene Lake Bonneville on the one hand and the nonhydrostatic bulge of the earth on the other—are appropriate to describe parameters of Newtonian viscosity in the mantle.

From observations of perturbations of satellite orbits, MacDonald (1963) has shown that the earth's surface is in a nonequilibrium shape for a fluid rotating with the present angular velocity of the earth. The nonequilibrium shape is presumably due to a fossil bulge representing the shape of the earth at some earlier period in its history when it was rotating faster. The viscosity of the earth's mantle can be estimated from observations on the rate of slowing of the earth's rotation (Munk and MacDonald, 1960). There is good evidence, from interpretation of earlier measurements of the rate of rebound of the Fenno-Scandian regions (Haskell, 1935–1937) following the removal of the glacial ice cap of the ice ages and of measurements by Crittenden (1963) on the rebound of Pleistocene Lake Bonneville, that the viscosity of the upper mantle is very much lower than that of the entire mantle. The viscosities obtained from the rebound observations are of the order of 10^{21}

to 10^{22} cgs units; the viscosity of the entire mantle is about 10^{26} cgs units. McConnell (1965), after a detailed analysis of the Fenno-Scandian rebound data, found that the viscosity of the upper mantle increases with depth (Fig. 9). Apparently the seat of the material with high viscosity, of the order of 10^{26} cgs units, is in the lower mantle. This is, strikingly, also the region with the low attenuation factor for seismic waves; Anderson (1966) has pointed out the correlation between viscosity and attenuation factors in the mantle. Perhaps it is not coincidental that a material with low efficiency in dissipating energy in the seismic-acoustic field also has a low efficiency in dissipating energy in the field of deformational flow. Thus, a material with low viscosity and a high seismic attenuation factor occupies the outermost part of the earth's mantle. It should be noted, in passing, that a high viscosity, of the order of 10^{26} cgs units, is sufficient to exclude Rayleigh convection as a possible state of motion for the lower mantle (Knopoff, 1964b).

Breaking Strength in Upper Mantle

In the regions of the mantle where the processes of deformation can be described by Newtonian viscous flow, probably associated with a process of diffusional creep as just outlined, the appropriate parameter is one of kinematic viscosity. This type of description is inappropriate for regions of the earth where fracture phenomena are prevalent. Here a different parameter must be used, and probably the appropriate one is the yield strength or fracture strength of the rocks.

Perhaps the easiest way to study this parameter is to investigate the ability of the rock to store energy before fracture, at least in those regions where earthquakes occur. The annual release of earthquake energy as a function of depth is shown in Figure 10. The only shocks which contribute significantly to this curve over a 50 year period are catastrophic ones; intermediate and small-sized shocks are not energetic enough to be observed on this curve.

Rocks with greater breaking strengths can store more energy of deformation before rupture occurs; hence, the greatest shocks occur in the regions of greatest yield strength. Thus, at least in regions where earthquakes occur, the crust is stronger than the mantle and the strength decreases with increasing depth in the mantle. Surprisingly, the upper mantle appears to have possibly two zones of low-

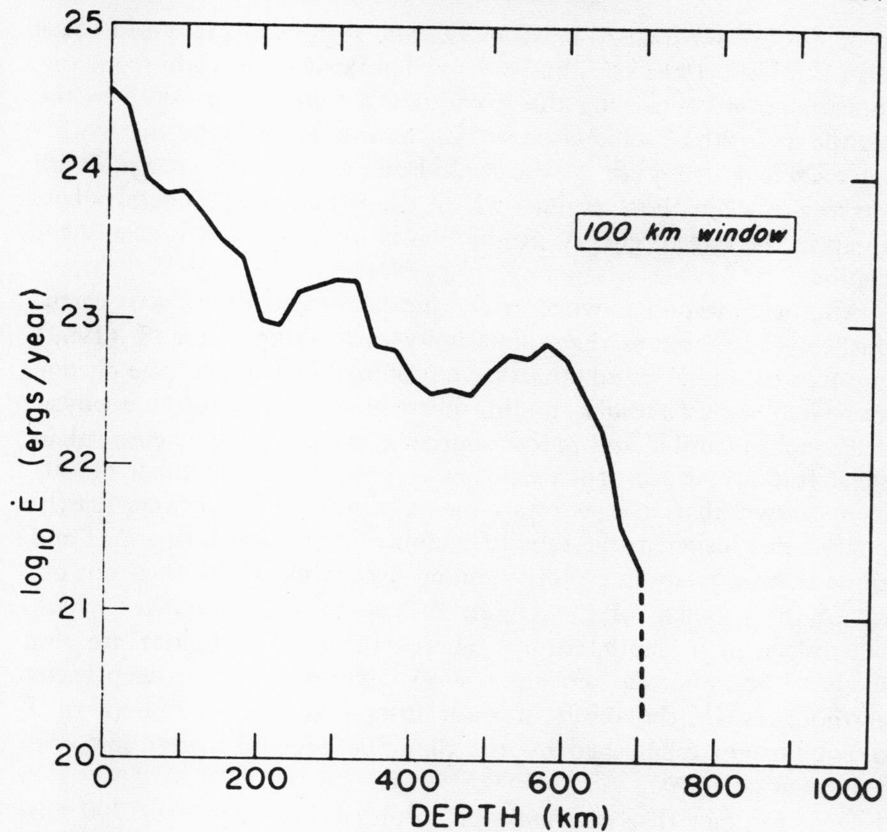

Fig. 10. Worldwide energy rate-depth function for earthquakes. After Knopoff (1964).

yield strength, each zone having a stronger zone beneath it. These zones are found below the low-velocity channel, in the same depth range as the low-viscosity zone under Fenno-Scandia, and in the zone of high attenuation of seismic waves, at least on my model.

Several assumptions are implicit in this interpretation, so that some of the results cited should not be taken literally. First, the curve of rate of energy release as a function of depth is obtained for worldwide catastrophic shocks. There is no reason to expect that the interpretation of this curve should apply to all strongly seismic regions. For example, no mantle shocks have been observed along the Pacific coast of the United States, the contributions to the curve from shocks at depths greater than 30 km coming from regions of the world other than the United States. Thus, no assumption can be made from this interpretation concerning the strength as a function of depth in the mantle under the Pacific coast of North America. Despite the world-wide average of Figure 10, the second minimum in the energy-depth curve is real; in those regions where deep-focus earthquakes occur, a gap in the rate of energy production is found at slightly shallower depths.

Another question is whether the mechanism of deep-focus earthquakes is the same as that for shallow earthquakes. Benioff (1963) and Randall (1964), and others have proposed that the volume change associated with a rapidly running phase transition might be a physically more plausible and potent source of seismic energy release than shear fracture, at least for deep shocks. Knopoff and Randall (1970) have shown that it is not possible to discriminate between earthquakes that originate in rapidly running phase transformations and those that originate in rapidly running dislocations. This does not get rid of the problem of the origin of deep focus earthquakes which take place in a depth regime where fracture instabilities are not likely to be important. Griggs (1969) has proposed that deep focus earthquakes are the result of shear instabilities due to the thermal disequilibrium established by the plunging of cold matter into the mantle at trenches.

Does the fact that no shocks are found below depths of 700 km imply that the lower mantle has a low breaking strength? The answer is that below a depth of 700 km thermal stresses are probably removed by diffusion creep. The only mechanism for accumulation of stresses is by convection, and convection is forbidden in this region because of the high viscosity (Anderson, 1966). In other words, it may be postulated that diffusion-creep phenomena, which led to the application of a kinematic viscosity for those parts of the upper mantle far from the surface and where earthquakes do not occur, may also apply globally to the earth at depths greater than 700 km.

Model Seismicity

The principal seismologic evidence that parts of the earth's interior are in relative motion is the fact that large earthquakes do exist. Significant fault scarps and other surface traces of rupture are associated with major earthquakes. Orogeny seems to occur in regions where earthquakes occur frequently.

The rate of occurrence of earthquakes and the associated energy release is probably relatively uniform over a period of a million years, but it is of considerable interest to construct a model for the occurrence of earthquakes on a short time scale. Until recently, no satisfactory model of the occurrence of earthquakes had been made, although a considerable body of statistical data associated with focal phenomena has been available. The mean seismic activity of different regions is known, as is the geographic distributions of regions in which earthquakes can occur. Major shocks are often followed by aftershocks. In a number of regions it has been found that earthquakes do not occur randomly in time; there seems to be a causal connection among earthquakes in a given region, so that the occurrence of a shock depends on the conditions resulting from a preceding shock.

Burridge and Knopoff (1967) have constructed a laboratory model and a numerical model of seismic focal phenomena based on the concept that friction at a fault surface is the principal generator of focal phenomenology. The model consists of a one-dimensional chain of discrete masses interconnected by springs. The masses rest upon an immovable tabletop with friction between the masses and the table. The masses are driven by "tectonic driving forces." With this model, frequency-magnitude relations have been observed for shocks of all sizes that are similar to those obtained in nature, and sequences of shocks are produced that are nonrandom in nature. The efficiency of converting potential energy into seismic energy was found to be independent of shock magnitude.

The principal role of the friction is to provide a mechanism whereby energy can be rapidly transferred from one part of the chain of masses to another part. Energy is concentrated at various spots along the chain of masses, due to stress concentrations at the termini of rupture surfaces of earlier shocks. A typical sequence of shocks along a fault is one in which small and intermediate sized shocks occur with relative regularity, producing stress concentrations and hence energy concentrations distributed along the fault. When enough energy has been concentrated along the fault, a catastrophic shock occurs in which a large fraction of the energy is released in seismic energy and in heat; not all the available potential energy is released for any shock, including the greatest ones. A catastrophic shock is accompanied by aftershocks due to the coupling of auxiliary faults to the principal fault by viscous elements; these auxiliary faults are also near a critical state. The viscous elements provide the time delays for the aftershocks; the time scale for aftershocks is much shorter than that for the small and intermediate shocks in the part of the cycle during which energy is being loaded into the fault. After the aftershocks have stopped, the system again starts on a cycle of small and intermediate shocks which load energy into the system.

Convection in Earth's Interior

Evidence in favor of a hypothesis of continental drift probably should not be reviewed in a chapter devoted to seismology and rheology. The pertinent evidence is reviewed elsewhere in this volume: the fit of the margins of the continents around the Atlantic; the paleomagnetic evidence for continental drift; the magnetic lineations in the sea floor and the geochronologic evidence for the spreading of the sea floor, the evidence for the disappearance of the lithosphere into the trenches, the focal mechanisms of earthquakes and the constructions of rigid plate tectonics, and the evidence of the drilling at sea. In addition to the present review, a review is given by Knopoff (1969). However, the discussion of a mechanism for drift is appropriate for this chapter.

Thermally driven convective motions in the earth's mantle have been suggested as a possible driving mechanism to account for continental drift, as well as for orogeny, earthquakes, etc. In fact, this is the only proposed mechanism which seems qualitatively potent enough to drive the continents apart. They are presumed to have moved apart by virtue of some process involving mass transport in the mantle.

If the mantle were homogeneous and with the viscosity determined from glacial uplift experiments, it would be quite unstable to Rayleigh-Bénard convection under trivially small radial temperature gradients. Thus the flows could operate as required by the theory. However, the mantle is not homogeneous; heterogeneity of the mantle places constraints on the allowable patterns of circulation of the mantle's material, considered here as a viscous fluid. For example, the arguments presented earlier indicate that, instead of the flow being required to move about 35 km or so of crustal matter, the outer parts of the earth, probably down to the low-velocity channel, must move as a unit. Thus, the thickness of the continental rafts is considerably greater than heretofore imagined.

If the viscosity of the lower mantle is higher by a factor of 10^5 than the upper mantle, this alone is sufficient to inhibit significant vertical convective flows in the lower mantle. Thus, mass transport of a great enough magnitude to account for the velocities of centimeters per year must, on this basis alone, be localized somewhere in the vicinity of the low-velocity channel.

The sharp discontinuity at 225 km apparently is an additional barrier to convection. Convection, basically a stirring mechanism, is incompatible with chemical discontinuities but may occur in the presence of a phase transition. An upwelling convection current traversing a phase transition is cooled if the reaction is endothermic. The cooling could be provided by the temperature gradient if the zone of transition is broad, but not if the transition takes place over a narrow interval of radius. The discontinuity at 360 km, already indicated as being relatively sharp and which may be associated with the bottom of the low-velocity channel and is most likely associated with a major phase transformation, may thus be a further barrier to convection unless the discontinuity lies in a zone of stagnancy. Mapping of this discontinuity in the future will shed additional light upon this problem.

If the range of depths in which convection becomes possible is more and more restricted, the magnitudes of the flows, and hence of the driving forces for continental drift, are significantly reduced on the Rayleigh-Bénard model.

The principal difficulties connected with theories of convection is that the horizontal scale of the convection is of the same order of magnitude as the vertical scale. If convection is confined to a layer approximately 600 km thick, or less as I have proposed, the horizontal scale must be of this order of magnitude, and it becomes difficult to imagine how the continents could have got as far apart as they are at present. An alternative theory to Rayleigh-Bénard convection has been proposed in which horizontal temperature gradients are used to drive the convection (Allan and co-workers, 1967). If the source of heat is confined to a small zone laterally, such as at the mid-Atlantic ridge or along a continental margin, it can be shown that the horizontal scale of the convection is once again of the order of the vertical scale. Thus, although it might be possible to generate earthquakes around the Pacific margin by taking advantage of the horizontal gradient of heat flow due to the differential mantle heat flow at shallow depths under oceans and continents, this gradient would not be potent enough to drive the continents apart to their present position.

How, then, can the continents have been separated by distances of the order of thousands of kilometers? In one possibility Howard and co-workers (1970) have proposed a model in which a continental raft of heat sources becomes self-propelled through a fluid once it acquired a sufficient momentum. The critical velocity is acquired through the mutual interaction an image continental raft receding from a common line of symmetry which may be thought of as the oceanic ridges. The critical velocities are of the correct magnitudes for the earth, namely of the order of cm/year.

Summary

The results of recent seismologic investigations have shown that the structure of the earth's crust, mantle, and core is more complex than hitherto suspected. These structural complexities have significant implications with regard to the chemical and physical constitution of the interior. For example, it is not clear whether the crust is the result of a simple differentiation of mantle material, or whether the crust and mantle are polymorphic phase changes of one another. Seismic discontinuities at depths of about 360 km and 600 km indicate that phase transformations may be important in determining the structure of Bullen's region "C." A complex inner core–outer core boundary indicates significant chemical complexity at those vast depths. That the lower mantle is significantly different from the upper one is borne out by the fact that, where the seismic velocities are higher in the lower mantle, the attenuation of elastic waves is significantly lower; the viscosity is much higher in the lower mantle than in the upper mantle.

Because the lower mantle is highly viscous, thermal convection is unlikely. On the other hand, the horizontal scale of thermal convection is probably of the same order as the vertical scale. If thermal convection is confined to the upper mantle, then convection due to vertical temperature gradients cannot have a great enough lateral extent to have driven the continents as far apart as their present locations. Seismic discontinuities at depths of the order of 360 km and 600 km in the mantle indicate further chemical differentiation and argue against convective mass transport across these boundaries. Convection appears to be constrained to thin layers, coupled to a highly mobile world-wide system of moving, rigid, interacting plates.

References

Adams, R. D., and M. J. Randall: The fine structure of the earth's core. *Bull. Seismol. Soc. Amer.* 54:1299–1313, 1964.

Aki, K.: Some problems in statistical seismology, *Zisin* 8:204–228, 1956.

Allan, D. W., N. O. Weiss, and W. B. Thompson: Convection in the earth's mantle, in *Mantles of the Earth and Terrestrial Planets* (S. K. Runcorn, ed. Interscience Pubs.), 507–512, 1967.

Anderson, D. L.: Densities of the mantle and core. *Trans. Amer. Geophys. Union* 45:101, 1964.

Anderson, D. L.: Universal Dispersion Tables 1. Love waves across oceans and continents on a spherical earth. *Bull. Seismol. Soc. Amer.* 54:681–726, 1964a.

Anderson, D. L., and C. B. Archambeau: The anelasticity of the earth. *J. Geophys. Res.* 69:2071–2084, 1964b.

Anderson, D. L.: Earth viscosity. *Science* 151:321–322, 1966.

Anderson, D. L.: Phase changes in the upper mantle. *Science* 157:1165–1173, 1967.

Anderson, D. L., and R. L. Kovach: Attenuation of shear waves in the upper and lower mantle. *Bull. Seismol. Soc. Amer.* 54:1855–1864, 1964.

Archambeau, C. B., E. A. Flinn, and D. G. Lambert: Fine structure of the upper mantle. *J. Geophys. Res.* 74:5825–5865, 1969.

Benioff, H.: Source wave forms of three earthquakes. *Bull. Seismol. Soc. Amer.* 53:893–903, 1963.

Berry, M. J., and L. Knopoff: Structure of the upper mantle under the western Mediterranean basin. *J. Geophys. Res.* 72:3613–3626, 1967.

Birch, F.: Elasticity and constitution of the Earth's Interior. *J. Geophys. Res.* 57:227–286, 1952.

Birch, F.: The velocity of compressional waves in rocks to 10 kilobars, Part 2. *J. Geophys. Res.* 66:2199–2229, 1961.

Bolt, B. A.: The velocity of seismic waves near the earth's center. *Bull. Seismol. Soc. Amer.* 54:191–208, 1964.

Brune, J. N., and H. J. Dorman: Seismic waves and earth structure in the Canadian shield. *Bull. Seismol. Soc. Amer.* 53:167–210, 1963.

Bullen, K. E.: *Introduction to the Theory of Seismology*, Cambridge University Press, 2nd Ed., 1953.

Burridge, R., and L. Knopoff: Model and Theoretical Seismicity. *Bull. Seismol. Soc. Amer.* 57:341–371, 1967.

Carder, D. S., D. W. Gordon, and J. N. Jordan: Analysis of surface-foci travel times. *Bull. Seismol. Soc. Amer.* 56:815–840, 1966.

Crittenden, M. D., Jr.: Viscosity of the earth. *J. Geophys. Res.* 68:5517–5530, 1963.

Engdahl, E. R., and E. A. Flinn: PKPPKP reflections from discontinuities within the earth's upper mantle. *Science* 163:177–179, 1969.

Ewing, M., W. Jardetzky, and F. Press: *Elastic Waves in Layered Media*, McGraw-Hill Book Co., New York, 1951.

Ewing, M., and F. Press: Determination of crustal structure from phase velocity of Rayleigh waves, III: The United States. *Bull. Geol. Soc. Amer.* 70:229–244, 1959.

Fahlquist, D.: Seismic refraction measurements in the western Mediterranean Sea, Ph.D. thesis, MIT, 1963.

Ferraes, S. G.: Test of Poisson process for earthquakes in Mexico City. *J. Geophys. Res.* 72:3741–3742, 1967.

Fuchs, K., S. Mueller, E. Peterschmitt, J. -P. Rothé, A. Stein, and K. Strobach: Krustenstruktur der Westalpen nach refraktionsseismischen messungen. *Gerl. Beitr. z Geophysik.* 72:149–169, 1963.

Griggs, D. T., personal communication.

Hannon, W. J., and R. L. Kovach: Velocity filtering of seismic core phases. *Bull. Seismol. Soc. Amer.* 56:441–454, 1966.

Haskell, N. A.: The motion of a viscous fluid under a surface load, 1. *Physics* 6:265–269, 1935.

Haskell, N. A.: The motion of a viscous fluid under a surface load, 2. *Physics* 7:56–61, 1936.

Haskell, N. A.: The viscosity of the asthenosphere. *Am. J. Sci.* 33:22–28, 1937.

Herrin, E., and J. Taggart: Regional variations in P_n velocity and their effect on the location of epicenters. *Bull. Seismol. Soc. Amer.* 52:1037–1046, 1966.

Howard, L. N., W. V. R. Malkus, and J. A. Whitehead: Self-convection in floating heat sources: a model for continental drift. *J. Fluid Mechanics,* in press, 1970.

Johnson, L. R.: Array measurements of *P* velocities in the upper mantle. *J. Geophys. Res.:* 6309–6325, 1967.

Knopoff, L.: The upper mantle of the earth. *Science* 163:1277–1287, 1969.

Knopoff L., and M. J. Randall: The compensated linear vector dipole: a possible mechanism for deep earthquakes. *J. Geophys. Res.* submitted 1970.

Knopoff, L.: Velocity of sound in a two-component system. *J. Geophys. Res.* 64:359–361, 1959.

Knopoff, L., and G. J. F. MacDonald: Models for acoustic loss in solids. *J. Geophys. Res.* 65:2191–2197, 1960.

Knopoff, L.: *Q. Reviews of Geophysics* 2:625–660, 1964a.

Knopoff, L.: The convection current hypothesis. *Reviews of Geophysics* 2:89–122, 1964b.

Knopoff, L.: The energy rate-depth function for earthquakes. *Proc. Nat. Acad. Sci.* 51:1–3, 1964.

Knopoff, L.: The statistics of earthquakes in southern California. *Bull. Seismol. Soc. Amer.* 54:1871–1873, 1964d.

Knopoff, L., S. Mueller, and W. L. Pilant: Structure of the crust and upper mantle in the Alps from the phase velocity of Rayleigh waves. *Bull. Seismol. Soc. Amer.* 56:1009–1044, 1966.

Knopoff, L.: Density-velocity relations for rocks. *Geophys. J. Roy. Astron. Soc.* 13:1–8, 1967.

Kosminskaya, I. P., and Y. V. Riznichenko: Chapter 4, *Research in Geophysics,* vol. 2, *Solid Earth and Interface Phenomena,* H. Odishaw, ed., MIT Press, 1964.

Landisman, M., Y. Sato, and J. Nafe: Free vibrations of the earth and the properties of its deep interior regions, I: Density. *Geophys. J. Roy. Astron. Soc.* 9:439–502, 1965.

Lehmann, I.: *P′*. Bur. Centr. Seismol. Intern. A-14:3–31, 1936.

Lehmann, I.: Velocities of longitudinal waves in the upper part of the earth's mantle. *Annales de Geophysique* 15:93–118, 1959.

Lehmann, I.: *S* and the structure of the upper mantle. *Geophys. J.* 4:124–138, 1961.

Lehmann, I.: The travel times of the longitudinal waves of the Logan and Blanca atomic explosions and their velocities in the upper mantle. *Bull. Seismol. Soc. Amer.* 52:519–526, 1962.

Lehmann, I.: On the travel times of *P* as determined from nuclear explosions. *Bull. Seismol. Soc. Amer.* 54:123–139, 1964a.

Lehmann, I.: On the velocity of *P* in the upper mantle. *Bull. Seismol. Soc. Amer.* 54:1097–1103, 1964b.

MacDonald, G. J. F.: The deep structure of the continents. *Reviews of Geophysics* 1:587–665, 1963.

McConnell, R. K., Jr.: Isostatic adjustment in a layered earth. *J. Geophys. Res.* 70:5171–5188, 1965.

McKenzie, D. P.: The viscosity of the lower mantle. *J. Geophys. Res.* 71:3995–4010, 1966.

Munk, W. H., and G. J. F. MacDonald: *The Rotation of the Earth,* Cambridge University Press, 1960.

Nafe, J. E., and C. L. Drake: In *Interpretation Theory in Applied Geophysics*, by F. S. Grant and G. F. West, McGraw-Hill Book Co., New York, 1951.

Oliver, J.: A summary of observed seismic surface wave dispersion. *Bull. Seismol. Soc. Amer.* 52:81–86, 1962.

Pakiser, L. C., and D. P. Hill: Crustal structure in Nevada and southern Idaho from nuclear explosions. *J. Geophys. Res.* 68:5757–5766, 1963.

Press, F.: Crustal structure in the California-Nevada region. *J. Geophys. Res.* 65:1039–1051, 1960.

Press, F.: Displacements, strains and tilts at teleseismic distances. *J. Geophys. Res.* 70:2395–2412, 1965.

Randall, M. J.: On the mechanism of earthquakes. *Bull. Seismol. Soc. Amer.* 54:1283–1289; Seismic energy generated by a sudden volume change. *Bull. Seismol. Soc. Amer.* 54:1291–1298, 1964.

Toksöz, M. N., and M. A. Chinnery: *P*-wave velocities in the mantle below 700 km. *Bull. Seismol. Soc. Amer.* 57:199–226, 1967.

Tozer, D. C.: Heat transfer and convection currents. *Phil. Trans. Roy. Soc. London* 258A:252–271, 1965.

Verhoogen, J.: Phase changes and convection in the earth's mantle. *Phil. Trans. Roy. Soc. London* 258A:276–283, 1965.

Warren, D. H., J. H. Healy, and W. H. Jackson: Crustal seismic measurements in southern Mississippi. *J. Geophys. Res.* 71:3437–3458, 1966.

Woollard, G. P.: Crustal structure from gravity and seismic measurements. *J. Geophys. Res.* 64:1521–1544, 1959.

7

Continental and Oceanic Geophysics

PAUL L. LYONS

Sinclair Oil & Gas Co.,
Tulsa, Oklahoma

The Conrad discontinuity roughly divides the entire crust of the earth into an upper and lower part. This separation, or interface, is detected by its velocity characteristics in seismic refraction profiles and contributes large anomalies to gravity and magnetic maps. Its study has some bearing on continental drifts, since the layer beneath the discontinuity provides a foundation for continents and must move with them if indeed they move as a whole. The study reported here suggests that large-scale crustal movements are confined to lateral displacements along faults.

Conrad Layer

The Conrad discontinuity was first described as such by V. Conrad in 1925. He defined it as a phase with compressional velocity equal to $\pm 6\frac{1}{2}$ km per second at its deepest point, and, correspondingly, with shear velocity equal to ± 3.6 km per second. Conrad thought of it as a continental layer, present only under the permanent continents. However, it has a counterpart under the oceans, where it is the so-called oceanic layer, defined by a compressional velocity range of 6 to 7 km per second.

The discontinuity is commonly referred to as marking the boundary between the overlying "granitic" and an underlying "gabbro" layer, although many observers deny the existence of an abrupt change. However, the velocity change from the overlying granitic layer on land, and from the overlying "layer 2" of the oceans is often quite distinct. The thesis of this report is that the observations of gradual change and abrupt change are all valid and not incompatible because of the irregular distribution of the Conrad layer beneath the discontinuity. For example, it is not present everywhere, it is sometimes at the surface of the crust, and its velocity may vary with depth of burial and rock composition. A further thesis is that it is next to impossible to differentiate the oceanic Conrad layer from the continental one, and that the two may indeed be continuous and much the same in original composition.

The oldest rocks of the earth that are widespread—let us call them Archean—apparently have a velocity in the Conrad range. Reported compressional velocities as high as 6.25 km per second are reported from the Archean rocks in the shield area of Canada. Since the rocks are not buried, a lesser velocity than the maximum is indicated. For comparison, the compressional velocity of Ordovician limestone is 5.5 km per second, and that of gabbro on the surface averages 5.4 km per second.

My principal thesis is that certain large interbasement gravity and magnetic anomalies appear to be the expression of masses which lie at depths at or below that of the Conrad discontinuity.

Primordial Rocks

A worldwide classification of Precambrian rocks by Hurley (1965) is shown in Figure 1. Of primary interest to our study are the rocks below the great Algoman unconformity, known as the Kenoran unconformity of Canada and the Karelian unconformity of Europe, and it is perhaps the greatest unconformity known. It is the consensus that it is the only Precambrian break continuous around the entire earth, and is dated, by clock minerals, as 2.5 billion, ±150 million years, old. Stille states: ". . . this interval represents the most profound break in the geotectonic history of the earth." In Canada, the Kenoran is defined as the last period of intensive folding, metamorphism, and intrusion throughout the major part of the Superior province.

It may also be postulated that the earth's ore mineral deposits were emplaced in the surface rocks in this Algoman time, and that all present emplacements of minerals are still in the rocks existing then or in younger rocks above the Algoman into which they have subsequently been dissipated or recycled.

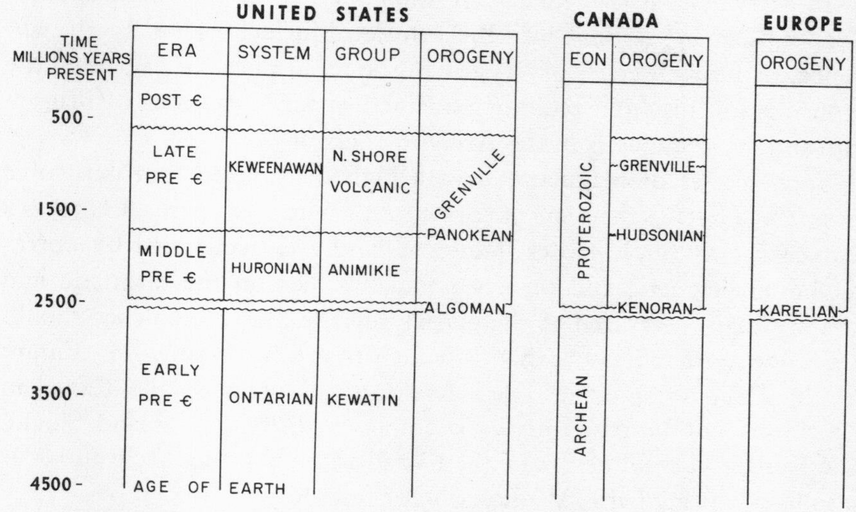

TIME MILLIONS YEARS PRESENT	UNITED STATES ERA	UNITED STATES SYSTEM	UNITED STATES GROUP	UNITED STATES OROGENY	CANADA EON	CANADA OROGENY	EUROPE OROGENY
	POST Є						
500 -							
	LATE PRE Є	KEWEENAWAN	N. SHORE VOLCANIC	GRENVILLE	PROTEROZOIC	GRENVILLE	
1500 -				PANOKEAN		HUDSONIAN	
	MIDDLE PRE Є	HURONIAN	ANIMIKIE				
2500 -				ALGOMAN		KENORAN	KARELIAN
	EARLY PRE Є	ONTARIAN	KEWATIN		ARCHEAN		
3500 -							
4500 -	AGE OF EARTH						

Fig. 1. Worldwide classification of Precambrian rocks. After Hurley (1965).

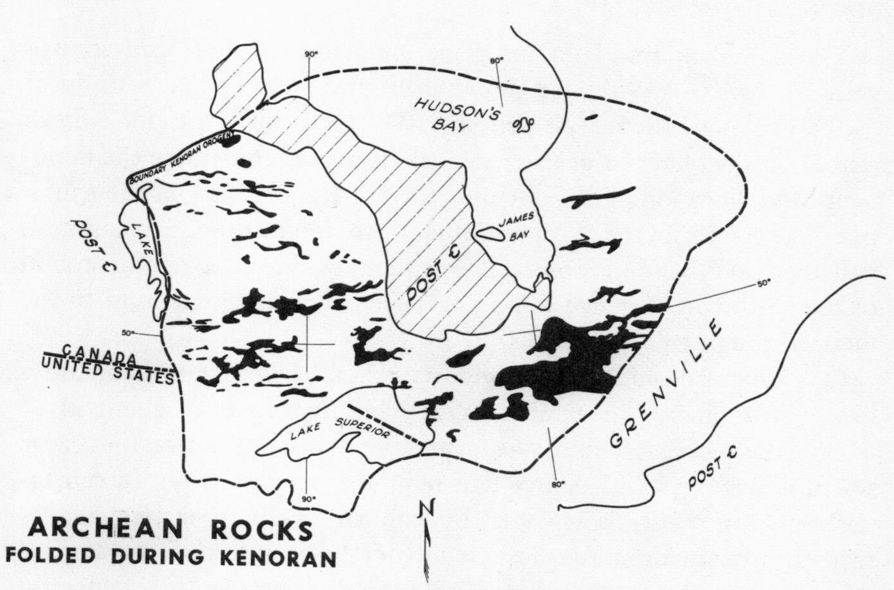

Fig. 2. Infolded Archean sediments of the Canadian Shield.

In particular, at that time large amounts of iron were available before subsequent settling into the mantle (Hurley, 1965). In the absence, or near absence, of oxygen in the atmosphere at the time, the iron was not disseminated, but remained as concentrations of magnetite and other minerals in the Archean rocks.

Such masses of magnetite have a high density, 5.2 gm per cubic cm. Of course, their magnetic susceptibility is high, 0.3 to 0.8 e.m.c.g.s/cm^3. Such masses should be, and are, manifested by corresponding magnetic and gravity anomalies seen on the magnetic and gravity maps measured at the earth's surface. Such anomalies should have one more tendency, that of being often E-W oriented. In Figure 2, the E-W oriented infolded Archean sediments in the Canadian Shield area of North America is emphasized. In the Finland Shield, the European counterpart of the Canadian Shield, the Archean rocks also have a principal E-W direction of folding.

Features of Magnetic and Gravity Maps

MAGNETIC MAPS

The dominant trends of magnetic anomalies as developed statistically by Affleck (1962) are shown in Figure 3. Note that a study of 34,000 magnetic trends yielded an E-W dominant lineation. Affleck states that the older, or deeper, anomalous masses contribute the dominant E-W lineation. This is in line with the hypothesis that the causative masses are in the Archean rocks. An interesting speculation is that the E-W banding observed was perhaps due to a rapid rotation of the earth before it had lost a part of its angular momentum to the moon through tidal action.

One other consideration is important here. In order to maintain a magnetic anomaly, a mass must have a limiting depth of about 21 to 25 km. Below that depth, the magnetite is heated above the Curie point, which is 585°C. Above that temperature, the strong ferromagnetism of magnetite is replaced by comparatively weak paramagnetism, and magnetic anomalies should not be manifest at the surface when causative masses are below that depth. That depth is generally below the Conrad discontinuity. Any gravity anomaly of deep origin, but without a magnetic twin, may therefore be caused by a mass below the 21 to 25 km limit for the corresponding magnetic anomaly.

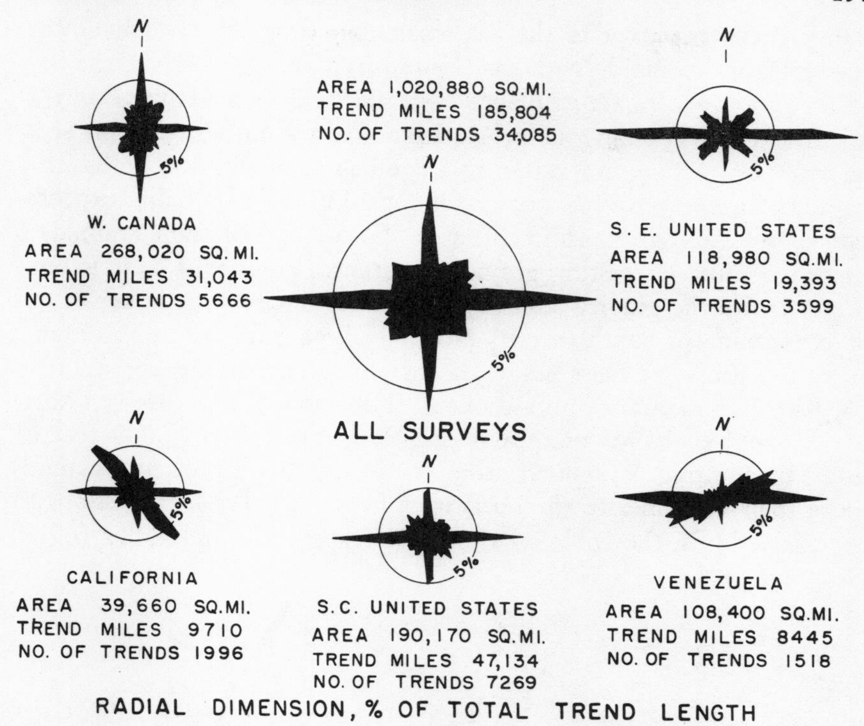

Fig. 3. Dominant trends of magnetic anomalies. From Affleck (1962).

GRAVITY MAPS

Gravity maps are seldom without large anomalies that appear to be expressions of masses at great depths. Of course, it is well known that unique depth determinations cannot be made on gravity or magnetic maps. However, a limiting lower depth can be established, and geologic and other considerations can often be used to fix a depth for the upper surface of the causative mass. Reference is here made only to those gravity anomalies which are deep in the crust. In general, such anomalies are only indirectly associated with the surface and subsurface geologic structure. In commercial gravity surveys, such anomalies are regarded as regional effects. An example of such an anomaly is the Crosbyton anomaly of north-central Texas. Drilled wells have established that the anomaly does not correspond to any density contrasts or folding in the sediments, and there is no uplift of the basement surface, so that the causative mass lies well within the crust. The thesis of our study is that the upper surface of this mass marks the Archean top, that is, the Algoman unconformity, and by inference it should coincide with the Conrad discontinuity.

Two such masses are emplaced beneath the Powder River Basin, in northeastern Wyoming (Fig. 4). Like Crosbyton, there is no sedimentary expression, nor is there a large basement uplift. The limiting depths of the gravity maxima are about 20 km, a reasonable expectation of the Conrad depth in this area. To the east of these anomalies, there is a most interesting gravity maximum, numbered 3. It is continuous with the gravity anomaly of the Cedar Creek or Baker Glendive anticline of southwestern North Dakota and eastern Montana. Here the causative mass lies at the top of the basement, and there is a marked sedimentary uplift above it. The basement feature is a horst of very old rocks, of unknown age, with more than 25,000 feet of uplift by faulting. The horst crops out in the Black Hills and extends more than 100 miles to the northwest from its gravity expression. In the Black Hills, the rocks of the horst feature are incredibly folded

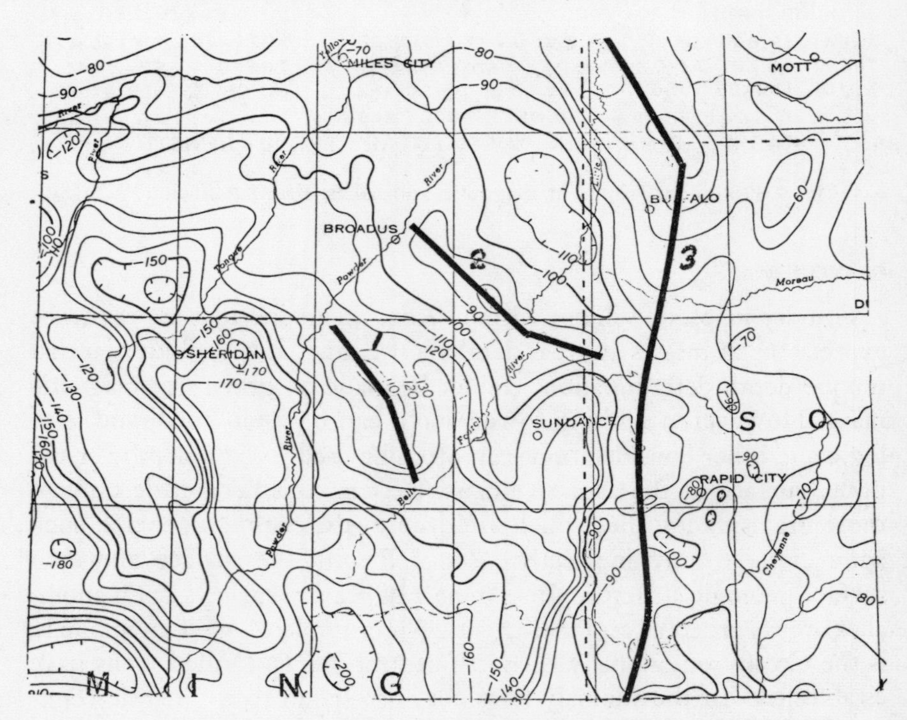

Fig. 4. Bouguer gravity anomalies in northeastern Wyoming and northwestern South Dakota. From USGS gravity map.

and contain a commercial iron formation. A folded conglomerate layer contains boulders the size of houses, evidence of tremendous elevation differences existing at the time it was deposited. The Baker-Glendive anticline is due to rejuvenation or later movements of the west-bounding fault.

Gravity and Magnetic Maps of Oklahoma

A gravity map of Oklahoma and a corresponding magnetic map of the state are shown in Figures 5 and 6, respectively. The southern part of the state shows, from left to right, the field expressions of the Wichita, Arbuckle, and Ouachita Mountains. Archean or Conrad features are also present. There are 20 gravity maximums and 12 gravity minimums with corresponding magnetic maximums and minimums, indicative of dense masses containing much magnetite. Proceeding by our criteria for the recognition of Archean masses on or below the Conrad surface, it has been possible to assign reasonable limiting depths to 5 gravity and twin magnetic anomalies. The average depths of these anomalies is 15 km.

Some confirmation is afforded by a seismic profile shot by the Uni-

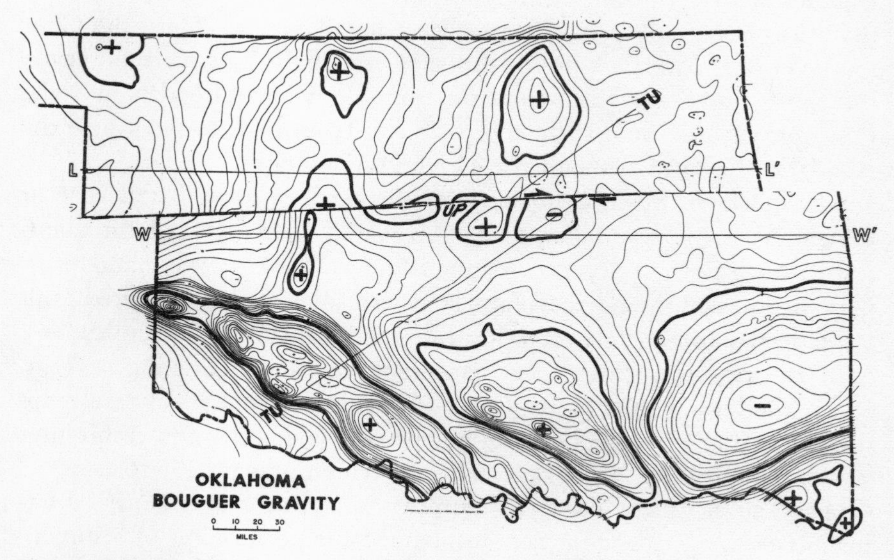

Fig. 5. Bouguer gravity map of Oklahoma, restored for hypothetical fault. Contouring based on data available through January, 1961.

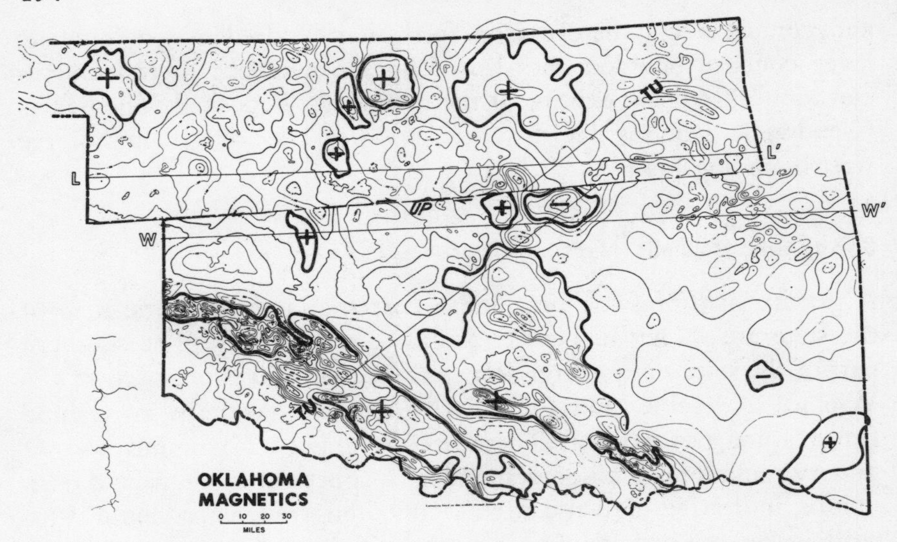

Fig. 6. Vertical magnetic map of Oklahoma, restored for hypothetical fault. Contouring based on data available through January, 1961.

versity of Tulsa from the northeastern to the southwestern portion of the state. The refraction profile shows an interface with the Conrad velocity at a depth of 15 km. The Moho discontinuity, or the base of the crust, is very deep on this profile; it lies at 51 km. Thus, the Conrad layer is 37 km thick in Oklahoma. Only slight dips were observed for the Conrad and the Moho on this seismic profile, although the Precambrian basement surface dips 13 km from N-E to S-W into the Anadarko Basin in front of the Wichita Mountains.

Also noteworthy in connection with the deep crustal structure of Oklahoma is the possibility of a deep-seated right-lateral fault manifest in both the gravity and magnetic maps. E-W gravity and magnetic profiles on the northern and southern sides of the suspected fault were drawn and correlated to indicate identical shifts of 35 miles.

In the restored gravity and magnetic maps of Oklahoma (Figs. 5, 6), the lateral shift along the E-W fault is removed. These maps show a rather striking N-S alignment of maximums and minimums which may indicate the simple relations which prevailed before lateral slippage along the fault. It also appears that later movements or adjustments along this fault have influenced the positioning of sedimentary structures above the Precambrian, as shown on the E-W profile correlation across the fault (Fig. 7).

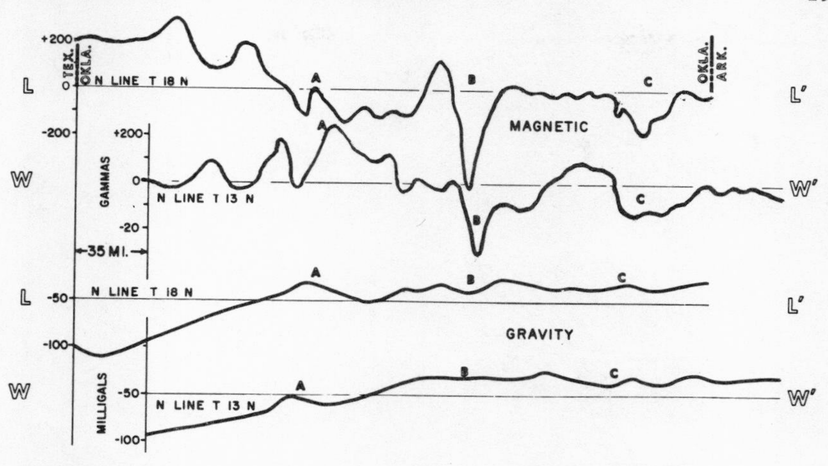

Fig. 7. E-W gravity and magnetic profile correlation for hypothetical fault shown in Figures 5 and 6.

The Greenleaf Anomaly

A remarkable gravity, and presumably magnetic, anomaly extending all the way from Oklahoma to Alabama by way of Lake Superior is shown in Figure 8. I have named it the Greenleaf anomaly from its occurrence in Kansas (Lyons, 1959). The portion from Iowa to Lake Superior has been named the Midcontinent anomaly by other workers. A thesis of the present study is that the Greenleaf and the Midcontinent anomalies are continuous, and that a further continuation extends through Michigan and Ohio on down into Alabama. All along this arcuate trend, gravity profiles drawn across the anomaly exhibit a "Mexican hat" profile; invariably, there is a central prominent maximum flanked by a prominent minimum on each side (Fig. 9). This is not a Conrad feature; in fact, it comes to the surface of the Precambrian. It is marked by ancient lava flows in the Lake Superior graben, and it is bounded in the area just south of Lake Superior by the great Keweenawan fault. Within this great arc, there are no large "regional" gravity anomalies which have a ready or apparent connotation with the tectonics of the basement rocks or of the sedimentary section until the area south of the continuous E-W fault shown in Figure 8 is reached. McGinnis (1965) has recently established an E-W fault in northern Illinois from gravity observations, and lavas are assumed to exist there at or near the basement surface.

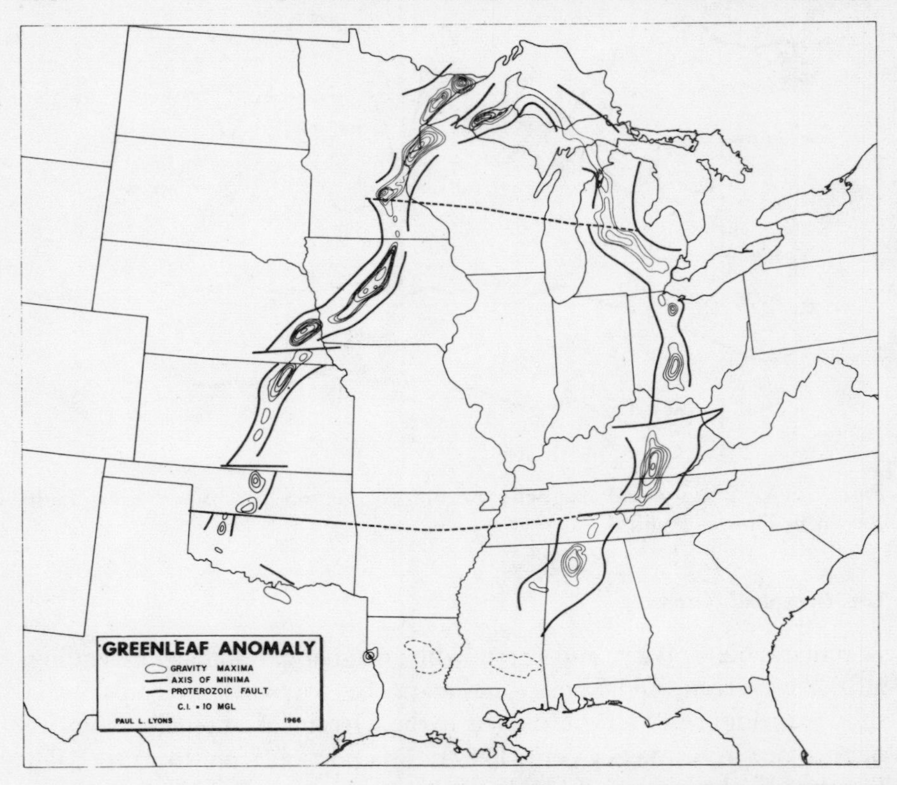

Fig. 8. The Greenleaf gravity anomaly of Kansas and its possible extensions.

Apparently, this great arc must have been a rift (at sea) during Keweenawan time, following the Grenville orogeny which added sediments to the southeastern flank of the existing shield area. We may speculate that volcanic and igneous intrusions were common within the area enclosed by this arc, and that the effect of these density contrasts comprise an overwhelming part of the present Bouguer gravity picture. This was a post-Algoman event, the effect of which is to mask the Conrad anomalies and to extend upward rocks with Conrad velocities as far as the basement surface. It is noteworthy that the central portion of this arc has a relatively high elevation of the Moho, a northward extension of the expected high Moho elevation existing in the Gulf of Mexico. The elevation of the central portion of the arc should therefore be relatively low, and indeed it is the great Mississippi River drainage area.

Calculations for the gravity anomaly in Iowa (Fig. 9) reveal that

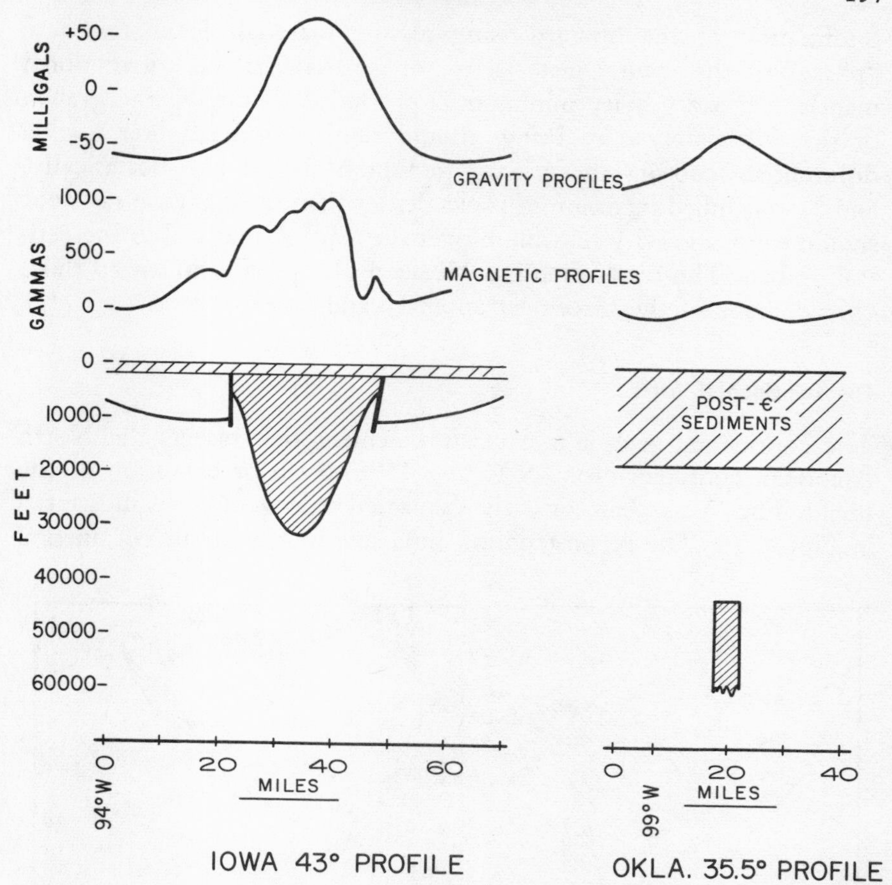

Fig. 9. "Mexican hat" profiles characteristic of the Greenleaf anomaly.

it is caused by a graben containing lavas at least 30,000 feet deep. Examination of gravity profiles over the Atlantic Rift shows a widely extended Bouguer minimum, with the rift centrally located in the minimum. No Conrad layer is seen in seismic profiles; further, mantle velocities in the range 8.0 km per second are not present. Instead, the velocity is about 7.2 km per second, intermediate between that of the Conrad and the Moho. Apparently, the rifting or spreading of the sea floor associated with the anomaly is accompanied by a phase differentiation of the upward-rising mantle material. Since the new "phase" of the mantle material is not as dense as the conventional mantle, a broad gravity minimum is produced, following the graben or rift. Ultimately, two things should result: 1) the lighter mantle differentiate causing the gravity minimum should rise isostatically, and 2) continued intrusion of rocks denser than those on the sea floor should cause a gravity maximum over the "dyke" material in the central graben. The result then is a Mexican Hat profile similar to those characteristic of the Greenleaf anomaly and its extensions.

The Continental Arch

The continental arch is a structural element that bisects the North American continent from N-E to S-W as a rib or extension of the shield. The Waucoban, or early Cambrian paleogeology is illustrated in Figure 10. The reconstruction indicates that a broad continental

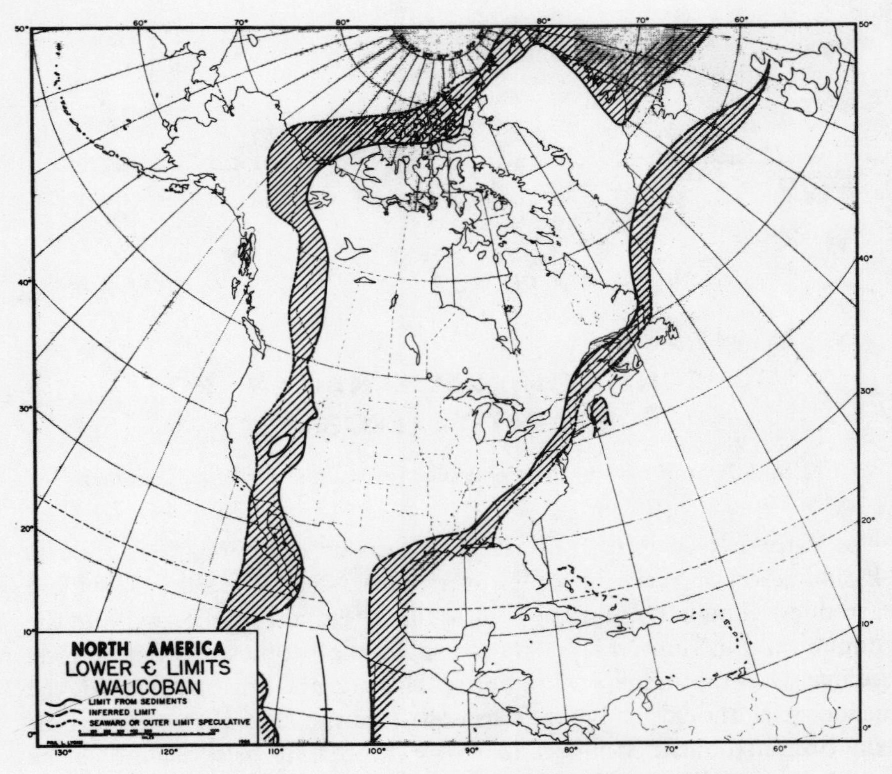

Fig. 10. Waucoban or Early Cambrian land and sediments.

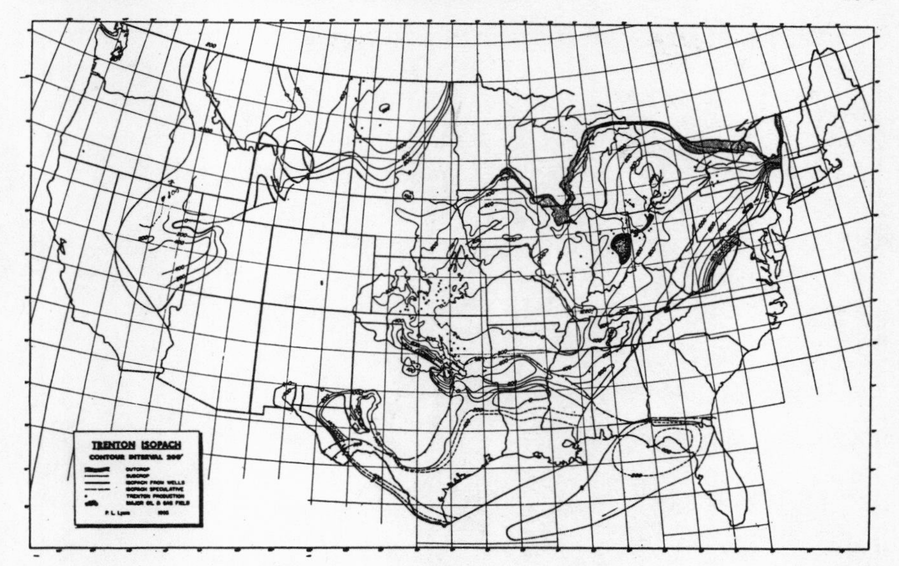

Fig. 11. Trenton isopach (Ordovician), showing continental arch extension of the shield area to the S-W.

arch was present, with the central position perhaps even then being occupied by uplifted Archean rocks. An isopach of rocks of Trenton age (Fig. 11) indicates that the arch was still present in Ordovician time, dividing the continent with basin areas on each side of the uplift.

Conrad Depths

An attempt to portray the depth of the Conrad discontinuity beneath and around the North American Continent by using all available seismic and gravity data is shown in Figure 12. The lightly outlined areas show the outcrop of rocks of Archean age; here the Conrad discontinuity reaches the surface if it indeed coincides with the top of the Archean or the Algoman unconformity. The continental arch is indicated as a high rib on the Conrad discontinuity. This ridge coincides with the site of the great metallic ore deposits of the continent, which are adjacent to the Archean rocks. Further, the continental arch appears to be a continuation of the East Pacific Rise. To the N-E, a connection with Europe through northern Scotland is indicated, with Iceland astride the arch as well as the Atlantic Rift.

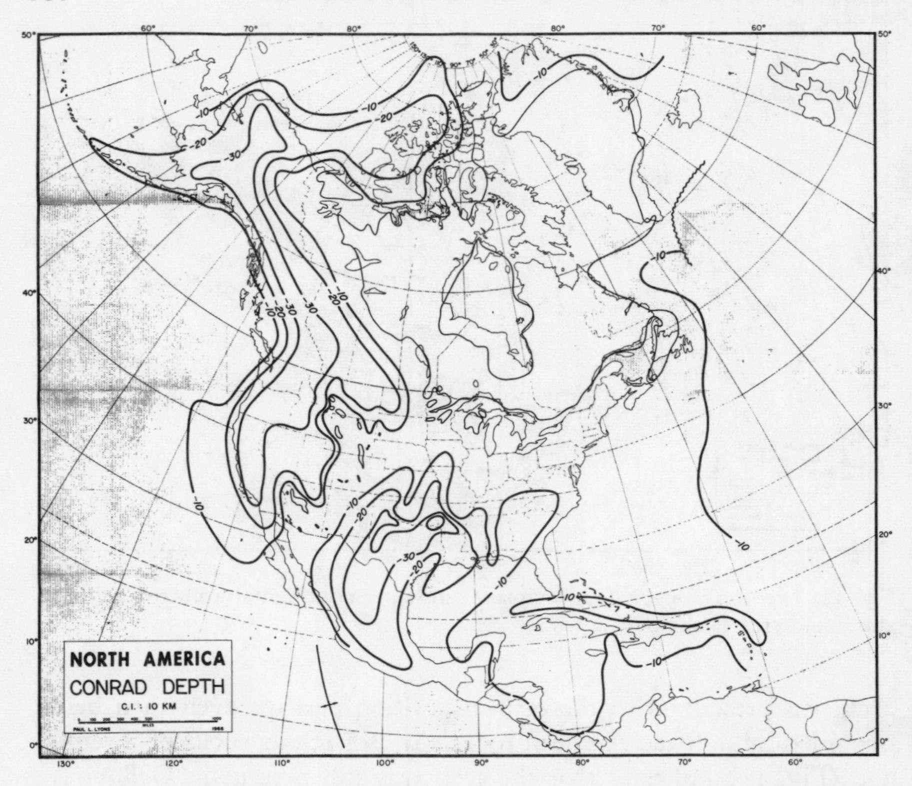

Fig. 12. Depth of Conrad discontinuity from seismic, gravity, and magnetic determinations. Light lines enclose Archean outcrops.

The question arises: Is the Pacific Rise an ancient feature, similar to the Atlantic Rift, which at one time resulted in continents as it girdled the earth?

Mantle Depths

An interpretation of the top of the mantle, or the Moho discontinuity, is portrayed in Figure 13. There is a high ridge on the mantle also, coinciding with the continental arch. Indeed, the mantle itself displays a "Mexican hat" profile with respect to this ridge, with profound deeps on either side of it. A recent seismic profile was obtained, running from west to east along the length of Lake Superior. The Moho was found at about twenty kilometers depth at the west

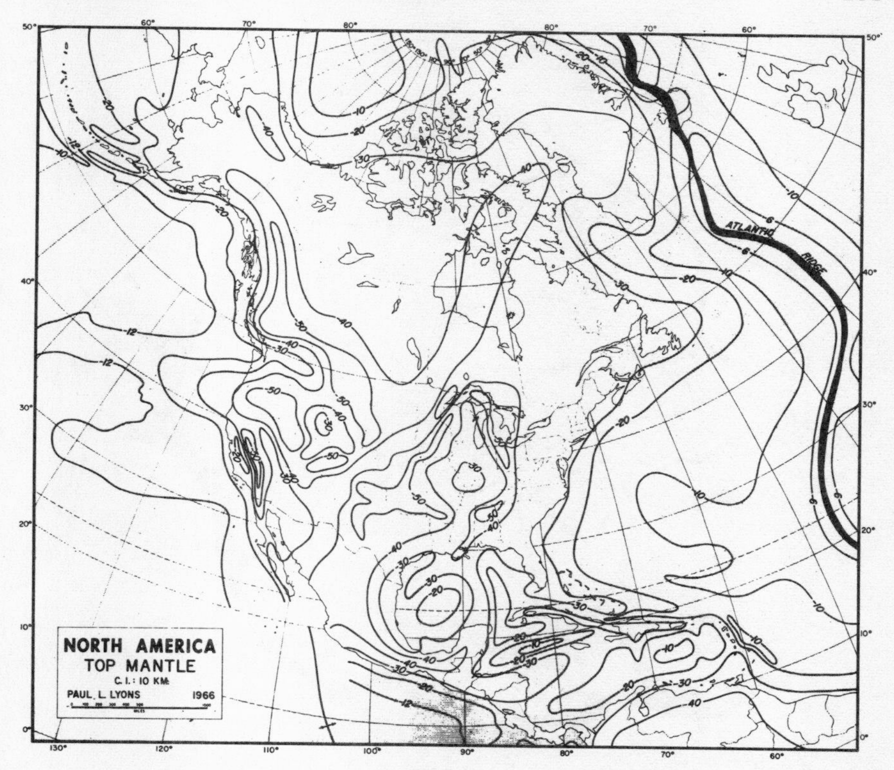

Fig. 13. Depth of the Moho discontinuity, or base of the crust, from seismic determinations.

end of the lake, and it dipped to more than fifty kilometers depth at the east end. This dip of thirty kilometers is off the east flank of the continental arch. The map as drawn suggests that the Greenleaf anomaly is consistently low at Moho depths relative to surrounding areas. Stable areas have low relief on the Moho surface, and areas of tectonic activity are sites of rather considerable relief on that surface.

Pacific Area

Selected features of the geology of the Pacific Ocean are shown in Figure 14. The Pacific Rise, if it does coincide with the continental arch of North America, extends over the earth's half portrayed here. On the other side, it is known to run up to India. Does it extend under Asia to the Finnish Shield and on to Iceland and North Amer-

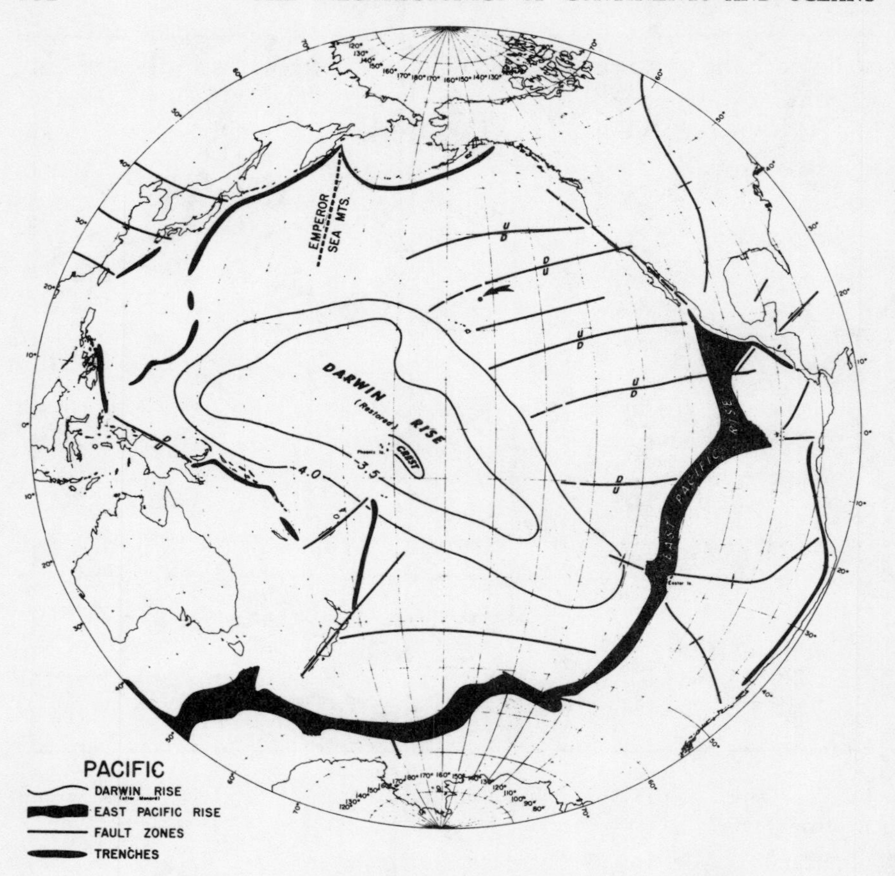

Fig. 14. Tectonic features of the Pacific Ocean, with restored Darwin Rise.

ica? The rise culminates at Easter Island, where it intersects the Darwin Rise. The rocks on Easter Island are reported as continental.

The most impressive feature of the Pacific is the Darwin Rise, as restored by Menard (1964). It was a vast swell which once existed in an area which is now the deepest part of the Pacific, excluding trenches. Restoration of the flat tops of sea mounts to sea level has enabled Menard to place 3.5 and 4.0 km depth contours, as shown on the map, as the one time elevation of this collapsed "blister." It was thus an incipient continent. The Conrad or oceanic layer is apparently uniform over the area of this restored rise, so that the rise was not a geologically ancient feature; it probably occurred in Early Cretaceous time and it collapsed in Late Cretaceous time. Did this have anything to do with the great unconformity which separates the Early and Late Cretaceous? The Moho appears to be relatively high in the area of the collapse, and its velocity is high compared to points of observation on the periphery. This situation is the same as in the Gulf of Mexico, another result of collapse, with the movement downward in response to higher densities of the mantle surface, as indicated by higher velocities. The thesis is that high Moho velocities and associated regional, high gravity values are characteristic of sinking elements of the earth's crust.

One other aspect of the ancient Darwin Rise is noteworthy: the feature marking the termination of the Mendecino family of lateral faults that trend roughly E-W off the west coast of the Americas. Another family of E-W faults has been reported in China. Were these faults obliterated or buried by the processes which at one time established a Darwin Rise? Perhaps gravity and magnetic studies will provide some answers. On Figure 14, the arrow marks the spot chosen for the well to be drilled to the Moho. To the N-E of this site, magnetic surveys of the sea have established a remarkable series of N-S trending magnetic anomalies. S-W of the site, in the vicinity of the Hawaiian Islands, surveys have established a similar series of E-W trending magnetic anomalies. Determination of the boundary between these divergent trends should be an interesting project.

Implications of Determining Age by Radioactivity

Use of clock minerals for radioactive dating of basement rocks poses another problem involving the Conrad layer. As is well known, there is a sort of concentric ring of basement rocks around the Canadian Shield of North America. By the radioactive method, the age of the central core is 2.5 billion years, and the rings around this nucleus consecutively younger. However, it seems that these ages are not those of the host rocks, but rather of invading materials from the mantle below; only the central shield area has been protected from this partial assimilation. This is doubtless a tensional phenomenon. Combined with the crustal spreading observed in some ocean areas, the assimilation procedure leads to the hypothesis that the density of the earth's crust is continually increasing. Further, the crust is being lost by this assimilation to the mantle, with the end result being a kind of geologic entropy in which the crust is utimately lost. If the thicker crust of the continents is reduced first, the oceans can readily cover the earth, and the processes of erosion producing low-density sediments will be absent. Only the processes of orogeny appear to resist and deter this tendency. The shields themselves are low lying and do not contribute greatly to sediment formation. The central portion of the Canadian Shield is still covered with a Paleozoic basin, so that the Archean rocks beneath the cover have not been eroded since Precambrian time.

East-West Faulting

There is growing evidence to support the presence of ancient E-W faulting in the continental crust. These faults are usually lateral ones. The Atlantic Rift itself is cut by a number of such faults. It is a simple exercise with a pair of scissors and a map to restore the Atlantic Rift to a straight N-S line. If this is done by cutting along the faults and extending them, the west coast of Africa also becomes a straight line, and South America moves westward to a position due south of North America. Under this hypothesis, the observed fit of the coastlines of South America and Africa is only a result of a shared right-lateral movement along the E-W fault system. It is much more difficult to doubt these great lateral faults and the implied movement along them than to doubt the hypothesis of continental drift. And it would seem that a choice must be made, for restorations of land and sea elements based on lateral faults invalidate the fitting of continental coastlines, or continental slopes, that are so important to the hypothesis of drift.

Possibly the E-W faults represent extremely ancient faults, persisting at least in the Conrad layer since the time when the earth had greater angular momentum, lost since then by tidal action to the moon. For example, E-W banding, truly faulting, is apparent on Jupiter. The equatorial regions of that planet rotate faster than the temperate zones. Faults of this type should be extremely long and persistent. There are indications that the Mendecino fault, at 40°N, may extend eastward into the deep crust through the United States. Magnetometer observations have found extensions in California, as well as E-W elements such as the Uinta Mountains of Utah and the Rough Creek fault zone of Kentucky. The Greenleaf gravity anomaly is cut by an E-W lateral fault just north of the Kansas border. Finally, a counterpart of the Mendecino fault has apparently been defined in the Atlantic Ocean. As for the westward trek of the Mendocino, it has already been observed that it is obliterated from ready observation as a seafloor feature by the collapsed Darwin Rise. Note, too, that there is an E-W fault at 40°N in China, just north of Korea. If all of this is indeed continuity of a single fault zone, then it has gone halfway around the earth.

The E-W fault pattern is important to oil prospecting. It is often possible to locate and extend E-W fault systems in the sediments, as well as, say, N-S systems. The "up" corners of fault blocks so formed are favored sites for the location of structures.

The E-W faults have another important bearing on the hypothesis of continental drift. The drift of continents implies rotation. If much drift has occurred, it is unlikely that the continuous patterns of E-W faulting would be preserved or maintained.

Conclusion

Large, usually coincident regional gravity and magnetic anomalies have been investigated to find whether they contribute to a study of the Conrad layer of the crust. They do provide reasonable depth estimates. The Conrad layer itself may in turn represent an ancient Algoman surface, which may be universally associated with E-W folding and faulting. These ancient lines of weaknesses, through rejuvenation, persist into the present and afford explanations, through lateral movements along faults, for observed shifts of the crust. Continued acquisition of regional, continental and oceanic geophysical data will ultimately provide real solutions to problems now just appreciated, but not solved.

References

Affleck, J.: Exploration for Petroleum by the Magnetic Method. In *Benedum Earth Magnetic Symposium.* Pittsburgh, University of Pittsburgh Press, 1962, pp. 159–175.

Hurley, P.: Lectures at Massachusetts Institute of Technology, Cambridge, Mass., 1965.

Lyons, P. L.: The Greenleaf Anomaly, a Significant Gravity Feature. In *Symposium on Geophysics in Kansas.* State Geological Survey of Kansas, Bulletin 137, 1959, pp. 105–120.

Lyons, P. L.: *Gravity Map of Oklahoma.* Geophysical Society of Tulsa Proceedings, vol. VIII, Pl. II. Oklahoma State Geological Survey, special printing, 1964.

Lyons, P. L., and V. L. Jones: *Magnetic Map of Oklahoma.* Geophysical Society of Tulsa Proceedings, vol. VIII, Pl. I. Oklahoma State Geological Survey, special printing, 1964.

Lyons, P. L.: Trenton extent in the United States: a regional study. *Tulsa Geological Society Digest* 34:99–109, 1966.

McGinnis, L. D.: *Crustal Movements in Northeastern Illinois,* University Microfilms, Ann Arbor, Mich., 1965.

Menard, H. W.: *Marine Geology of the Pacific,* McGraw-Hill, New York, N.Y., 1964.

8

The Mediterranean, Ophiolites, and Continental Drift

JOHN C. MAXWELL *

Department of Geology
Princeton University

Just as in the past the concepts of certain men have dominated geologic thought—we speak of the Huttonian and Wernerian eras, for example–so the last half century might well be designated the Wegenerian era. The concept of continental drift, first conceived by this German meteorologist, led to expansive speculations which in turn inspired energetic and fruitful scientific studies. Unfortunately, the concept, so far resisting proof or disproof, has assumed the character of dogma. The common question: "Do you believe in continental drift?" indicates clearly that the matter is still in the realm of faith. This is a statement of fact, not of criticism. The mind boggles at the complexity of the problem if rigorous proof or disproof is

* It is impossible to acknowledge properly everyone who has contributed significant opinions and ideas to a summary report such as this. Over the past 15 years of work related to the Mediterranean I have had particularly fruitful discussions with G. Merla, A. Azzaroli, R. Selli, L. Trevisan, B. D'Argenio, J. H. Brunn, and K. Zachos. My Princeton colleagues H. H. Hess, W. Elsasser, E. Moore, and F. J. Vine, and J. Rodgers of Yale University have done their best to educate me, while A. G. Fischer offered comments on the manuscript as well as stimulating discussions during field trips in the Apennines and Alps.

Studies in the Mediterranean area extended over two sabbatical years and three additional summers. For thesis opportunities I am indebted to the Fulbright Commission, which provided a fellowship covering the first year in Italy, and to the National Science Foundation for a Senior Post-Doctoral Fellowship, for making possible the International Field Institute of the American Geological Institute, held in Italy in 1964, and for research grant G.P. 4535 to Princeton University.

Finally I am indebted to the Department of Geology of Rutgers–The State University, for providing the impetus for this report.

sought. Perhaps for the time being we can best serve science by characterizing the kinds of relative motion that may have occurred and simultaneously ruling out certain suggested movements which are highly improbable on geologic grounds. I will therefore not attempt to assess the "truth" of the concept of continental drift, but rather restrict myself to an appraisal of the limitations imposed on drift by the geology of the Mediterranean area.

As a corollary of the hypothesis of the drifting of rigid continents, the concept of a Mediterranean Sea residual from an ancient and more extensive oceanic area has come to be widely accepted. However, the nature and distribution of Paleozoic and younger rocks argues strongly for the continuity of sialic rocks between Africa and Eurasia, from the Paleozoic to the present. Furthermore, the Betic-Rif-Atlas mountain arc, encircling the west end of the Mediterranean, also extends northward through Sicily and Italy to join the Alpine chain, thus linking the western Mediterranean area into a unit which has not been affected by lateral displacements of global extent, at least since the mid-Paleozoic. There is convincing evidence that the entire Mediterranean was a sialic, continental area after the Hercynian episode. This reconstruction results from the distribution of stratigraphic sequences and from the inferred presence of now foundered sources of late Mesozoic and Tertiary sediments and gravity slides. A kind of "oceanization" of the present Mediterranean area began in the Late Jurassic with the emplacement of mantle material as intercrustal plutons and as rocks of the ophiolite sequence.

Continental Drift and Tethys

The observation which apparently suggested the idea of continental drift to Wegener, and for which no other reasonable explanation has been advanced, is the remarkable "fit" of the continents and the matching of many major geologic features if the Atlantic ocean is "closed" by moving continents together (Fig. 1*A*). For the best fit, a rotation of Europe with respect to Africa is indicated. The rotation would open a wide oceanic area between the continents. It is this hypothetical oceanic area which is commonly identified with the Tethyan sea in which the sedimentary rocks of the present Alpine (or more properly Tethyan) chains were deposited.

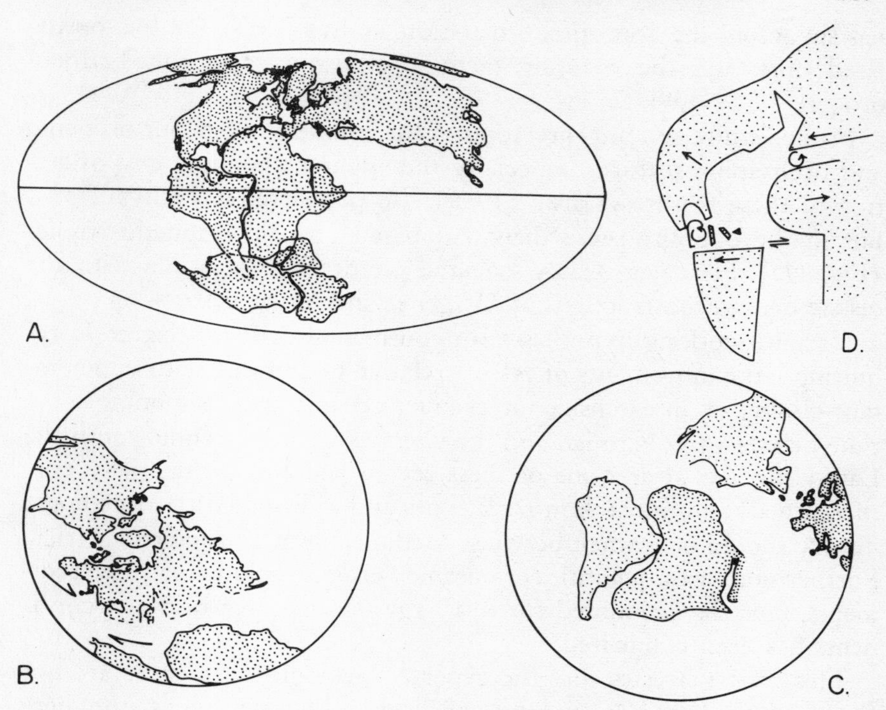

Fig. 1. Continental drift and the Mediterranean area. *A*, after Wilson, 1963. *B*, *C*, arrangement of continents based on interpretation of paleomagnetic data; *B*, in Triassic (Van Hilten, 1964); *C*, in Late Paleozoic (Van Hilten, 1965, after Creer). *D*, after Carey, 1958.

In the past decade, paleomagnetism, a new and potentially powerful geologic tool, has provided the impetus for a renewed interest and reevaluation of continental drift (Irving, 1964; Runcorn, 1962). Results to date do indeed suggest large relative movements between continents, but these movements do not necessarily match those suggested by the physical fit of continents, especially for the South Atlantic (Fig. 1*B*, *C*). Africa, rather than drifting eastward relative to South America in post-Permian time, has, in one possibly extreme interpretation, drifted northwestward from a position in the present Pacific Ocean, while Eurasia rotated through nearly 90° and drifted relatively eastward (DeBoer, 1963; Van Hilten and Zijderveld, 1966). The wide oceanic area south of Eurasia was simultaneously narrowed to the dimensions of the present Mediterranean. Van Hilten (1964) has suggested the appropriate name of Tethys Twist for this postulated shift, and the resulting tectonic adjustments in the Tethyan orogen.

From a study and interpretation of the geometry of major orogenic and orographic features, especially the apparent bending and offsetting of these features, Carey (1958) also postulated a Tethys Twist, but in the opposite sense than that based on paleomagnetic studies (Fig. 1*D*). However, Carey's overall picture is compatible with the classic drift reconstructions of Wegener and his successors.

Various working hypotheses for continental drift thus agree in requiring large movements of Africa relative to Eurasia, with concomitant closing of an extensive intervening oceanic area, beginning some time between the Permian and the Cretaceous and extending into the Late Cenozoic. If any one of these reconstructions is essentially correct, a marked mismatch in rock types and a dissimilarity in geologic history should be evident between Mediterranean Europe and Africa. Furthermore, vast quantities of oceanic "crust," including abyssal sediments, must be accounted for if a large oceanic area between continents has been eliminated.

This report stresses that the expected geologic differences are not found, nor are the large volumes of displaced oceanic rocks anywhere identifiable. Instead, as pointed out by European geologists actively working in the Mediterranean area, there is convincing evidence of a continuity of sialic basement between Europe and Africa in late Carboniferous (post-Hercynian) time, and probably much earlier. Furthermore, the Mediterranean Sea itself developed later, beginning with widespread deposition of shallow-water Permo-Triassic sediments which are everywhere disconformable on older sialic rocks. Marine sediments of the Alpine system seem to have been deposited everywhere on continental crust, beginning late in the Middle Triassic. There is no evidence of the former presence of thousands of square kilometers of oceanic rocks which is implied by the continental drift reconstructions, nor will density relations permit such rocks to be swallowed without trace as is commonly postulated (Fig. 2).

Deeply ingrained in American geologic thought is the concept that mountain systems develop on the margins of continents. The geology of North America so satisfyingly displays this order that it is tempting to assign to it worldwide significance. On this premise the great Tethyan orogenic system was assumed to have developed on the

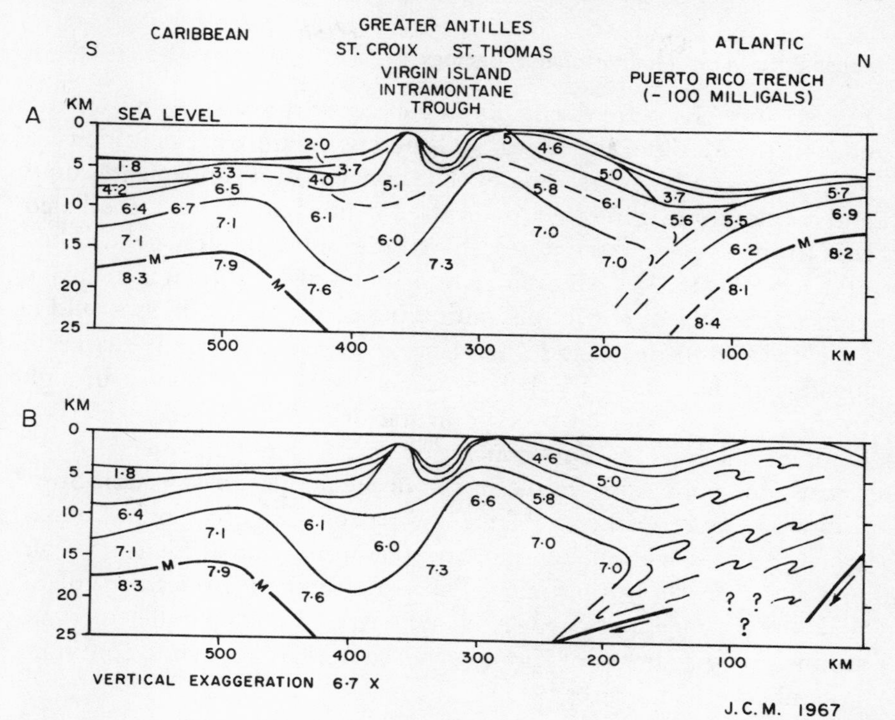

Fig. 2. Problem of disposal of old oceanic crust, to accommodate formation of new crust above mid-oceanic ridges. *A*. Suggestion that crustal material is dragged down into the mantle (modified from section by Officer, et al., 1959); *B*. Piling up of relatively light oceanic crustal rocks which would be expected if island arc- or continent-ocean boundary were loci of down-dragged crust. This situation is not known to occur.

margins of preexisting, separate continents—Africa and Eurasia—and Tethys was regarded as truly "oceanic." As pointed out, this premise is at variance with Eurasian geology. Smirnow (1964) and Aubouin (1965), among others, demonstrate convincingly that the Alpine, Hercynian, and Caledonide mountain systems in each case developed across older orogenic belts, all within continental crust which was locally subjected alternately to erosion and sedimentation, interrupted by orogeny. In Eurasian experience, then, orogenic systems characteristically occur within and across continental masses. The Tethyan seaway and the Mediterranean Sea are conceived as having been initiated on continental crust, by local rifting and development of a trough and ridge topography.

Gross Pattern of Mediterranean Geology

Eurasia and Africa are essentially in contact at the eastern and western ends of the Mediterranean Sea, and there is an apparent continuity of geologic features at these points. The concept of a fundamental unity and continuity of geology around the Mediterranean, from the Paleozoic to the present, is supported by stratigraphic studies such as those of Klemme (1958). An examination of his paleolithologic maps reveals an absence of abrupt discontinuities of the type which would be expected were Europe and Africa far apart, either laterally or meridionally, at the end of the Paleozoic and were subsequently brought to their present position by continental drift over distances of thousands of kilometers. This conclusion is further emphasized by the remarkable continuity visible in the orogenic belt of southern Spain, Gibraltar, and northern Africa (Morocco, Algeria, and Tunisia) which extends without significant interruption across Sicily to Calabria and then along the Italian Peninsula to the European mainland. This great orogenic system is characterized by subparallel tectonic-sedimentologic belts of impressive lateral extent and continuity (Choubert, 1956; Caire, 1965).

These gross observations, suggesting essential stability and continuity, do not support the concept of independent development of geologic histories and of great intervening distances, as required by the various drift reconstructions. A more detailed examination of circum-Mediterranean geology further emphasizes similarity and continuity rather than a divergence of geologic histories between Europe and Africa.

A striking pattern repeated many times in the circum-Mediterranean orogenic complex is the bilateral symmetry of the mountain systems, reflecting an outward-directed tectonic transport (Fig. 3). The symmetry centers are of three types: 1) depressed parts of the Mediterranean Sea with bordering mountains, such as the western Mediterranean and the Corsica-Elba-Italy-Tyrrhenian Sea system; 2) central depressions on land, possibly homologous to type 1, such as the Hungarian depression or Pannonian Basin between the Carpathian and Dinaride chains; and 3) central crystalline massifs, such as the Rhodope and Mendres massifs with the associated Hellenides, the Greek Islands, and the mountains of Turkey. The recent International Tectonic Map

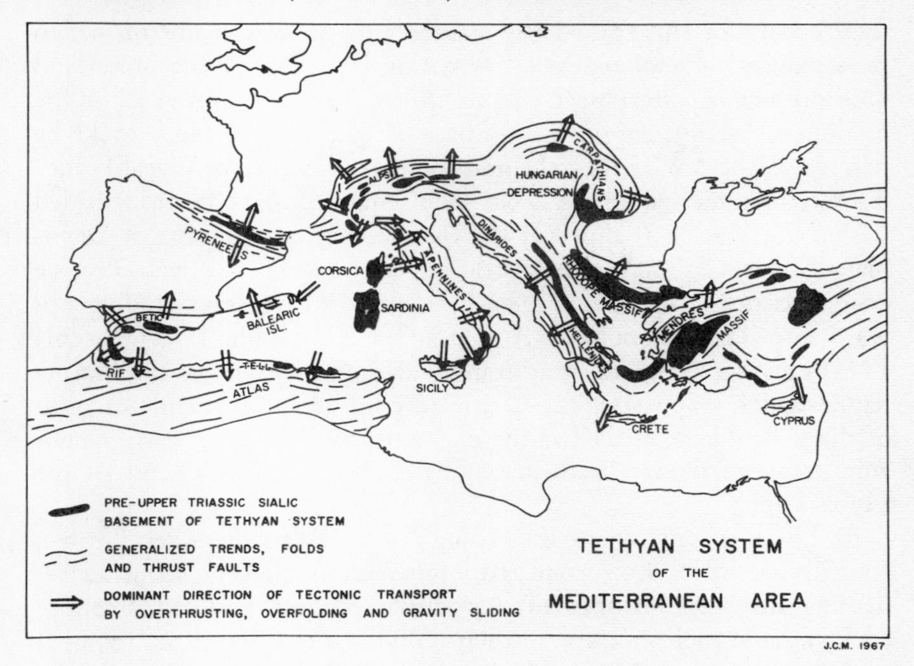

Fig. 3. Bilateral symmetry in Tethyan System, Mediterranean area.

of Europe (published in Moscow for the International Geological Congress, 1964) strikingly illustrates the symmetry within the Mediterranean area.

WESTERN END OF MEDITERRANEAN

*Structure and Stratigraphy.** The western end of the Mediterranean is a particularly obvious example of bilateral symmetry (Fig. 3). The Betic system extends from the Island of Menorca in the Balearic group westward along the southern coast of Spain to the vicinity of Gibraltar. From there, the Rif system swings southward and then eastward through the great Atlas chain and the coastal Tell. These systems are geologically similar and all are characterized by outward transport of rocks from the present Mediterranean toward the continents.

The western Mediterranean mountains are of Alpine type; that is,

* Our knowledge of western Mediterranean geology stems largely from the work of M. Blumenthal, P. Fallot, M. Durand-Delga and their students. Particularly useful summaries and bibliographies are found in Fallot, 1948, and in Durand-Delga, ed. (1962).

they consist of superposed thrust sheets or nappes, many of which reflect differing geologic histories within the parts of the hinterland (the present Mediterranean) from which they came. In each of the mountain systems, nappes with a base of Paleozoic or older rocks lie near or at the top of the pile, and these are themselves divisible into discrete tectonic sequences. Two such sequences have been identified within the Malaga "nappe" of the Betic chain: one consists of Devonian–Lower Carboniferous clastics overlain by red Permo-Triassic coarse clastics of Germanic facies; the other, of phyllites grading upward into Triassic rocks of Alpine affinities (Boulin, 1962). Azema (1961) gives a schematic stratigraphic column for rocks of the Malaga nappe which emphasizes the neritic to continental character of most of this assemblage, as well as the characteristic Mesozoic–early Cenozoic Alpine carbonate sequence and the absence of sediments of the Flysch type.

In the Rif system, where the geology of the Paleozoic is best known, four distinct slices are recognized, each with a Paleozoic sequence reflecting different histories of deposition, erosion, and deformation; rock types in each slice are, in their individual characteristics, comparable to rocks found in the Malaga nappe of the Betic chain to the north (Durand-Delga and co-workers, 1962). In the Rif, and indeed throughout the western Mediterranean chains, the Paleozoic rocks are overlain by a transgressive Permo-Triassic sequence, characteristically of continental clastic rocks. Ancient Tethys was epicontinental, not oceanic. The Alpine cycle itself is usually considered to have started with the deposition of late Middle or Upper Triassic sediments which are everywhere discordant on older sialic rocks (Aubouin, 1965).

The similarity in Paleozoic section between the Betic and Rif mountain systems is said to extend eastward to the Kabyle massifs of the Algerian Tell; furthermore, the Mesozoic and Cenozoic rocks and the structural units in which they are involved are remarkably similar and continuous through the same western Mediterranean arc (Durand-Delga, 1966; Durand-Delga and co-workers, 1962; Fallot, 1948). It also seems certain that the Betic-Rif arc at Gibraltar essentially marks the westward termination of the "Alpine" deformation (Fallot, 1948; Durand-Delga, 1966).

On the continental or northern side of the Betic chain, Triassic sediments of Germano-Andalusian facies extend from the massifs of central Spain southward well into the deformed zone. During the

Jurassic, the area of the present mountain system subsided; the sediments deposited in the resulting trough grade from neritic in the north to pelagic in the south. The pre-Betic sequence of neritic sedimentary rocks is overthrust by the more pelagic rocks of the sub-Betic sequence, originally deposited to the south. On top of the sub-Betic rocks reposes the large Alpujarride nappe with a basement of crystalline schists overlain by a thick Triassic dolomite-limestone sequence. Paleozoic rocks and an associated peridotite pluton of the Malaga nappe rest on the Alpujarride nappe. Both nappes (actually composite) were derived from the present Mediterranean area to the south.

A cross section prepared by Caire (1963) for the Tellian Atlas illustrates essentially the same sequence of offshore domains reflecting parallel belts of uplift and erosion, and of downwarping and filling, throughout the period of the Alpine cycle; reversals of direction of vertical movements are evident. The picture that emerges, then, is one of a complex range and trough topography which developed in the westernmost part of the present Mediterranean Sea, beginning in Late Triassic time. A similar sequence of active blocks seems to have characterized the Mediterranean from Sicily and western Italy to Gibraltar. The cross sections (Fig. 4) summarize this concept of the development of the Mediterranean within a preexisting sialic basement.

The stratigraphy of the more interior nappes, involving Paleozoic rocks, shows clearly a history of shallow water deposition and intermittent emergence, apparently of ridges paralleling the Flysch troughs. Azema and co-workers (1960) compare the unusually thin Mesozoic-Eocene sedimentary cover of the Malaga nappe with that of the Briançonais geanticline of the French Alps. Caire (1965) similarly compares the Kabyles of the Algerian Tell with the Briançonais structure.

Extensive Flysch deposits appeared perhaps as early as the Cretaceous, or even Late Jurassic (Durand-Delga, 1961), deposited in elongated troughs, some of great length. Internal Flysch trough development continued into the Oligocene.

The transport of rocks from the interior zones of the present Mediterranean basin onto the adjacent continents seems to have occurred mainly during the lower Miocene. Caire (1963, p. 307) states that the Tellian nappes lie on lower Miocene, contain lower Miocene, and are overlain—discordantly—by lower Miocene. In the Betic, nappes of the internal zone involve rocks as young as late Oligocene. Fallot's (1948)

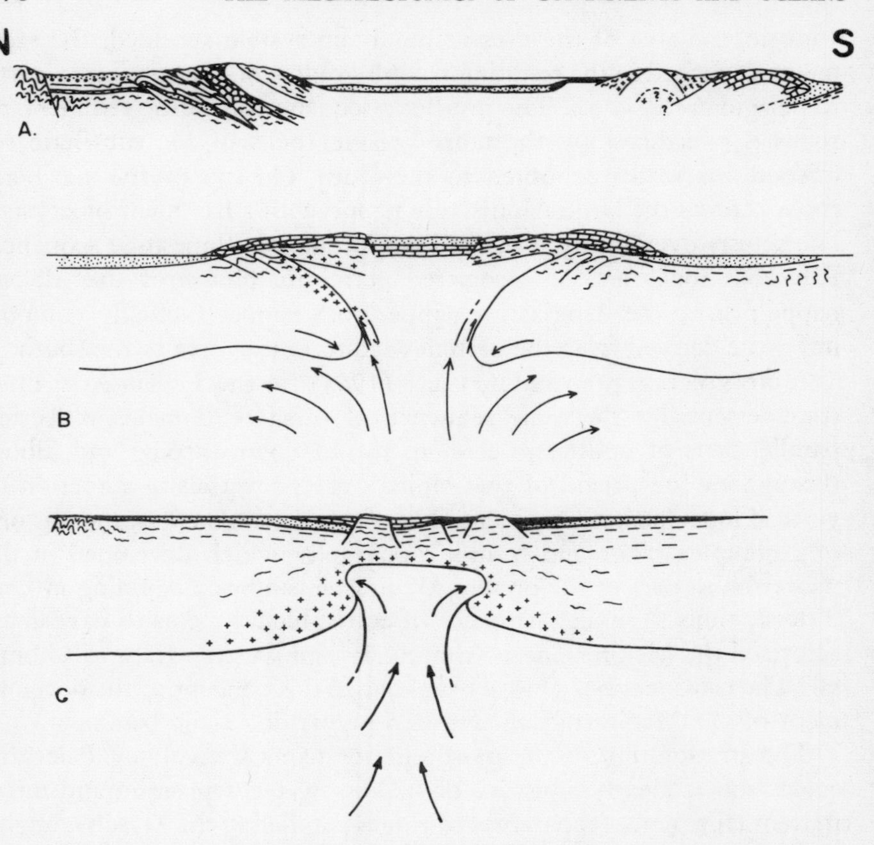

Fig. 4. Origin of western Mediterranean, modified from Glangeaud, 1957. *A*, present N-S section, Spain to Morocco. *B*, same in Miocene. *C*, same in latest Jurassic–earliest Cretaceous.

attribution of an Eocene age for emplacement of these nappes is erroneous. The movements are now thought to have occurred in early Miocene, while movements in the external zone occurred between middle and late Miocene (Chauve and co-workers, 1964; Durand-Delga and Fontbote, 1960). Movement of the nappes involve extensive flowage of Flysch type sediments and production of great subaqueous mud flows or olistostromes (Caire, 1963; Perconig, 1962), strongly suggesting that the emplacement occurred predominantly by outward movements under the force of gravity, and perhaps largely in a submarine environment. The present western Mediterranean basin is thus no older than the late Miocene.

Igneous Rocks and Metamorphism. The western Mediterranean mountains are typically "Alpine" in that metamorphism and igneous activity are only locally conspicuous. Metamorphism which affects rocks as young as Miocene tends to be restricted to the more interior zones and is mostly of greenschist facies. Garnet-sillimanite-graphite gneiss occurs in the Beni-Bouchera area of the Rif (Milliard, 1959), and gneisses have been described in the Kabyles of the Tellian Atlas (Durand-Delga, 1962). The gneisses, with interbedded marbles of possible Early Cambrian age, grade upward to mica schists and phyllites in which Ordovician fossils have been found. The metamorphics are overlain unconformably by unmetamorphosed fossiliferous Silurian and younger Paleozoic rocks. Young granites are shown to cut the metamorphic rocks in the "Petite Kabylie" (Keiken, 1962, Fig. 7). Miocene andesites and trachytes and Pliocene and Quaternary basalts occur in northeastern Morocco near Melilla, and to the north across the Mediterranean Sea at Cabo de Gata.

Of fundamental tectonic interest is the presence of large masses of peridotite in the Betic and Rif Mountains, associated with metamorphosed Paleozoic rocks. In the Rif, and possibly also in the Betic, the peridotite occupies the core of anticlines. As Kornprobst (1966) points out, these peridotites are particularly interesting in that they contain significant amounts of lherzolite, pyroxenite, garnetiferous pyroxenite, and, more rarely, garnetiferous peridotite, approximating the theoretical upper mantle rock "pyrolite" (Ringwood and co-workers, 1964). The age of these peridotite masses has not been fixed precisely, but on structural grounds, apparent emplacement before metamorphism, and regional considerations they are thought to be of "Alpine" affinities, the equivalent of the ophiolites which are characteristic of the alpine milieu (Durand-Delga, and co-workers, 1962, p. 401).

TYRRHENIAN AREA

The Tyrrhenian Sea is a second tethyan area exhibiting bilaterally symmetrical outward thrusting from a region presently occupied by deep water, and characterized by "oceanic" crust having seismic velocities of 7.2 to 7.4 km per second (Berckhemer and Hersey, 1965). On its southern and eastward sides, the Sicilian-Calabrian arc recalls the western Mediterranean arc and is similarly characterized by mid-Miocene radial displacements involving gravity tectonics (Caire, 1965, p. 165; Merla, 1964; Selli, 1964).

To the north, where the Tyrrhenian Sea narrows and becomes shallower between Corsica and Elba, a particularly interesting symmetrical pattern is found (Kraus, 1962; Trevisan and co-workers, 1964). Outward thrusting or large-scale gravity sliding has occurred from the center of the present sea, towards Corsica on the west and Elba and Italy on the east. Corsica is made up largely of Hercynian granitic and metamorphic rocks, but in the northeastern coastal area there is a slightly metamorphosed, chaotic ophiolite-bearing melange resembling similar rocks to the north in Liguria and to the northeast and east in Tuscany (Bortolotti and Passerini, 1963). The map pattern suggests that the ophiolitic melange was emplaced on an erosion surface developed on the crystalline rocks.

In Elba, similar rocks appear as components of overlapping and generally less chaotic tectonic slices (Fig. 5). These slices include schists of probable Carboniferous age at the base, Triassic-Jurassic calcareous rocks of the Tuscan autochthonous facies, and Cretaceous-Eocene Flysch (Trevisan and co-workers, 1964; Trevisan, 1950). In both Corsica and Elba, the tectonic style and evidence of "soft rock" deformation points to emplacement primarily by gravity sliding away from a N-S axis which approximately bisects the Tyrrhenian Sea. The symmetry axis extends southward toward Calabria and northward toward the serpentine-bearing metamorphic rocks of the Voltri zone west of Genova. As indicated by Kraus (1962), the Pennide-like rocks of the Voltri zone may mark the southern extension of the Alpine-Briançonais geanticlinal zone.

Beginning late in the Cenozoic, the axial zone was a local hot spot. The Monte Capanne granite batholith of western Elba has been dated radiometrically at 7 million years, while granites and ignimbrites of Campiglia on the adjacent mainland are even younger (about 4 million years, according to Trevisan, 1964, and remarks during 1964 International Field Institute field trip). Granitic rocks are also exposed on the islands of Montecristo and Giglio south of Elba. A belt of Quaternary calderas and cones lies east of Campiglia and extends southward past Rome to the latitude of Naples. Recent activity is restricted to the Sicilian-Vesuvian portion of the arc.

Although evidence of bilateral tectonic symmetry has not been noted in the regions between Liguria and the northern tip of Corsica, nor between southern Corsica and Sicily, there is conclusive evidence that much of the Italian Peninsula is covered by allochthonous rock

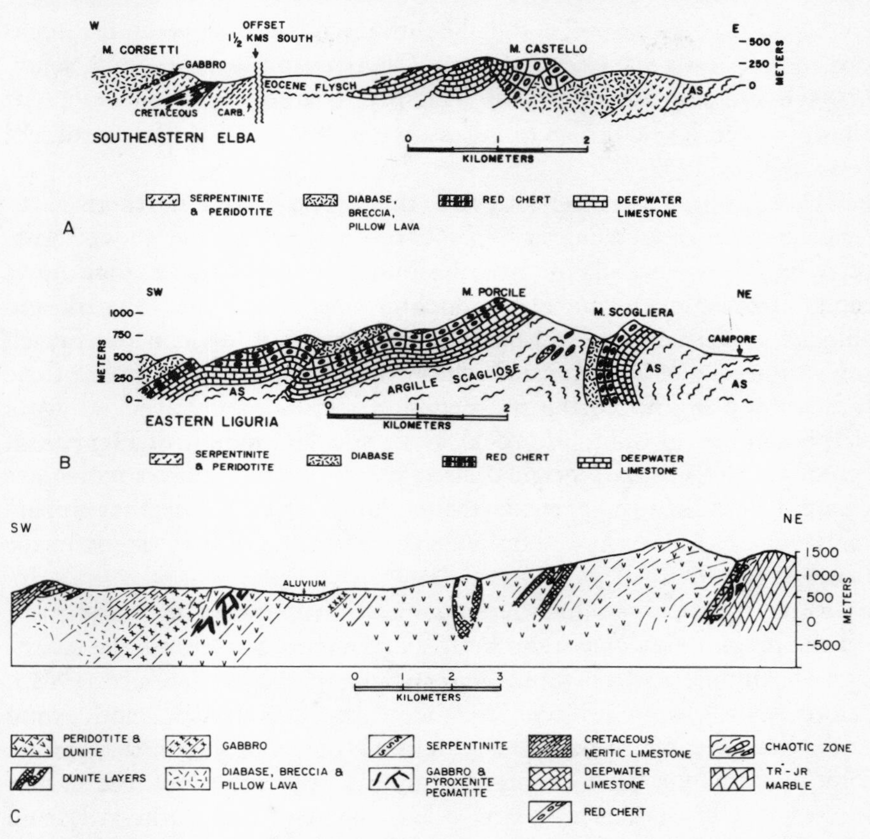

Fig. 5. Ophiolite terranes. *A*, western Elba (modified from Trevisan, 1950); *B*, Liguria (after Mostardini, 1957); *C*, Vourinos complex, Northern Greece, modified from Moores (in press). Vourinos and Elba terranes have normal sequences; the Ligurian sequence lies inverted in chaotic Flysch assemblage.

derived by gravity movements from the present area of the Tyrrhenian and Ligurian Seas (Merla, 1951, 1964; Trevisan, 1964; Maxwell, 1959). In the northern Apennines, for example, the great Tuscan nappe, a slab of Late Triassic to Oligocene marine sedimentary rocks 2 km thick and at least 40 km broad, came from the southwest to cover a sequence of similar but presumably autochthonous rocks of the Alpi Apuane. Above the Tuscan nappe is the thick ophiolite-bearing melange which has been called the Ligurian nappe and also *argille scagliose*. This broken to chaotic allochthonous debris was likewise derived from the direction of the present Ligurian Sea. In addition to the exotic blocks of ophiolitic rocks, the allochthonous melange contains blocks and great slabs of Cretaceous to Oligocene calcareous and sandy Flysch and less numerous but widespread large and small blocks of highly sheared Hercynian granites (Merla, 1933, 1951; Eberhardt and co-workers, 1962).

Thus, from the present area of the Ligurian and northern Tyrrhenian Seas vast quantities of allochthonous rocks have moved easterly onto the peninsula by gravity sliding and as submarine mud flows and related movements. The allochthonous rocks, both matrix and blocks, consist mostly of Lower Cretaceous to mid-Cenozoic Flysch of various types, but Mesozoic carbonates comparable to those of the autochthon of the northern Apennines were also involved, as were large volumes of ophiolitic rocks and a smaller amount of Hercynian granite. The inferred original areal extent of the rocks involved is much larger than the present area of the sea from which they apparently came. Late Alpine compressive deformation may account for the loss of area.

Equally extensive allochthonous masses have probably moved out of the present area of the southern Tyrrhenian basin across the southern Apennines and northern Sicily (Selli, 1962, 1964; Caire, 1965). Here again chaotic Flysch sequences are characteristic, and exotic ophiolites are present though not so numerous as in the north. Granite blocks have not been reported from the allochthonous rocks, but Hercynian crystalline rocks may have been thrust southward from the Tyrrhenian area in northern Sicily (Caire, 1965), and perhaps eastward at the northern end of the Calabrian massif (Selli, 1962). The allochthonous lower Miocene San Giorgio formation contains conglomerates with boulders of granites and metamorphic rocks up to 45 cm in diameter. The only possible source seems to be in the Tyrrhenian area (Selli, 1964, p. IV-18 and remarks during 1964, I.F.I. field trip). Recently, Sarpi (1967) has described coarse Miocene conglomerates in the Cilento region, carrying boulders and pebbles of granitic and metamorphic rocks for which the only likely source is the Tyrrhenian area. It is thus clear that areas which are now deep Mediterranean Sea floor were source areas for the materials which make up much of the Apennines and were denuded mainly by gravitational mass movements, but also partly by subaerial erosion. They must have subsided rapidly in Neogene times. The mechanism of this subsidence remains uncertain, though it probably was related in part to adjustments near the surface above a column of cooling and differentiating mantle material. Rapid subsidence also occurred earlier in parts of the Mediterranean Sea area, associated in space and time with emplacement of the ophiolites in latest Jurassic-earliest Cretaceous times.

Ophiolite Sequence

The assemblage of igneous rocks which Europeans call ophiolites seems incompatible with the dominantly sedimentary melange in which it typically occurs. In the melange environment the ophiolite sequence is disarticulated into discrete blocks which, however, may range in size to many cubic kilometers. The original sequence has been preserved on the island of Elba (Trevisan, 1951), where the distance of travel from the point of origin to the west is apparently small (Fig. 5*A*).

Ophiolites in nonmetamorphic terrains have a characteristic stratigraphy (Dubertret, 1955; Gansser, 1959; Brunn, 1960; Maxwell and Azzaroli, 1962; Passerini, 1965). Dense, black serpentinite commonly forms the base; the serpentinite grades upward into partially serpentinized peridotite which, in the thicker masses, may have interlayered dunite and pyroxenic peridotite. Above the peridotite, especially in the thicker bodies, there is a layer of very coarse gabbro including both dark and light varieties. The gabbro gives way abruptly upward to coarse diabase, which in turn grades upward to finer grained diabase, then to microlitic lava with breccia and pillow structure characteristic of an extrusive environment. Bedded, usually red, radiolarian cherts overlie the extrusive lavas, though locally the lavas and cherts may be interbedded. On Elba, I found angular fragments of red chert in the cores of pillows. The base of the chert sequence is commonly a chippy chloritic mudstone similar to the filling between pillows; it frequently contains fragments of variolites, derived from adjacent pillows, and appears to have been produced by the reaction of hot lava to sea water. In many areas, the radiolarian cherts pass upward through a thin transition zone of gray, calcareous, radiolarian chert to pale gray or cream-colored finely grained limestone with abundant calpionellids of Tithonian to earliest Cretaceous age.

Although individual units may be thin or locally absent, the stratigraphic order within most ophiolite complexes seems to be everywhere essentially the same. Never is there a preexisting roof rock above the igneous complex; internal unconformities and igneous contacts are rarely found; and successive multiple intrusions are essentially restricted to coarse pegmatitic pyroxenite and gabbro which are common in the thicker peridotite masses and may penetrate the gabbro zone.

The problems posed by the invariably sedimentary environment of the ophiolites has long been recognized. For more than a century geologists working in the Apennines have recognized their submarine origin. Indeed, certain kinds of evidence indicate a deep marine environment of emplacement, perhaps at depths great enough to exceed the critical pressure of water (Glangeaud, 1952). The volcanics are strikingly free of vesicles and amygdules, in contrast to most extrusive basaltic and andesitic lavas, and instead are characterized by variolitic crusts around pillows. The constant association of coarse igneous rocks —peridotite, gabbro, and diabase—with obviously extrusive lavas also poses problems. Steinman (1926) recognized the essential unity of the ophiolitic suite, but invoked a complicated mechanism involving multiple intrusion into middle Mesozoic deep sea sediments beneath already formed radiolarite. The mechanism does not account for the ubiquitous extrusive lavas, nor refute the abundant evidence that the radiolarites, though partly contemporaneous, were for the most part deposited on top of the extrusive lavas. In fact, field evidence convincingly demonstrates that the formation of the ophiolites sequence involved only a single, but complex, event which took place largely on top of deep water sediments and in contact with sea water.

Ophiolites of the type described characterize the entire Alpine belt from the Alps and Apennines eastward to the Himalayas (Gansser, 1959; Dubertret, 1955; Bailey and McCallien, 1953). They are particularly well developed in the mountains of Yugoslavia and Greece. Aubouin (1959), in his reconstruction of the geology of peninsular Greece, depicts a ridge and basin environment comparable to that inferred for the western Mediterranean, and suggests that the ophiolites were extruded on the sea bottom from the margin of the rising internal Pelagonian massif. My colleagues * and I are currently studying the Vourinos ophiolite mass in northern Greece (Fig. 5*C*). Brunn (1960) and Zachos (1953), who brought this important area to the attention of geologists, demonstrated a gross stratigraphic sequence

* H. H. Hess, E. Moores, and J. Zimmerman.

within the ophiolite mass, in general conforming to the sequence in ophiolites elsewhere. Detailed work by Moores (in press) largely substantiates the conclusions of Brunn and Zachos, but reveals a much more complex internal structure. It also indicates, to our satisfaction at least, that the complex was formed not by fractional crystallization of a gabbro magma, as suggested by Brunn, but directly from ultramafic material (Fig. 6) which differentiated by partial melting as it neared the sea bottom. There is excellent evidence that much of this material was extruded on the sea bottom, but it is quite likely that vast amounts were also intruded at some levels within the crust, perhaps giving rise to great peridotite plutons such as those of the Betic and Rif chains (Fig. 4).

The enormous masses of ultramafic rocks which have appeared at moderate depths in the crust and in the sea bottom most probably represent an upwelling of upper mantle material. No other source seems compatible with the large volumes of ultramafic rocks involved and the regularity of the phenomenon which occurred along a quarter of the earth's circumference. The mechanism of emplacement undoubtedly is complex (Maxwell and Azzaroli, 1962), but currently the best guess is that mantle material rose as a plastic mass, probably facilitated by partial melting and the formation of intergranular liquid (Fig. 7). The intergranular melt might be expected to coalesce and rise faster than the ultramafic mass, perhaps breaking out on the sea floor as shield volcanoes (Dubertret, 1939).[1] Where a rising mass was bordered by a deep marginal depression, the ultramafic part of the column locally reached the level of the sea bottom and flowed plastically beneath the carapace of diabase and lavas, at the same time intensely deforming and extruding the unconsolidated Flysch sediments from beneath the growing mass (Fig. 7). Gabbroic differentiates formed pegmatite masses and great sill-like lenses between ultramafic rocks and already solid overlying diabase. In the meantime the liberation of silica into the sea water, and presumably also the water turbulence related to heating, initiated a bloom of radiolarians above and far beyond the limits of the igneous extrusion.

Ophiolite emplacement involves the rise of large masses of mantle well above its normal level, into and above lighter crustal rocks. The mechanism can hardly be other than the diapiric rise of plastic rocks,

[1] The inferred process recalls that suggested by Nayudu, 1962, for the origin of guyots and seamounts.

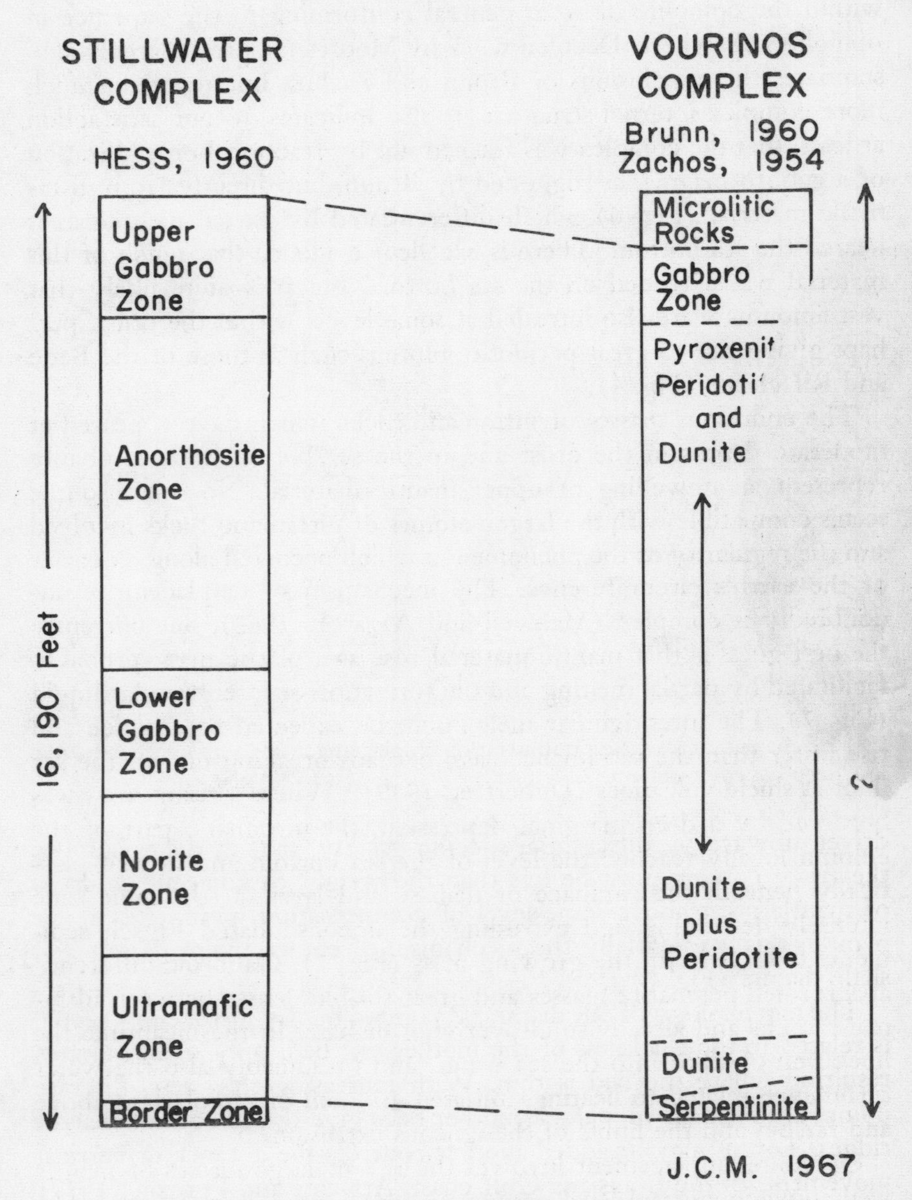

Fig. 6. Comparison of differentiated gabbro complex (Stillwater) with Vourinos ultramafic complex.

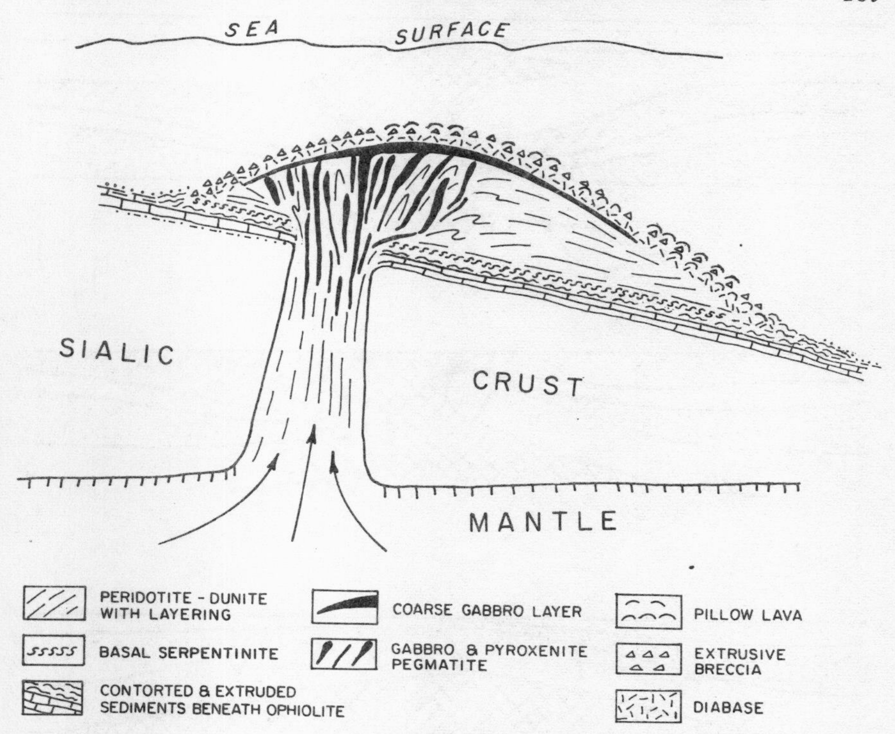

Fig. 7. Schematic cross section illustrating emplacement of an ophiolite sequence in a deep sea environment.

driven upward by differential specific gravity; that is, the weight of the rising column must be less than the weight of adjacent columns of mantle-crust above some level of compensation within the upper mantle; this is essentially the mechanism of emplacement of salt and shale diapirs (Fig. 8).

The morphology of salt diapirs, as pointed out by Trusheim (1960), is related to the thickness of the mother salt bed, only small mounds resulting where the bed is thin. With increasing thickness, discrete domes are formed; very thick mother beds give rise to continuous salt ridges. To this may be added another control—the distance of vertical movement. In the Louisiana Gulf coast, Atwater and Forman (1959) have noted the presence of "salt massifs," continuous ridges of salt at great depth from which rise individual salt domes. Gravity surveys of the Texas-Louisiana Gulf coast reveal a pattern of elongated minima, approximately parallel to regional strike of formations, which have

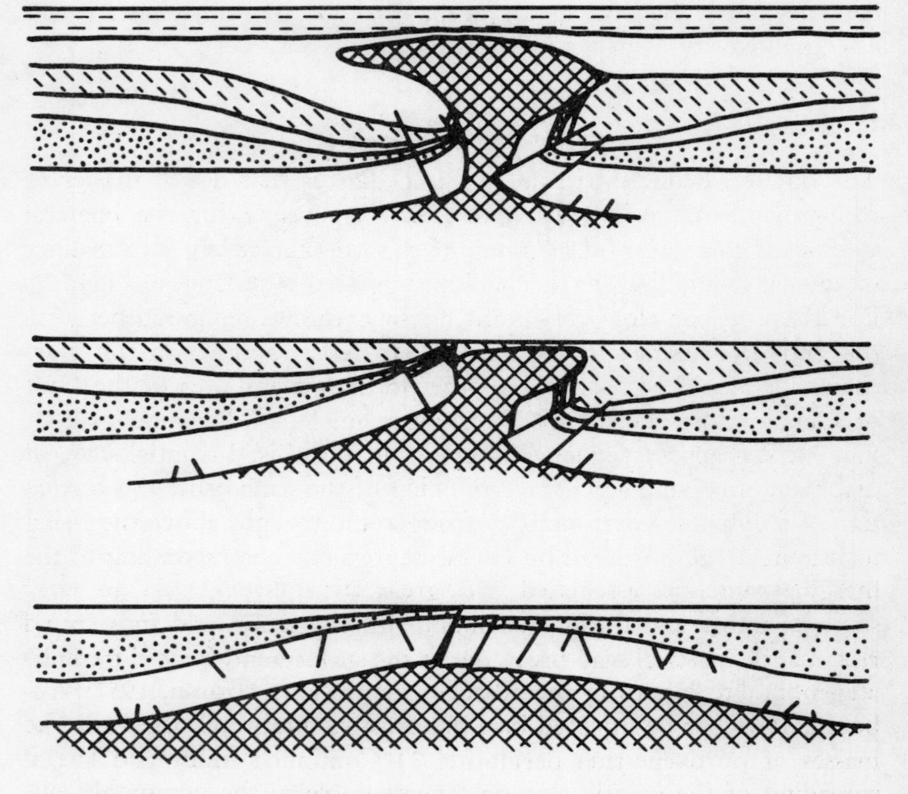

Fig. 8. Initiation, emplacement, and lateral spreading of salt diapirs. After Trusheim (1960).

long been interpreted as representing great, deep salt ridges from which rise individual domes.

Probably, the thickness of the plastic upper mantle zone is everywhere great enough to permit the formation of diapiric ridges. However, by analogy with the behavior of salt, it might be postulated that the distance of vertical travel also will control the form of diapiric masses of mantle rocks. In the deep ocean, where the crust of light rocks above the mantle is thin, long ridges would perhaps be the dominant form, whereas in areas of thicker crust, such as the Mediterranean, domal masses might be expected to form above rising mantle ridges. Perhaps, too, the formation of domal masses is facilitated by the progressive differentiation of the upward-moving plastic ultramafic mass and faster rise of the differentiate, because the column contains relatively more gabbroic material and therefore has a lower effective density as it nears the surface.

Tectonism

The obvious deduction from the preceding is that domal masses of rising mantle material may have been responsible for the bilateral symmetry characteristic of much of the Mediterranean, in a manner somewhat resembling the mechanism suggested by Glangeaud (1957). For example, one elongated uplift might have been centered between the Betic cordillera and the Rif–Tellian Atlas; another in the Tyrrhenian Sea; and perhaps still another in the general area of the Gulf of Genoa (note the mid-Mediterranean nuclei of Klemme (1958), Fig. 7). Perhaps the sequence of events involved is: 1) Initial stage of diapirism, involving a general stretching of the sialic crust and formation of a dynamic environment of ridges and troughs above the rising mantle material, bordered by lateral depressions corresponding to the marginal synclines associated with great salt uplifts. 2) At an early stage of uplift and during maximum differentiation and most rapid rise, mantle material may break out at the surface and form differentiated ophiolite masses in the adjacent depressions (Brunn, 1957; Aubouin, 1965) or be injected into the crust as undifferentiated sill-like masses of pyroxene-rich peridotite. 3) Continued uplift and lateral spreading of the mantle material (comparable to the commonly observed lateral spreading of the tops of salt domes, but more extensive) to permit long-continuing rise of mantle material, with resulting thrusting and gravity sliding of nappes and chaotic masses landward from the present sea areas. The lateral spreading of the tops of the rising mantle columns and consequent underthrusting of adjacent crust may then account for the modest crustal compression found, for example, in the autochthon of the northern Apennines estimated at about 20 percent shortening (Merla, 1951, p. 365). 4) Finally, during the later stages, the gradual heating up of crust overlying and adjacent to the rising mantle column and associated satellitic plutons could well have resulted in regional metamorphism and the production of the young silicic intrusive and volcanic rocks of Elba, Corsica, the adjacent Italian Peninsula, the Rif, and elsewhere. Geothermal steam of the Larderello area, south of Volterra, used for decades to generate electric power, is a remnant of this mantle heat. Perhaps also the long-continued presence of hotter and more plastic mantle beneath the Tethyan orogen would facilitate not only the well-documented vertical movements within the "mobile belt" itself, but also the release of crustal stress by compressive movements directed perpendicular, or shear movements directed parallel, to the orogen and its various parts. A certain independence in behavior of local areas within a broad orogenic pattern is implied.

Gravity data for the Mediterranean area (Bruyn, 1955) are compatible with the hypothesis developed here. The entire western Mediterranean is an area of modest but persistent positive anomalies. The small amount of available seismic data (Hersey, 1965; Berckhemer and Hersey, 1965) indicates the presence of high velocity rocks at relatively shallow depths which, in turn, has led many workers to regard the western Mediterranean as being an oceanic area. The reported velocities (7.0 to 7.7 km per second at 6 to 12 km depth, with one area in the northern Balearic basin reaching 7.9 to 8.0 at 12 km depth), however, are not so high as those expected in "true" oceanic areas. Moreover, Knopoff, in his presentation at this conference, points out that the minimum depth of a zone with a velocity as high as 8.0 must be at least 50 km in this area.

The available geologic evidence and the geophysical results are compatible with the hypothesis of a dynamic ridge and trough environment established on sialic crust and injected by mantle material which locally differentiated and extruded on the surface as ophiolites.

It is interesting that the larger ophiolite masses have precisely the structure postulated for oceanic crust on the basis of seismic data. The obvious inference is that some parts of "oceanic areas" determined seismically may, in fact, contain ophiolites overlying sedimentary rocks, and false depths of the Moho discontinuity may have been reported.

The eastern Mediterranean and the Ionian Sea differ in structure from the western Mediterranean. Whereas the western Mediterranean is a symmetrical axis of bilateral orogenic systems, to the east of Sicily the bilaterally symmetrical orogenic system lies entirely on the north side of the Mediterranean Sea. The south side appears to be developed on little deformed African shield. The differences between east and west are further emphasized by the gravity data. In contrast to the dominantly positive anomalies in the west, the eastern Mediterranean is characterized by a large negative anomaly arcing along the southern, outer margin of the Greek Islands, paralleled on the north by large positive gravity anomalies. One of these gravity positives is related to a large, rootless, ultramafic-mafic assemblage resembling ophiolite rocks on Cyprus (Gass and Masson-Smith, 1963). The present topography of the eastern Mediterranean is a series of parallel ridges and troughs (Ryan and Heezen, 1965). The regional history and the large negative gravity anomaly suggest that these ridges and troughs developed on sialic crust. A reasonable postulate is that the present eastern Mediterranean closely resembles the western Mediterranean as it was developing in late Mesozoic and early Cenozoic time.

Conclusion

If Europe and Africa were not close together now, the "geologic fit" of the two continents is so striking that we would certainly postulate they had once been together and had drifted apart. No lateral displacement of oceanic dimensions can reasonably be postulated between Africa and Eurasia, at least since mid-Paleozoic. Instead, the Mediterranean is best interpreted as a strip at one time of continental crust in which massive injections of mantle material accompanied the generation of a dynamic ridge and basin environment, with resulting parallel belts of erosion and sedimentation, extensive thrusting, gravity sliding, and tectonic denudation. The present pattern of Mediterranean basins largely reflects subsidence since the late Miocene.

References

Atwater, G. I., and M. J. Forman: Nature of growth of southern Louisiana salt domes and its effect on petroleum accumulation. *Bull. Am. Ass. Petrol. Geol.* 43:2592–2622, 1959.

Aubouin, J.: *Contribution a l'étude géologique de la Grèce septentrionale.* Ann. Géol. Pays Helléniques vol. 10, 1959.

Aubouin, J.: *Geosynclines.* Amsterdam, Elsevier Publishing Company, Inc., 1965.

Azema, J.: *Etude géologique des abords de Malaga (Espagne).* Invest. Geol. "Lucas Mallada" (Madrid); Est. Geol. Inst., vol. 17, 1961.

Azema, J., M. Durand-Delga, and Y. Peyre: *Corte del Mesozoico y del Eoceno en el Palo de Malaga* (Andalucia). Inst. Geol. Minero España, Notas y Comunicaciones No. 59, 1960.

Bailey, E. B., and W. J. McCallien: Serpentine lavas, the Ankara melange and the Anatolian thrust. *Trans. Roy. Soc. Edinburgh* 62, P. II:403–442, 1953.

Berckhemer, H., and J. B. Hersey: Some Features of the Alpine-Mediterranean Orogenesis. In *Continental Margins and Island Arcs.* Geol. Surv. Canada Paper No. 66–15:114–123, 1965.

Bortolotti, V., and P. Passerini: Sulla presenza di depositi da frane sottomarine nella parte settentrionale della Corsica. *Boll. Soc. Geol. Ital.* 82:167–172, 1963.

Boulin, J.: Sur la serie metamorphique de Velez-Malaga. *Bull. Soc. Géol. France* s. 7, 4:165–169, 1962.

Brunn, J. H.: Mouvements verticaux et translations dans le couple axe ancient-sillon orogene de la Grece septentrionale. *Bull. Soc. Géol. France,* 13, s. 6, t. 7, f. 4–5, pp. 305–325, 1957.

Brunn, J. H.: Mise en place et différenciation de l'association pluto-volcanique du cortège ophiolitique. *Rev. Géog. Phys. Geol. Dyn. (Paris)* 3, n. 3:115–132, 1960.

Bruyn, J. W., de: Isogram maps of Europe and North Africa. *Geophys. Prospecting* 3:1–14, 1955.

Caire, A.: Phénomènes tectonique de biseautage et de rabotage dans le Tell algérien. *Rev. Géog. Phys. Geol. Dyn.* 5, 4:299–325, 1963.

Caire, A.: Comparison entre les orogénese berbère et apénninique. *Ann. Soc. Géol. Nord.* 48:163–174, 1965.

Carey, S. W.: The Tectonic Approach to Continental Drift. In *Continental Drift: A Symposium.* Hobart, University of Tasmania, Geology Department, 1958, pp. 177–355.

Chauve, P., et al.: Mise au point sur l'âge des phénomenes tectoniques majeurs dans les Cordillères Bétique Occidentales. *Geol. Mijnbouw* 43:273–276, 1964.

Choubert, G.: *Les grands traits de la Geologie du Maroc.* Lexique Strat. Int., v. 4, Afrique, f. la Maroc., 1956.

DeBoer, J.: *The Geology of the Vicentinian Alps, N.E. Italy.* Geol. Ultraiectina No. 11, 1963 (doctoral thesis, University of Utrecht).

Dubertret, L.: Sur la genèse et l'âge des roches vertes syriennes. *Comp. Rend. Acad. Sci. (Paris)* 209:763–764, 1939.

Dubertret, L.: Géologie des roches vertes du nord-ouest de la Syrie et du Hatay (Turquie). *Notes et Mém. sur le Moyen-Orient, Mus. Natl. d'Histoire Naturelle (Paris)* 6:1–179, 1955.

Durand-Delga, M.: I. Le sillon flysch Eocene en Mediterranée occidentale: II. Le sillon geosynclinal des flyschs oligocenes en Mediterranée Occidentale. *Compt. Rend. Acad. Sci.* 296–298, 431–433, 1961.

Durand-Delga, M., ed.: *Livre à la Mémoire du Professeur Paul Fallot.* Mém. Soc. Geol. France, vol. 1.

Durand-Delga, M.: Esquisse structurale de l'Algérie en 1961. *Lexique Strat. Int.*, v. 4, *Afrique*, flb, *Algerie*, 1962, pp. 3–37.

Durand-Delga, M.: *Titres et Travaux Scientifiques.* Paris, Priester, 1966.

Durand-Delga, M., and J. M. Fontbote: Le problème de l'âge des nappes alpujarrides d'Andalousie. *Rev. Géog. Phys. Geol. Dyn.* 3:181–187, 1960.

Durand-Delga, M., et al.: Données Actuelles sur la Structure du Rif. In *Livre à la Mémoire du Professeur Paul Fallot*, vol. 1, M. Durand-Delga, ed. Mem. Soc. Geol. France, 1962, pp. 399–422.

Eberhardt, P., G. Ferrara, and E. Tongiorgi: Détermination de l'âge des granites allochthones de l'Apennin septentrional. *Bull. Soc. Geol. France* s. 7, 4:666–667, 1962.

Fallot, P.: Les Cordillères betique. Est. Geol. Inst. Invest. Geol. "Lucas Mallada" (Madrid) no. 8, pp. 83–172, 1948.

Gansser, A.: Ausseralpine Ophiolithprobleme. *Eclog. Geol. Helvetiae* 52: 659–680, 1959.

Gass, I. G., and D. Masson-Smith: The geology and gravity anomalies of the Troodos massif, Cyprus. *Phil. Trans. Roy. Soc. London*, ser. A. 255: 417–467, 1963.

Glangeaud, L.: Reflexions sur les travaux de la XVe section (Paleovolcanologie et Tectonique) au Congres International d'Alger. *Comp. Rend. Int. Geol. Cong.*, 19th session, sec. XV, fasc. 17, pp. 235–239, 1952 (published 1954).

Glangeaud, L.: Essai de classification geodynamique des chaines et des phenomenes orogeniques. *Rev. Geog. Phys. Geol. Dynam.* 1, f. 4:214, 1957.

Hersey, J. B.: Sedimentary Basins of the Mediterranean Sea. In *Submarine Geology and Geophysics* (Proc. 17th Symp. Colston Res. Soc.). London, Butterworths, 1965, p. 78.

Hess, H. H.: *Stillwater Igneous Complex, Montana.* Geol. Soc. America, Mem. 80, 1960.

Irving, E.: *Paleomagnetism and Its Application to Geological and Geophysical Problems.* New York, John Wiley & Sons, Inc., 1964.

Keiken, M.: Les Traits Essentiels de la Géologie Algèrienne. In *Livre à la Mémoire du Professeur Paul Fallot*, M. Durand-Delga, ed. Mém. Soc. Geol. France, vol. 1, 1962, pp. 543–614.

Klemme, H. D.: Regional geology of circum-Mediterranean region. *Bull. Am. Ass. Pet. Geol.* 42:477–512, 1958.

Kornprobst, J.: À propos des peridotites du massif des Beni-Bouchera (Rif septentrional, Maroc). *Bull. Soc. franç. Miner. Crist.* 89:399–404, 1966.

Kraus, E.: Le problème de l'espace en tectonique dans la region méditerranéenne. In *Livre à la Mémoire du Professeur Paul Fallot,* M. Durand-Delga, ed., vol. 1, 1962, pp. 117–124.

Maxwell, J. C.: Turbidibe, tectonic and gravity transport, Northern Apennine Mountains, Italy. *Bull. Am. Ass. Pet. Geol.* 43:2701–2719, 1959.

Maxwell, J. C., and A. Azzaroli: Submarine extrusion of ultramafic magma (abstract). Geol. Soc. America, Special Paper 73, 1962, pp. 203–204.

Merla, G.: *I graniti della formazione ofiolitica appennina.* Boll. R. Uff. Geol. d'Italia, v. 58, no. 6, 1933.

Merla, G.: Geologia dell' Appennino Settentrionale. *Boll. Soc. Geol. Italiana,* 1951, pp. 95–382.

Merla, G.: *The Apennines.* Guidebook, Int. Field Inst. Italy; Am. Geol. Inst., Sec. I, 1964.

Milliard, Y.: Les massifs métamorphiques et ultrabasiques de la zone paliozoique interne du Rif. *Notes Mém. Serv. geol. Morocco* 147:125–160, 1959.

Moores, E.: Petrology and structure of the Vourinos ophiolite complex, northern Greece. *Bull. Geol. Soc. Am.* Sp. paper.

Mostardini, F.: Unpublished thesis, University of Florence, Italy, 1957.

Nayudu, Y. R.: A new hypothesis for origin of guyots and seamount terraces; Crust of the Pacific Basin. *Geophys. Mon.* 6, Am. Geophys. Un., p. 171–180, 1962.

Officer, C. B., et al.: Geophysical investigations in the Eastern Caribbean. *Physics and Chemistry of Earth,* vol. 3. New York, Pergamon Press, 1959, Fig. 17.

Passerini, P.: Rapporti fra le ofiolite le formazioni sedimentarie fra piacenza e il Mare Tirreno. *Boll. Soc. Geol. Italiana,* 84:3, 1965.

Perconig, E.: Sur la constitution géologique de l'Andalousie occidentale, en particulier du bassin du Guadalquivir. In *Livre à la Mémoire du Professeur Paul Fallot.* M. Durand-Delga, ed., vol. 1, 1962, pp. 229–256.

Runcorn, S. K., ed.: *Continental Drift.* New York, Academic Press, 1962.

Ringwood, A. E., I. D. MacGregor, and F. R. Boyd: Petrological constitution of the upper mantle. *Yearbook Carnegie Inst.,* 63:147–152, 1964.

Ryan, W. B. F., and B. C. Heezen: Ionian Sea submarine canyons and the 1908 Messina turbidity current. *Bull. Geol. Soc. Am.* 76:915–932, 1965.

Sarpi, E.: Stratigraphy of the Flysch Sequence in the Cilento Area, Southern Apennines, Italy. Unpublished Ph.D. thesis. Columbia University, New York, Faculty of Pure Science, 1967.

Selli, R.: Il paleogene nel quadro della geologia dell' Italia meridionale. *Mem. Soc. Geol. Italiana* 3:737–789, 1962.

Selli, R.: *Southern Apennines and Umbria.* Guidebook Int. Field Inst., Italy; Am. Geol. Inst., sec. IV, 1964.

Smirnow. L.: *An Introduction to the Geological Structure of Europe and Asia.* Univ. Toronto, Inst. Earth Sciences, Scientific Report no. 5, 1964.

Steinman, G.: Die ophiolithischen Zonen in den mediterranen Kettengebirgen. 14th International Geological Congress, vol. 2, pp. 637–667, 1926.

Trevisan, L., G. Giglia, and F. Barberi: *Elba.* Guidebook, Int. Field Inst. Italy; Am. Geol. Inst., sec. V, 1964.

Trevisan, L.: *L'Elba orientale e la sua tettonica di scivolamento per gravita. Mem.* Ist. Geol. Univ. Padova, No. 16, 1950.

Trevisan, L.: Isola d'Elba. *Boll. Soc. Geol. Italiana* 70:435–470, 1951.

Trevisan, L., et al.: *General Description of the Apuane.* Guidebook, Int. Field Inst., Italy; Am. Geol. Inst., sec. VIII, 1964.

Trusheim, F.: Mechanism of salt migration in northern Germany. *Bull. Am. Ass. Petrol. Geologists* 44:1519–1540, 1960.

Van Hilten, D.: Evaluation of some geotectonic hypotheses by paleomagnetism. *Tectonophysics* 1:3–71, 1964.

Van Hilten, D., and J. D. A. Zijderveld: Paleomagnetism and the Alpine tectonics of Eurasia. II. The magnetism of Permiam porphyries near Lugano. *Tectonophysics* 3:429–446, 1966.

Van Hilten, D.: Discussion of some recent papers on ancient continental configurations reconstructed from paleomagnetic evidence. *Tectonophysics* 1:523, 1965.

Wilson, J. T.: Continental drift. *Scient. Am.*, April, 1963.

Zachos, K.: Chromite Deposits of Vourinos (Kozani) Area. In *The Mineral Wealth of Greece. Pub. Inst. Geol. Subsurface Res.* 3:47–82, 1953.

9

The Axial Valley: A Steady-State Feature of the Terrain *

KENNETH S. DEFFEYES

Department of Geology,
Princeton University, Princeton, New Jersey

From the time of its initial discovery by Heezen (1960) the axial valley of the mid-ocean ridge has been widely regarded as an extensional feature. After Hess's introduction of the broader concept of sea-floor spreading, the axial valley assumed special importance as the site of generation of new oceanic crust. The author was one of many who therefore wanted to study the axial valley, probably because of the old uniformitarian urge to study an important process in action today. Although a fully documented review paper bringing together all that has been written about ridge axes would be valuable, I prefer to take this opportunity to attempt a synthesis. Enough is now known about the geology and geophysics of the mid-ocean

* Virtually every ship that has worked near the ridge axis has contributed information that a successful model cannot ignore. Fortunately, the flow of information and of ideas has been sufficiently open and sufficiently international to make the study of the ridge axes more successful than it would have been under a single program. Unfortunately, it is possible occasionally to lose track of the genesis of an idea, and in a synthesis of this type it is particularly difficult to ascribe each fragment to its originator. To colleagues who recognize an unattributed fragment, my apologies and my thanks.

Since the flow of money is easier to trace than the flow of ideas, I can thank the Office of Naval Research and the National Science Foundation for supporting the operation of the R/V *Yaquina* of Oregon State University. Five cruises, spaced over several years, have furnished the concrete involvement to keep alive an interest in the more abstract parts of the problem.

I wish to thank Fred Vine for reading the manuscript and for an education on the geology of Cyprus.

ridge to place severe constraints on hypothetical models. Of course we do not now have, and probably will not ever have, enough information to prove the uniqueness of one particular model. It is the purpose of this paper to summarize those constraints which seem to restrict my choice of models, and then to show that at least one model is possible that does not violate the constraints.

Constraints

Before attempting to put together a model, it will be helpful to bring together the boundary conditions. With hundreds of ship crossings and dozens of kinds of observations this is necessarily a selected list. In part the purpose of the list is to encourage others to nominate additional constraints that further restrict the choice of models.

UNIFORMITARIANISM

Since most geologists (including marine geologists) were originally trained on land, our initial reaction to the discovery of the axial valley was to think of the Rift Valleys of East Africa, or of the Great Basin of the western United States. Both of these continental features bear the imprint of a few tens of kilometers of extension. But neither the Rift Valleys nor the Great Basin could be steady-state features of the terrain. If additional hundreds of kilometers of extension were to happen to either continental area, the existing materials of the Rift Valleys would be transported far away, with no provision for replacing continental crust along the axis. In contrast, the mid-ocean ridge can be in steady-state. Ten million years from now the materials now in the axial valley will be a portion of the ordinary sea floor at least 100 kilometers from the axis, yet I fully expect the axial valley to exist in essentially its present form. This steady-state is a special case under uniformitarianism. We are accustomed to seeing geologic processes today extending the evolution of geologic features, but we usually think an evolution toward an end point, not toward a steady-state. Geosynclines become mountain ranges which in turn become shield areas, but the steady-state has been not much more popular in geology than it was in cosmology. John Hack's model of weathering in the Appalachian terrain is the outstanding steady-state concept in land geology. For areas in and near the axis of the mid-ocean ridge, we can state the steady-state condition explicitly: the rate of acquisition of material at each location must equal the transport of material away from that location.

Before committing the following discussion permanently to the steady-state restriction, possible exceptions to that restriction should be explored. Of course, it would be much simpler if one could take a pre-existing oceanic crust, give it about 10 kilometers of extension in very recent geologic time, and produce an axial valley. As will be discussed in detail in the section on magnetics, the discovery of the "tape recorder" by Vine and Matthews (1963) has required that new sea floor be created during every 100,000 years of time. If one had a brief extensional episode create the axial valley as a unique event, it would have to happen within the last half million years and it would leave the magnetic anomalies unexplained and unexplainable.

There has been some discussion of discontinuities in spreading rate (Ewing and Ewing, 1967). Fortunately, the time scale relevant to the axial valley, the last 3 million years, is solidly established by potassium-argon dating and the arguments about discontinuities concern times of 10 million years and older. The only possibility that I can see for any present-day change in spreading rate is a change in the equatorial bulge of the earth associated with the Pleistocene ice. Taking off 100 meters of ocean and storing it as polar ice increases the rotation rate, and presumably increases the equatorial bulge. If this is the cause of the smooth, emergent coasts equatorwards of 45 degrees latitude and drowned valleys polewards of 45 degrees, then it is just possible that the increased equatorial radius would exhibit itself as a jump in spreading rate on the mid-ocean ridges. We can show quickly that this is a small effect. At most, the equatorial emergence is a half kilometer in elevation. The increase in circumference is 2π times that, or about three kilometers spread over at least two million years. Even a slowly spreading ridge at 1 centimeter per year would move 20 kilometers in two million years. So at most, if all of the maximum increase in equatorial circumference were to appear on one side of the mid-Atlantic ridge, one would still be hard pressed to locate an increase of 3 parts in twenty. Therefore, until some unexpected new data appear, there seems to be every reason to adopt a firmly uniformitarian view of the ridge axis.

BATHYMETRY

We must recognize first that along almost half of its length, the mid-ocean ridge lacks an axial valley. That this lack was not simply the result of inadequate surveying was resolved when Menard (1967) pointed out the restriction of axial valleys to half spreading rates less than 3 cm/yr. (It has become customary to quote spreading rates in terms of the amount of surface appearing annually on each side of the ridge. The full annual extension between two points on opposite sides well away from the ridge axis is twice the quoted rate. Hereafter "spreading rate" will refer to the customary one-sided or half spreading rate.)

Spreading rates greater than 3 cm/yr typically produce a ridge with less roughness at the top of the basaltic layer and with either no axial valley at the ridge crest or a small axial valley of the type shown in Figure 1. Certainly one of the important constraints on a model that produces axial valleys is some method of turning the model off for higher spreading rates.

Several detailed surveys have been made over axial valleys. Before satellite navigation was available, a sizeable effort involving anchored navigational buoys was required for a detailed survey. All work had to be done in one episode, because the buoys would be lost before a return expedition was possible. A delightful exception is available on the Gorda Ridge, about 120 miles off of the Oregon-California coast. Loran A navigation, with an east-west accuracy of about a half mile and north-south accuracy of a few hundred yards is available over the Gorda axial valley. Figure 2 shows a segment of the valley, based on a survey of 45 lines at half-mile spacing. Besides the general form of the valley, which is about the depth and width of the Grand Canyon, this survey shows a series of steps in the valley

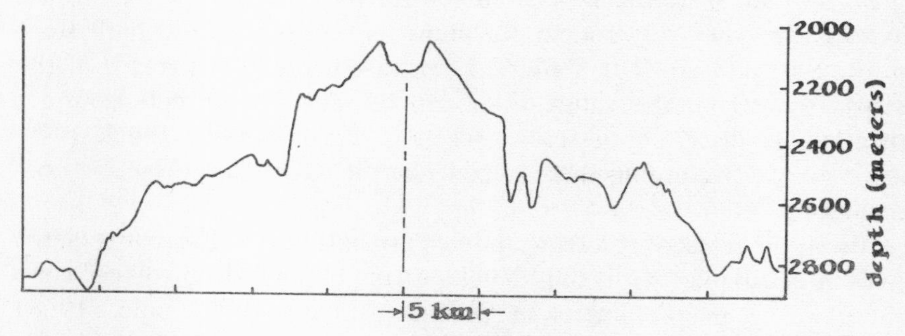

Fig. 1. Bathymetry of the axial zone of the Juan de Fuca Ridge, showing symmetrical topography and a small axial valley at the crest of the central peak (Melson, 1969).

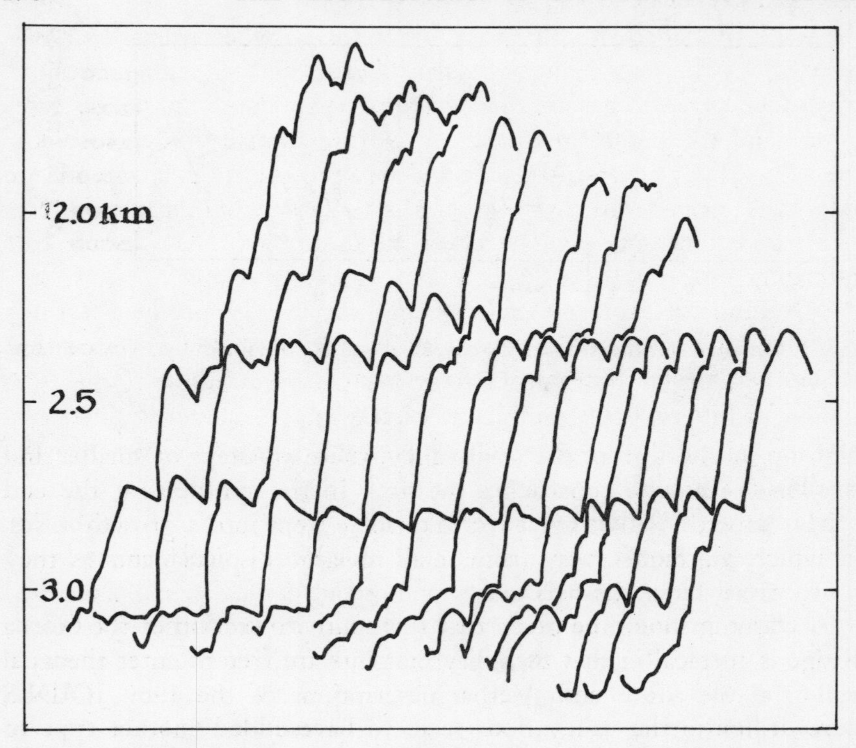

Fig. 2. Wide-beam echo-sounder profiles at half-mile spacing on the east wall of the Gorda axial valley at 41°20′N and 127°15′W. Prominent steps between fault scarps occur at 2.4 and 2.7 km depth. For their fortitude in gathering these east-west profiles in a strong south wind, I wish to thank the crew of R/V *Yaquina* and K. L. Russell. Vertical exaggeration is about 15 times.

wall. Individual steps can be followed for tens of miles parallel to the axis of the valley. Of course, the steps remind one of the fault steps in the East African Rift Valleys. Even the tilting of the treads of the steps away from the valley floor can be seen on the echo-sounder profiles. To date, every reason seems to favor making the leap by analogy and presuming these steps to be the expression of active normal faults.

Vastly increased resolving power in examining the bathymetry became available with the development of the deep-towed echo sounder and magnetometer at Scripps. Atwater and Mudie (1968) made a deep tow over the Gorda axial valley, and their profile (reproduced here as Figure 3) shows that in addition to the steps

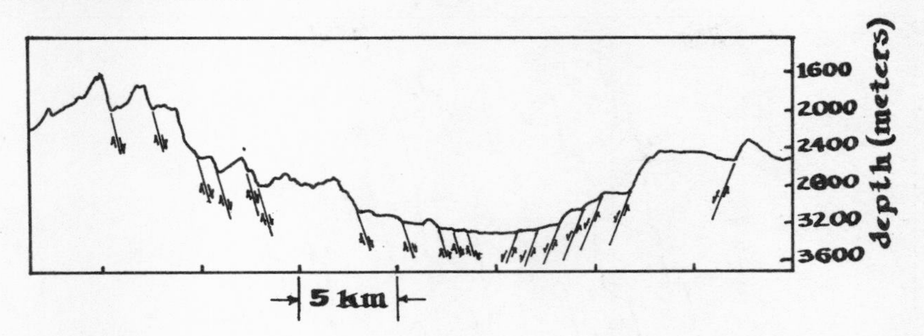

Fig. 3. Detailed profile of the Gorda axial valley obtained by Atwater and Mudie (1968) with a deep-towed echo sounder.

high up on the wall of the axial valley, there are steps of smaller and smaller size as one approaches the axis. In the synthesis at the end of this paper, I will attempt to make these steps into a growth series, in which the faults grow from small to large displacements as they move away from the axis.

A warning should be given that this southern section of the Gorda Ridge is atypical in that turbidity currents are free to enter the axial valley at the south end. Verbal descriptions of the 1969 JOIDES core drilled in the valley floor seem to have added another type of geosyncline to Marshall Kay's list. The Pleistocene sediments in the axial valley are being treated to instant metamorphism.

The sea floor away from the axial valley contains very few additional bathymetric hints about the processes in the valley. From the survey shown in Figure 2, one could recognize a saw-tooth form, with the steeper side of the teeth facing the axial valley. The scarps between the normal faults high up on the wall of the axial valley are somehow diminished in magnitude before they travel out across the sea floor. This reduction of scarp height is about the least convenient condition to adapt to of all the constraints in the list. Some sort of unfaulting, possibly due to isostasy, will be called upon later; I do not pretend to be satisfied with this aspect of the problem.

MAGNETICS

The Vine-Matthews magnetic model played a central role in confirming the sea-floor spreading hypothesis. In addition it is possible to squeeze several important constraints out of the magnetic data. The first constraint initially seems obvious; the age of the sea floor

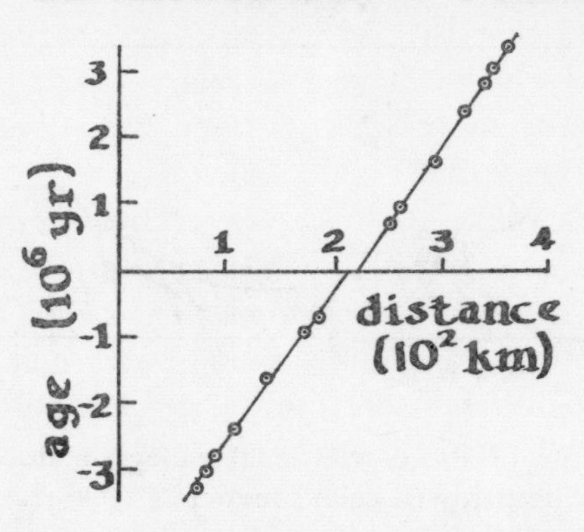

Fig. 4. The ages of different magnetic events can be plotted against their locations in the magnetic profile. Lines fitted separately to the two flanks of the ridge are extrapolated to zero age. This profile (Eltanin-19) suggests that there can be a measurable difference from zero age at the center.

extrapolates to zero on the axis of the ridge. As shown in Figure 4, we can artificially consider one half of the ridge to represent negative distances and negative time, and estimate whether there is a need to draw an intercept at other than the origin. Several attempts at this seem to indicate that the intercept at zero time is within a few kilometers of the center of symmetry. In the early mathematical models of the ridge one was slipping in thin dikes only at the origin; therefore zero distance at zero time was a trivial consequence of the model. We now know several reasons why we cannot build the sea floor entirely out of vertically intruded dikes:

1. The overwhelming majority of the rocks dredged near the axis are pillow lavas, which implies flows of some horizontal extent.
2. If the magnetic record were being written on thin vertical sheets originating at the center of symmetry, the resolving power of the magnetometer tows (especially the deep tows) would be even higher than observed.
3. The growth of faults from small steps on the axis to large steps on the valley walls (Figure 3) implies that the "spreading" takes place over a zone a few kilometers wide, rather than at a singularity at the center.

One is left with both the extension and the emplacement of basalt distributed over several kilometers to either side of the axis. Now it is no longer required that the age extrapolate exactly to zero on the axis, but quantitatively we have to balance basalt emplacement and the zone of extension to get the observed near-coincidence between zero distance and zero time.

The resolving power of the magnetic record is an important input. Although pillow lavas, dikes, and intrusives probably exist, it is the pillow lavas which contribute most heavily to the magnetic record. Probably this is due to their fine grain size which produces single-domain iron oxide grains. Therefore the resolving power of the magnetic record is determined by the width over which the pillow lavas are extruded. Both Matthews and Bath (1967) and Vine and Morgan (1967) have independently simulated magnetic profiles resulting from a Gaussian scatter of units of emplacement of the magnetic record. Vine and Morgan's conclusion was unexpected: the faster the spreading rates, the narrower the width over which the magnetic record was written. Quantitatively, Vine (personal communication) has generalized the relationship to state that the half width of the magnetic resolving power (in kilometers) is equal to 3.5 over the square root of the spreading rate (in cm/yr). For ridges with axial valleys, this restricts the writing of the magnetic history to the floor of the axial valley. Very little, if any, of the magnetic record originates on or under the walls of the valley.

One of the first aspects noted about the magnetic records over mid-ocean ridges was the increased magnetic intensities near the axis. In addition, the axial anomaly itself was not only largest of all, there was no diminution of the anomaly exactly on the axis. If the magnetic record originated even two or three kilometers from the axis, a very noticeable drop in the anomaly would occur over the axis.

HEAT FLOW

The discovery of abnormally large heat flows near the axis of the ridge played an important role in the early history of the sea-floor spreading hypothesis. At first, it seemed obvious enough that if convection were required to transport heat out of the upper mantle, the convected heat would appear at the site of the convective upwelling. A second look was disappointing. The amount of excess heat flow appearing in the mid-ocean ridge anomalies was only a small fraction of the total heat flow appearing at the earth's surface. Specifically, the 200 kilometer width of the heat flow anomaly near the ridge covers about two percent of the surface of the earth (Lee and Uyeda, 1965). If half of the heat were transported out by simple mantle convection, one would expect the two percent of the area to have a heat flow 50 times normal. The average measured heat flow anomaly is only about 4 times normal.

There are two possible explanations for the apparently small size of the heat flow anomaly. Either the conventional heat flow technique does not measure all of the forms of heat transfer in the near-ridge environment, or the hypothesis that the ridge is the outcrop of an immense convective system is incorrect.

The first possibility is that some heat transfer does not appear in the conventional measurements. It is possible that extrusive and shallow intrusive igneous rocks deliver their heat to the sea water before they are transported far enough from the axis to be accessible to the sedimentary heat probe. In a later section I will discuss convection of sea water through the rocks as a possible mode of heat removal near the ridge axis. Regardless of how we proceed, the total anomalous heat flow at the ridge cannot be as much as 10 percent of the total heat liberated from the earth.

The second possibility is that the ridge is not simply the outcrop of a major convection cell. One can examine the closely spaced transform faults in the equatorial Atlantic and quickly convince oneself that asking each segment between two transform faults to be a simple convective cell, requires the cells to be narrower north-south than they are deep. Although intuition is not always a reliable guide, the pattern in the equatorial Atlantic is hardly what one expects from the top of a cell involving the whole upper mantle. With Morgan's (1968) discovery of the plate-like behavior of the earth's surface, the segments of ridge and transform faults become simply alternations between shear and tension fractures between two coherent plates. Wherever the tension fractures appear, at those sites material from the low-viscosity zone must be drawn up to fill in the gap. In this model, the plates respond to the vector sum of all the forces at their undersides and margins. Convection can still exist, it can still drive the plates around; but what happens at the ridge crest is due to the local divergence of the plates. Orowan (1969) reached a similar conclusion recently. Elsasser (1967) has shown that "convection" in the earth takes on a specific thin-skinned form and that the equality of heat flow on land and at sea is an expression of the mass transfer and heat flux. For this present purpose, it suffices to know that the mid-ocean ridge is very probably the driven rather than the driving phenomenon in the global plate tectonics. Everything we see at the ridge crest in the way of vulcanism, heat flow, and active faulting can be the consequence of forces applied elsewhere to the plates.

The remaining notable point about the heat flow observations is their very large scatter, even in observations spaced a few tens of kilometers apart (Figure 5). Although instrumental errors, natural disturbances, and underlying inhomogeneities in conductivity have been postulated, the large variance in the near-ridge observations tempts one to introduce an additional hypothesis. In the interpretive section of this paper, the water convection hypothesis suggested by Elder's Figure 1 (1965) will be employed to explain the inhomogeneities in heat flow.

SEISMIC OBSERVATIONS

In verifying J. Tuzo Wilson's (1965) hypothesized transform faults, Sykes (1967) showed that most of the seismic events from the mid-ocean ridges come from the transform fault segments between ridge crests. The ridge segments themselves produce a lesser number of detectable earthquakes. First motion studies of these few ridge earthquakes confirm the normal fault behavior that was inferred from the bathymetry. For the purpose of this synthesis, the important

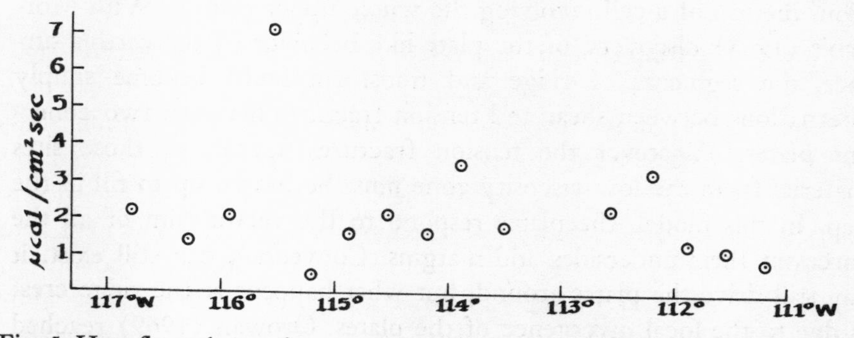

Fig. 5. Heat flow observations near the ridge axis show abnormally low values even among the spectacularly high heat flows. These data are for the East Pacific Rise at 17°S. (Vema-19 data as reported by Lee and Uyeda, 1965.)

result from Sykes' work is the restriction of the normal fault events to a zone no wider than 20 kilometers. Unfortunately, the 20 kilometer width is approximately the resolving power of the method, so all we have is an upper limit to the width of the seismically active zone. However, the knowledge of this upper bound is sufficient to exclude models like the one presented by Vogt and Ostenso (1967), in which the extension is distributed over a width of 1,500 kilometers.

Refraction seismology with explosive sources has rarely been reported from the axial valley. Some of the early work in the Atlantic by Ewing and Ewing (1959) suggests layer thicknesses on the ridge axis to be about half the normal thicknesses. Away from the ridge axis, Menard (1967) has noted that slow spreading rates tend to be associated with a thicker second layer.

GRAVITY

Shipborne gravimeter observations over the ridge have produced a useful null result: the anomalies are small, therefore the terrain is very close to isostatic compensation. Of course, with very hot rock very close to the surface one would not expect measurable deviations from isostasy. As a result, I will hold the proposed model to exact isostasy.

The large-scale gravity observations available from satellite tracking appear frustratingly unrelated to the presence of mid-ocean ridges. McKenzie (1969) discussed some probable reasons for the lack of correlation between ridges and the large-scale gravity anomalies.

SEDIMENTATION

Accommodating the pelagic sedimentation into a steady-state spreading model is the one straightforward pleasure in the whole procedure. Pelagic sedimentation commences in the axial region, sediment thicknesses build up with distance exactly as predicted by dividing the sedimentation rate by the spreading rate, and the paleontologic age of the bottom of the sediments equals the magnetic age inferred for the underlying basalt. Since this result worked out so beautifully on JOIDES leg 3 in the South Atlantic, there is one additional inference: once a piece of sea floor is clear of the axial region, it is quite rare to have any additional lava flow out on top of the sediments.

PETROLOGY

Since the mid-ocean ridge is the only area on the sea floor, aside from occasional seamounts, accessible to hard-rock petrologists, a considerable effort has been concentrated on the ridge. The overwhelming majority of the rocks dredged from ridge axes are basalts, specifically they are the oceanic tholeiites so carefully characterized by Engel and Engel (1965). The two common textural varieties are pillows (Figure 6) and intrusive rocks having grain sizes of a few millimeters. My own estimate is that 95 percent of the rocks dredged from the ridges are basalts or metabasalts. Because anyone makes a quite justifiable fuss whenever anything other than a basalt comes up in the dredge, the literature gives the impression that a fair amount of other rocks exist on the sea floor. It is statistically certain that basalts dominate the top of the oceanic crust, but it is presently impossible to give convincing proof of the composition of the third layer from dredging. From those dredge hauls specifically aimed at

Fig. 6. Photograph of basalt pillows dredged from the Gorda axial valley.

the bottom of the largest scarps in the northeast Pacific, it is my own opinion that the third layer is largely composed of intrusives of tholeiitic composition.

GREENSTONES

In the Atlantic (Melson and Van Andel, 1966), Pacific (this report and Duncan, 1968), and Indian (Cann and Vine, 1966) Oceans, basalts metamorphosed to greenstones have been dredged. Most students of these rocks infer them to be normal oceanic tholeiites altered at a few hundred degrees Centigrade to an assemblage of albite, actinolite, epidote, and chlorite. Typically these rocks come from fault scarps, usually high up on the wall of the axial valley. Part of the evidence for the alkali metasomatism of these rocks is shown later in Figure 11.

SERPENTINE

Rocks dominated by serpentine minerals have been dredged at many localities, although their absolute abundance is in doubt. Serpentine occurrences are of special interest because in the paper proposing sea-floor spreading, Hess postulated a third layer dominated by serpentine. In a paper with F. J. Vine, scheduled to appear in The Sea, volume IV, Hess brought together the known occurrences of serpentine on the sea floor. In recent years, a serpentine third layer has become increasingly hard to defend, and in that final paper, Hess admits that there are some arguments for a basaltic third layer without abandoning the case for serpentine. The objections to a serpentine third layer, in addition to the dredging results, are an increase in the total amount of sea-floor spreading which upsets the calculation about the delivery of juvenile water (given in a later section of this paper), and the difficulty in maintaining the 500°C isotherm at constant depth in the face of variations in spreading rate and variations in heat flow. A basaltic third layer has the simplistic advantage that upwelling a given amount of upper mantle from the low velocity zone always fractionates a certain amount of basalt.

PERIDOTITES

Although rare, oceanic occurrences of peridotite are treasured as possible samples of the upper mantle. There is not a site in the typical axial valley that consistently exposes peridotite. Therefore we must see that an acceptable model keeps the earth's mantle modestly veiled with a layer of basalt. The very few exposures of peridotite are probably associated with unusually deep faulting.

Within these constraints, one could construct a large number of models. Future work will make the constraints even more rigorous, but for now a unique model would not be derivable from these observations, and one would have to await either a dramatic improvement in data-gathering or deep drilling to establish a model. This impasse may have been broken with the probable discovery of the oceanic crust cropping out on Cyprus (Gass, 1968) and the Macquarie Islands (Varne and co-workers, 1969).

The Cyprus Analogy

Although the ophiolite sequences in the Alps had been attributed to a deep-sea origin by several authors, the variations among the different ophiolite occurrences did not tempt one to base a detailed sea-floor model on them. The Troodos Massif of Cyprus exhibits a sequence thicker and possibly more complete than the usual ophiolites. After a considerable amount of detailed mapping that is reported in the Cyprus Geological Survey Memoirs, Gass suggested in 1968 that the Cyprus outcrop represents a complete section of typical oceanic crust, including the Mohorovicic discontinuity and an outcrop of the mantle. Field work directed towards testing the validity of the Cyprus outcrop as a model of the sea floor was already underway by Eldridge Moores and Fred Vine when Gass's paper appeared. After two summers' field work in Cyprus, Moores and Vine see no objections to the validity of accepting Cyprus as a complete section of the sea floor (Moores and Vine, in press). The reporting of similar exposures on the Macquarie Islands south of New Zealand has reinforced the probability that Cyprus is a valid sample (Varne and co-workers, 1969).

A section through the critical exposure on Cyprus, the Troodos Massif, is shown in Figure 7. Although the reader is referred to the

Fig. 7. A cross section of the Troodos Massif, Cyprus. Sketched from Gass, 1968, and Moores and Vine, in press.

Cyprus Geological Survey Memoirs for the details of the geology, it is impossible to resist commenting on the "sheeted intrusive complex." This remarkable feature which consists almost entirely of nearly vertical dikes with apparently nothing between them, reminds one of the early sea-floor spreading models in which one keeps inserting dikes at the origin. More than any other feature, the sheeted complex would be difficult to emplace by any other mechanism besides sea-floor spreading.

Other features of Cyprus which were once regarded as anomalous are the extensive hydrothermal alteration, especially in the sheeted complex and the massive iron-copper sulfide deposits at the top of the pillow lavas. The remarkable chemical similarity between the hydrothermally altered rocks on Cyprus (Gass, 1968) and the greenstones dredged on the mid-ocean ridges (Melson and Van Andel, 1966) suggests the view that the hydrothermal alteration on Cyprus is not necessarily atypical of the sea floor. One hesitates to claim that there are massive iron-copper sulfide deposits littering the sea floor. However, some limited evidence of hydrothermal activity is available from the deep sea in the form of euhedral quartz crystals (Cann and Vine, 1966; Duncan, 1968; and from my own dredging on the Gorda Ridge). Minor amounts of sulfides are present in some of the dredged rocks.

As yet there has been no way to establish what kind of sea-floor spreading produced the Cyprus outcrop. The possibilities must include a slow-spreading ridge of the mid-Atlantic type, a fast-spreading ridge like the equatorial East Pacific Rise, and even the new suggestion from JOIDES Leg 6 of spreading on the concave side of the island arcs. For this paper I will make the double assumption first that the Cyprus outcrops are a valid sample of the sea floor and second that a spreading rate of less than 3 cm/yr would produce the Cyprus pattern. There are abundant grounds for the first assumption, the second assumption rests solely on my inability to find anything contradicting the Cyprus model while studying the slow-spreading ridges.

A Model for the Axial Valley

It is unlikely, at least to my intuition, that either the supply of basaltic magma or the extension happens at a singularity one meter wide at the axis. It will be shown that allowing both magma supply and extension to take place over widths of tens of kilometers can generate an axial valley. In such a model, the delivery of basalts and the extension are set by deeper processes and the presence or absence of an axial valley is dependent on the relative width of extension versus basalt supply. The mechanics of the model will be discussed first, then the chemistry and heat flow will be considered.

MECHANICS: BASALT AND EXTENSION

For the moment I will use "basalt" to mean all intrusive and extrusive igneous rocks of basaltic composition; the equivalent rocks on Cyprus would be the pillow lavas, the sheeted complex, and the gabbroic intrusives. At sea, I intend this "basalt" to represent the combined second and third layers. (If there are important amounts of serpentine in the third layer, one would want to modify this model.) The other component of the mechanics, "extension" is the fractional increase of length ($\Delta l/l$) that would be measured between two markers glued to the bedrock. Of course this motion is discontinuous in time (earthquakes) and discontinuous in space (faults), but continuous motions give us an overall description and breaking the continuity into individual faults will be postponed until later.

Both the rate of delivery of basalt and the extension rate will be assumed to be described by Gaussian curves centered on the axis. Someday it will be desirable to replace the Gaussian assumption with a curve derived from observation, but for now Gaussian curves give a fair reproduction of the shape of the axial valley and lead to no obvious contradictions. Figure 8 (a and b) shows the starting assumptions, with the extension taking place over a narrower width than the basalt is supplied. The integral of the extension rate (8c) is the velocity with respect to the axis. Well away from the axis this integral becomes identical with the conventionally observed spreading rate. The integral of the basalt supply rate (8d) is the number of square centimeters of basalt (in the plane of a vertical cross section) passing a given point each year as a function of that point's distance from the axis. Then, in the crucial step, we divide the integrated basalt curve (8d) by the integrated velocity curve (8c) to get the thickness of basalt present at each distance from the axis (Figure 8e). We now have an axial valley. There are four parameters in the model: the

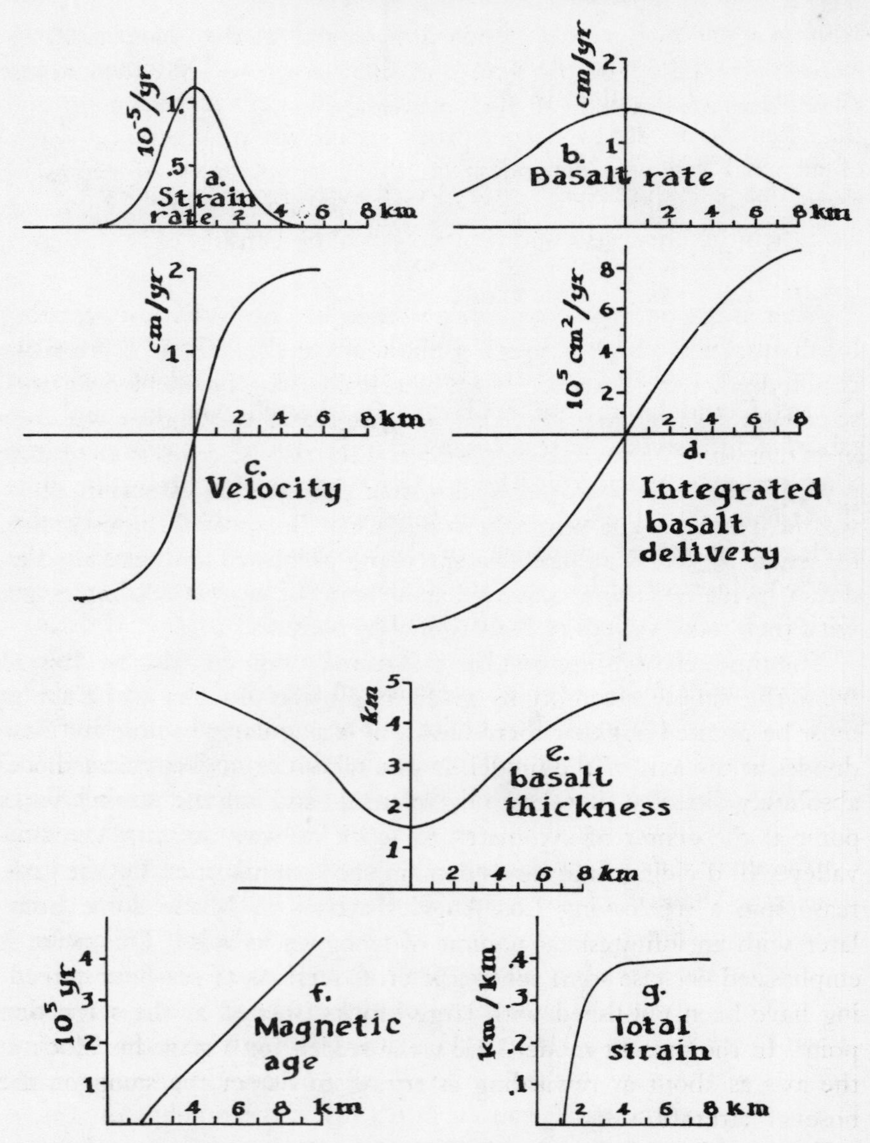

Fig. 8. Gaussian distributions of strain rate (a) and basalt "sedimentation" rate (b) which explain the existence of an axial valley. Integrating (a) gives the velocity (c), and integrating (b) gives the area of basalt passing a point each year (d). Dividing (d) by (c) gives the basalt thickness (e). Integrating the reciprocal of the velocity (c) gives the mean magnetic age (f) and the total strain (g).

heights at the axis and the standard deviations of the Gaussian distributions for basalt supply rate and extension rate. We can adjust these four parameters to fit four quantities:

1. The width of the axial valley
2. The spreading rate measured well away from the valley
3. The basalt thickness measured well away from the valley
4. The basalt thickness on the axis.

After assigning appropriate magnitudes, one finds that the resulting basalt thickness passes through a minimum at the axis. We therefore conclude that the presence of an axial valley can be the simple consequence of supplying the basalt over a greater width than the zone of extension. The reverse condition should also be true: if the basalt is supplied over a narrower width than the zone of extension, there will be a peak at the origin. It is tempting, but probably premature, to assign the form of the fast-spreading ridges to the reverse condition (wide extension, narrow basalt) and the slow-spreading ridges with their axial valleys to the original hypothesis.

The time relationships and the integrated strain can also be derived from the simple assumptions given in Figures 8a and 8b. First, it must be pointed out that there has to be a singularity where the flow divides at the axis of the model. If one takes the mathematical model absolutely literally, it takes infinite time (and infinite strain) for a point at the center of symmetry to work its way out past the axial valley. All dividing fluid flows contain such singularities. By the same reasoning, a jet leaving Los Angeles arrives in Miami some hours later with an infinitesimal volume of smog on its nose. This point is emphasized because some schematic cross sections of sea-floor spreading have been published with large blocks isolated at the stagnation point. In the present model I view the search for a stagnant block at the axis as about as rewarding as trying to detect the smog on the nose of aircraft.

Fortunately, the time required to leave the axis is no more meaningful operationally than it is mathematically. The time-versus-distance information observed for the mid-ocean ridges is derived by picking the inflection points on the magnetometer profiles. Since we know from Vine and Morgan's (1967) random injection models that the magnetic history is written over a width of a few kilometers, the inflection point on the magnetometer record is roughly the point at which half of the basalt beneath is reversely magnetized and half is normally magnetized. However, the magnetic record resides almost entirely in the shallow basalts, probably because pillow lavas have single-domain oxide grains as opposed to multi-domain grains in intrusives. Therefore the critical distance for the magnetic record is the distance from the axis that includes one half of the area under a pillow basalt delivery curve. Vine's result would place the width of the magnetic record at about $3.5/\sqrt{2}$ kilometers for a 2 cm/year spreading rate. We therefore conclude that most of the pillow lavas are emplaced within 2 kilometers of the axis. An important consequence is that the pillow lavas are all emplaced on the floor of the axial valley. Hydrodynamically, this makes some sense; the valley floor is the site that requires the least pressure to reach the surface.

The time-distance relationship can be obtained readily from the velocity plot (Figure 8c). The reciprocal of the velocity is the number of years the surface rock takes to pass each centimeter, and integrating with respect to distance from the axis gives the age in years. If one starts with Gaussian curves initially, it is easiest to do the velocity integration numerically.

The time-distance plot, Figure 8f, is not only consistent with the observed data, but also clarifies the reason why the apparent age extrapolates almost exactly to the origin. Two offsetting effects are present. The slower spreading rate near the axis and the writing of the magnetic record over a zone having a finite width combine to make the more distant record appear to point toward the origin. If there were a prominent magnetic reversal about 100,000 years ago, it would have been obvious much earlier that the change of spreading rate and basalt supply were interlinked near the axis.

SUMMARY

If one utilizes this model and the Cyprus analogy, a cross section of the axial valley resembles Figure 9. With six adjustable parameters (width and height of three curves representing total basalt, pillow basalt, and strain rate), one can account for:

1. The existence or absence of an axial valley
2. A completely steady-state terrain

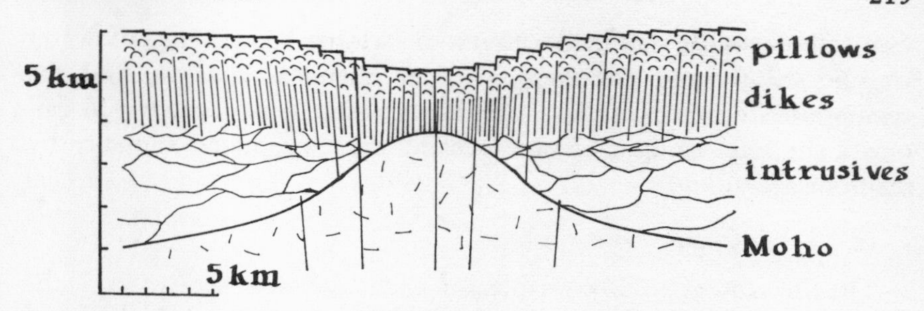

Fig. 9. A cross section under the axial valley constructed using the curves of Figure 8 and following the Cyprus outcrop as a guide to the lithologies. The vertical scale is exaggerated by a factor of two.

3. The shape of the valley
4. The pattern of normal fault steps of increasing amplitude
5. The resolution of the magnetic record
6. The time versus distance properties of the magnetic record
7. The narrow zone of seismicity
8. The absence of isostatic anomalies, and
9. The abundance of tholeiitic rocks and the rarity of other rock types.

To be explained in a later section are the paradoxes concerning the heat flow and the presence of greenstones. In addition, it is only fair to point out a deficiency in the model. The troublesome point is the change from the top of the axial valley onto the adjacent sea floor. Either the throw on some of the normal faults is reversed or possibly the viscous decay suggested by Cramer (1970) removes some of the relief.

Hydrothermal Alteration and Heat Flow

Elder (1965) recognized that the high heat flow at the mid-ocean ridges could lead to active water convection within the rocks, similar to the well-known geothermal areas on land. If water convection is active, there will be large effects on the heat flow and possibly large chemical effects as well. I would like to examine the consequences of water convection through fractures in the rock for the chemistry and heat flow before deciding on the plausibility of a water convection hypothesis. It will be necessary to estimate the amount of water transported through the rock, the net chemical changes, and the temperature differences. As a preliminary step, it is necessary to calculate the total amount of new oceanic crust produced annually by sea-floor spreading.

RATE OF SEA-FLOOR SPREADING

The amount of rock available for exchange with sea water is the product of the rate of creation of new sea floor and the thickness of the basaltic layer. Morgan's (1968) observation that the crust behaves as a set of rigid plates has served to organize the many measurements of spreading rates into a compact form. Morgan referred the relative motion between two blocks to a rotation about an axis, and Le Pichon (1968) later compiled a table of the axis locations and rotation rates. As shown in Figure 10, we can determine the rate of production of new sea floor by integrating the spreading rates between individual blocks. Table 1 gives the angles from the rotation axes to the ends of

Table 1. Summation of the Total New Oceanic Crust Created Annually by Sea-Floor Spreading *

Location	Se Equatorial Half Rate	θ_2	$\sin\theta_2$	θ_1	$\sin\theta_1$	Half Area rSe $(\sin\theta_2 - \sin\theta_1)$
N Atlantic	1.1	90	1.000	30	.500	.035
Pacific-Cocos	10.8	71	.945	50	.766	.123
Pacific-Nazca	8.9	11	.191	−24	−.407	.339
Galapagos Rise	7.3	72	.951	57	.839	.052
Chile Rise	2.9	24	.407	30	.500	.017
Pacific-Antarctic	6.0	80	.984	20	.342	.245
Indian Ocean	2.2	76	.970	−45	−.707	.213
Antarctic-Africa	1.5	63	.891	18	.309	.055
S Atlantic	2.0	70	.939	−30	−.500	.182
N Atlantic	1.1	90	1.000	30	.500	.035
					Sum of half areas	1.326
					Total new area (km^2/yr)	2.65

* The angles are measured from the pole describing the relative motion between the plates (Morgan, 1968). The equatorial half rate is the spreading rate that would be observed 90 degrees away from that same pole. Some of these data are known only indirectly, therefore a final estimate of 2½ km^2/yr will be used.

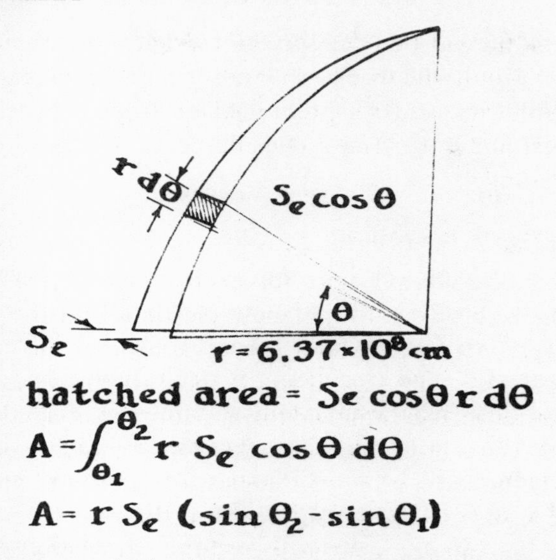

Fig. 10. Integration to obtain the new area created on a segment of the ridge. The vertical axis is a line from the center of the earth to the pole describing the relative motion between the plates (Morgan, 1968).

the segments of the mid-ocean ridge and their rotation rates. The resulting rate of creation of new sea floor is 2.5 ± 0.2 km^2 per year.

SERPENTINE

It is now possible to repeat Hess's (1962, p. 613) calculation for the rate of water released if juvenile water causes the serpentinization:

$$4.7 \text{ km} \times 2.5 \text{ km}^2/\text{yr} \times .25 = 2.9 \text{ km}^3/\text{yr}.$$

This is seven times Hess's estimate of 0.4 km^3/yr. The revision is entirely due to an improved knowledge of spreading rates. Since the 0.4 km^3/yr would produce the present volume of sea water in 4 billion years, the new rate would produce the ocean in post-Cambrian time. Therefore either serpentinization uses recycled water instead of juvenile water or the third layer (as in the present model) is mostly basalt.

The thickness of the rocks being produced are less well known. Using an observation that the second layer thickness was spreading-rate dependent, Menard (1967) computed a world basaltic output of 5 to 6 km^3 per year. If the Cyprus analogy has any merit, the entire second and third layers are basaltic and the volume is about 12 km^3 per year. For the moment we will simply examine the hypothesis that a 1 km thick layer of tholeiitic basalt is converted to greenstone, giving 2.5 km^3 per year of new metamorphic rocks.

COMPOSITIONAL CHANGE ON METAMORPHISM

In general, the metamorphic rocks have gained sodium and magnesium and lost calcium and potassium when compared to fresh tholeiitic basalts. Data on the alkaline earth contents of metamorphosed basalts from Cyprus and from the Mid-Atlantic Ridge are shown in Figure 11. The circulating water responsible for the metasomatism must lose magnesium and sodium while gaining calcium and potassium. If sea water is the water responsible for the exchange, one limiting factor will be the amount of magnesium and sodium in the incoming water. Because magnesium in the sea is a factor of ten less abundant than sodium, we will base the calculation on a magnesium balance, a parallel calculation for sodium would result in a lesser water requirement. The mass of magnesium required is 2.5 $km^3/yr \times 10^{15}$ $cm^3/km^3 \times 3$ grams/$cm^3 \times 0.006$ g Mg/g = 4.5×10^{13} g/yr. Surprisingly, this flux amounts to 55 percent of the 8×10^{13} grams/year of magnesium which rivers deliver to the sea. The amount of calcium released from the tholeiites upon conversion of the same one kilometer thickness to greenstone is equivalent to 30 percent of the river flux for calcium.

It is suggestive that the two most abundant cations in sea water, sodium and magnesium, are the ones absorbed by the greenstones and the cations released (calcium and potassium) are those underrepresented in sea water. It has long been difficult to explain why marine sediments are so low in sodium and high in calcium when compared with the igenous rocks from which the sediments were presumably weathered. Producing a substantial amount of oceanic greenstones helps balance the stubborn budgets for sodium and calcium.

HEAT FLOW

Sea-floor spreading has as an inescapable consequence the presence of steep thermal gradients beneath the locus of spreading. Although the exact gradients are not known, they are at least two orders of magnitude greater than the adiabatic gradient for sea water. There-

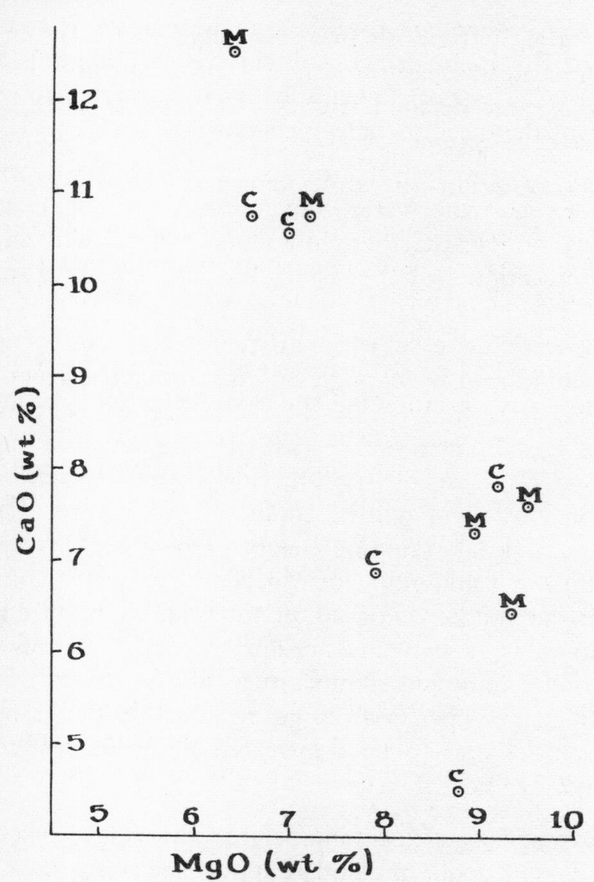

Fig. 11. Calcium and magnesium content of greenstones. Mid-Atlantic Ridge samples (Melson and Van Andel, 1966) are marked M and Cyprus samples (marked C) are taken from the various Cyprus Geological Survey Bulletins. The most altered specimens are those on the lower right.

fore water in a vertical crack would break into upward- and downward-moving fingers. Convective exchange tends to warm (and to cool) the wall rock, and the thermal diffusion into the wall rock sets a minimum horizontal distance between upward-flowing and downward-flowing fingers. This distance increases as $\sqrt{Kt}$ and reaches a kilometer in 30,000 years, and 6 km in a million years. Therefore one would expect the heat flow over the mid-ocean ridge to be extremely erratic, with a scale of a few kilometers between highs and lows. The heat flow measurements are always wildly variable near the ridge, the most detailed studies (Lee and Uyeda, 1965) with spacings between stations of about 20 kilometers, show the heat flow to vary by almost a factor of 10 between adjacent stations (Figure 5). Even if the water convection is no longer active away from the axial valley, the thermal anomalies impressed on the rocks by water convection may be detectable in the heat flow measurements.

It is comparatively easy to show that water convection of this magnitude would transfer enough calories to make a major difference in the crustal heat budget. In the following calculations we will consider the heat budget for one square centimeter of surface area. If there is one kilometer of hydrothermally altered rock beneath this square centimeter, there will be about 3×10^5 grams of rock to be altered. If the rock takes up one weight percent MgO, we need 3,000 grams of MgO or 1,800 grams of Mg. Since sea water contains 1.272 grams of magnesium per kilogram of solution, we require a minimum of 1,400 kilograms of sea water per cm^2 of surface. If this water is to be raised to the minimum temperature of the greenstone facies, it will take 300 calories per gram to get to 300°C, giving a minimum of 4.2×10^8 calories/cm^2 to be transferred by water convection. For comparison, other heat fluxes are:

1. The average heat flow anomaly at the ridge, as observed by the sediment probe method is about 3 μcal/sec cm^2. Integrated over a 10 km width of axial valley at 1 cm/year, this would amount to 10^8 calories/cm^2.
2. The heat of fusion of 3 km of intrusive basalt in the third layer at a density of 3 gm/cm^3 and a heat of fusion of 72 cal/gm amounts to 7×10^7 cal/cm^2.
3. Cooling the same 3 km of basalt from its melting point to 400°C requires 175 calories per gram, giving a total of 1.5×10^8 calories/cm^2.

It seems, therefore, that the water required to alter the greenstones would remove a major portion of the heat produced by the creation of the new oceanic crust. In fact, if it were not for this water convection, an intrusive third layer would cause unacceptably large increases in the observed heat flow.

PERMEABILITY

Finally it must be asked if there is sufficient permeability in the newly formed second and third layers to make water convection possible. Because of the presence of normal faults, even near the axis, at the very minimum there will be fractures in which water can move. In order to get a conservative estimate, we will take fractures one kilometer apart and a temperature difference of 100°C between the upgoing and downgoing water streams. This temperature difference, at 500 bars total pressure, will give a density difference of .04 g/cm^3. If the depth from top to bottom of the fracture is one kilometer, this density difference will result in a head of 4,000 cm of water. The amount of water is the 1,400 kg/cm^2 computed in the section on metamorphism. Since there is a kilometer (10^5 cm) between fractures, it is necessary to have $1{,}400 \times 10^3 \times 10^5$ cm^3 of water pass along each cm of the fracture. Giving 300,000 years for the convection to take place, the required flow rate is .014 cm^3/sec per cm width of the fracture. Now one can use the Pouiselle equation for flow between parallel plates:

$$h = \frac{12\mu Vl}{\rho g t^2}$$

where h is the fluid head, μ is the viscosity, V is the velocity, l is the length along the flow path, ρ is the density, g is the acceleration of gravity, and t is the spacing between the walls of the fracture. The equation can be rearranged to give the thickness required for a certain delivery rate:

$$t^3 = \frac{12\mu(Vt)l}{\rho g h} = \frac{12 \times .01 \times (.014) \times 10^5}{1.0 \times 970 \times 4{,}000}$$

Solving for t, the required fracture thickness turns out to be .035 cm, about the thickness of 4 sheets of book paper. Although this calculation is inexact, I have taken conservative magnitudes. Further, the dependence of the flow on the cube of the thickness means that increases in the required flow will mean only modest increases in the fracture thickness.

From this calculation, it would seem very difficult to keep sea water from circulating into a hot, normal faulted terrain. The effect on the heat flow cannot avoid being large, and the chemical exchange seems to require about the same amount of sea water as the heat exchange.

Preliminary calculations suggest that heat and chemical fluxes of this magnitude will not be easy to detect when streamed into the overlying sea water. For some years Arrhenius (1966) and Bonatti and Nayudu (1965) have been suggesting that the volcanics from the mid-ocean ridge make a dissolved chemical contribution that appears in the surrounding sediments. As long as the pillow basalts dredged at the surface showed only weak signs of chemical exchange, it was difficult to accept their hypothesis. It seems plausible now that the greenstone metasomatism could eject a flux of dissolved ions that might be recognizable in the nearby sediments.

Conclusions

A model of a slow-spreading ridge consistent with the Cyprus outcrop can be created that conforms to all of the original constraints. At this time, some aspects of the model are speculative and some of the quantitative inputs are inexact. However, only two entities are postulated: a distribution for basalt delivery and a distribution of strain rate. The model then attempts to account for all of the observations made near the ridge axis. As a possible future step, one would like to have enough data to work the problem in the other direction and to extract the distributions of basalt delivery and strain rate from field observations.

References

Arrhenius, Gustav: Sedimentary record of long-period phenomena. Pp. 155–174 in *Advances in Earth Science*, M.I.T. Press, Cambridge, Mass. P. M. Hurley, editor, 1966.

Atwater, Tanya and J. D. Mudie: Block faulting on the Gorda Rise. *Science*, v. 159, pp. 729–731, 1968.

Bonatti, Enrico and Y. R. Nayudy: The origin of manganese nodules on the ocean floor. *Amer. Jour. Sci.*, v. 263, pp. 17–39, 1965.

Cann, J. R., and F. J. Vine: An area on the crest of the Carlsberg Ridge: petrology and magnetic survey. *Proc. Roy. Soc. London*, v. A-259, pp. 198–217, 1966.

Cramer, C. H.; Viscosity of the Atlantic Ocean bottom. *Science*, v. 167, pp. 1123–1124, 1970.

Gass, I. G.: Is the Troodos Massif of Cyprus a fragment of the Mesozoic ocean floor? *Nature*, v. 220, pp. 39–42, 1968.

Elder, J. W.: Physical processes in geothermal areas. Pp. 211–239 in *Terrestrial Heat Flow*, Amer. Geophysical Union, Geophysical Monograph Series, No. 8, W. H. K. Lee, editor, 1965.

Elsasser, W. M.: Interpretation of heat flow equality. *Jour. Geophysical Research*, v. 72, pp. 4768–4770, 1967.

Engle, A. E. M., C. G. Engel and R. G. Havens: Chemical characteristics of oceanic basalts and the upper mantle. *Bulletin Geol. Soc. Amer.*, v. 76, pp. 719–734, 1965.

Ewing, J., and M. Ewing: Seismic refraction measurements in the Atlantic Ocean basins, in the Mediterranean Sea, on the Mid-Atlantic Ridge and in the Norwegian Sea. *Bulletin Geological Soc. Amer.*, v. 70, pp. 291–318, 1959.

Ewing, J., and M. Ewing: Sediment distribution on the mid-ocean ridges with respect to spreading of the sea floor. *Science*, v. 156, pp. 1590–1592, 1967.

Hack, J. T.: Interpretation of erosional topography in humid temperate regions. *Amer. Jour. Science*, Bradley Volume, v. 258-A, pp. 80–97, 1960.

Heezen, B. C.: The rift in the ocean floor. *Scientific American*, v. 203, pp. 98–110, 1960.

Lee, W. H. K., and Seiya Uyeda: Review of heat flow data. Pp. 87–190 in *Terrestrial Heat Flow*, Amer. Geophysical Union, Geophysical Monograph Series, No. 8, W. H. K. Lee, editor, 1965.

Matthews, D. H., and J. Bath: Formation of magnetic anomaly pattern of Mid-Atlantic Ridge. *Geophysical Journal*, v. 13, pp. 349–357, 1967.

McKenzie, D. P.: Speculations on the consequences and causes of plate motions. *Geophysical Journal*, v. 18, pp. 1–32, 1969.

Melson, W. G., and T. H. Van Audel: Metamorphism in the Mid-Atlantic Ridge, 22°N Latitude. *Marine Geology*, v. 4, pp. 165–186, 1966.

Menard, H. W.: Sea floor spreading, topography, and the second layer. *Science*, v. 157, pp. 923–924, 1967.

Moores, E. M., and F. J. Vine: The Troodos Massif, Cyprus and other ophiolites as oceanic crust, evaluation and implications. Submitted to *Transactions*, Royal Society at London, in press.

Morgan, W. J.: Rises, trenches, great faults, and crustal blocks. *Jour. Geophysical Research*, v. 73, pp. 1959–1982, 1968.

Orowan, E.: The origin of the oceanic ridges. *Scientific American*, v. 221, pp. 102–119, 1968.

Sykes, L. R.: Mechanism of earthquakes and nature of faulting on the mid-ocean ridges. *Jour. Geophysical Research*, v. 72, pp. 2131–2153, 1967.

Varne, Richard, R. D. Gee and P. G. J. Quilty: Macquarie Island and the cause of oceanic linear anomalies. *Science*, v. 166, pp. 230–233, 1969.

Vine, F. J., and D. H. Matthews: Magnetic anomalies over ocean ridges. *Nature*, v. 199, pp. 947–949, 1963.

Vine, F. J., and W. J. Morgan: Simulation of mid-ocean ridge magnetic anomalies using a random injection model. Geol. Soc. Amer. Annual Meeting *Abstracts*, pp. 228, 1967.

Vogt, P. R., and N. A. Ostenso: Comments on mantle convection and mid-ocean ridges. *Jour. Geophys. Research*, v. 72, pp. 2077–2085, 1967.

Wilson, J. T.: A new class of faults and their bearing on continental drift. *Nature*, v. 207, pp. 343–347, 1965.

10

The Large-scale Tectonic Stress Field in the Earth

ADRIAN E. SCHEIDEGGER *

University of Illinois

The earth presents many striking and interesting surface features, such as continents, mountains, and plains, which clearly must be the result of the action of stress. Obviously, the earth could not have always been in a state of equilibrium. At one time during its history, the earth was presumably in a molten state; as a rotating celestial body, it should exhibit a regularly layered structure had it always been in equilibrium. The densest materials would form the core; going outward from it, there would be layers of progressively less dense substances, with the interfaces between the layers forming perfect equipotential surfaces of the gravity field. In a state of perfect equilibrium, therefore, the earth's solid surface would be level and covered by a layer of water of constant depth. Under such conditions land areas could not exist, and human life, as we know it, would be impossible. The presence of mountainous regions on the earth's surface is thus a positive indication that forces other than gravity must have been active at some time during its history.

Many geotectonic hypotheses attempt to explain the surface features of the earth (Scheidegger, 1963b). All of them must show that a stress field had occurred which could in turn produce the major features observed. Obviously, the stress in the earth could not invariably have been entirely isotropic, and modern geophysics is attempting to find out more about these stresses. The question whether anisotropic stresses are still present, or whether they were active only during periods of tectonic activity, is particularly interesting.

* I am extremely grateful for stimulating discussions with Dr. Rhodes Fairbridge of Columbia University, which helped me to clarify my ideas of Alpine tectonics.

Among the indications of continued activity of anisotropic stresses are their manifestations in mine shafts, in the origin of earthquakes, and in measurable geodetic movements. Studies of the earth's stress field, past and present, in the same way as studies of other types of fields (e.g., the gravity field or the magnetic field), therefore seems worthwhile.

General Characteristics of Tectonic Stress Field

In physics, stress is expressed in the form of a symmetric tensor, denoted by the symbol σ_{ik}, in which $ik = 1, 2, 3$. Any three-dimensional symmetric tensor has three principal orthogonal directions, unless the tensor is degenerate in which case three orthogonal directions can be chosen. In the case of stress, these directions represent the axes of the least, the intermediate, and the highest principal pressures; the T, B, and P axes, respectively. A symmetric tensor may be regarded as the sum of two parts: the isotropic and the deviational. The former is expressed by the equation

$$\sigma_{ik}^{\text{isotr}} = \frac{1}{3}\left(\sum_{j=1}^{3} \sigma_{jj}\delta_{ik}\right) \quad (1)$$

where σ_{ik} is the Kronecker tensor. The equation for the latter is

$$\sigma_{ik}^{\text{dev}} = \sigma_{ik} - \frac{1}{3}\left(\sum_{j=1}^{3} \sigma_{jj}\delta_{ik}\right) \quad (2)$$

An assumption that the underground stress field is simply the isotropic compression due to overburden would be naive. Investigation of the density of the material in the earth's crust would be expressed as

$$\sigma_{ik}^{\text{isotr}} = (1\ \text{lb/in}^2/\text{foot depth}) \times \delta_{ik} \quad (3)$$

But this theory is too simple. There are significant indications that the earth's stress field is not isotropic. The deviations from isotropy are generally ascribed to the action of tectonic stresses; however, the mere fact that the crust is a solid substance induces lateral stresses which differ from the vertical ones, even under the sole effect of the weight of the overburden (Isaacson, 1962). Presumably, a solid substance may be regarded as elastic in the first approximation. Since, under conditions of complete confinement, the lateral strain must be zero, the condition (based on the well-known stress-strain relation for isotropic elastic bodies) can be expressed as

$$0 = \frac{1}{E}\,[\sigma_{11} - m(\sigma_{22} + \sigma_{33})] \qquad (4)$$

where E is Young's modulus; m is Poisson's number; 1 and 2 are the horizontal direction, and 3 the vertical one. Since $\sigma_{11} = \sigma_{22}$, equation 4 immediately yields

$$\sigma_{11} = \sigma_{22} = \left(\frac{m}{1 - m}\right)\sigma_{33} \qquad (5)$$

The lateral stresses are therefore less than the pressure of the overburden; the solidity of the material takes up some of the weight.

The relation expressed by equation 5 is commonly used, but its accuracy is questionable, for creep in due course equalizes the strains in all directions. It is a reasonable assumption that stress differences must be continuously regenerated if they are to exist for any length of time. On the other hand, the stresses at any one moment must satisfy the equilibrium conditions of elasticity theory. This concept is the basis of Anderson's (1942) theory of the standard tectonic stress state. He noted that near the earth's surface (to a depth of about 10 miles) the effect of a boundary condition is present, but that at the surface itself a principal stress direction must be the normal one. Near the surface, therefore, the vertical must still be a principal stress direction, so that one principal stress (σ_v) is vertical, the other two are horizontal (σ_{H1}, σ_{H2}). Evidently, three orientations of the P, T, B trihedron are possible with regard to the three principal stresses which Anderson, who assumed that faults are only fractures in a brittle material, related to geologic faulting phenomena. The usual engineering theory (Mohr, 1928) holds that the potential fracture surfaces in brittle materials are conjugate, contain the intermediate principal stress, and are inclined at an angle of $\leqq 45°$ towards the greatest principal pressure. Based on this theory, three types of standard stress states can be distinguished: B is σ_v, or state of incipient transcurrent faulting; P is σ_V, or state of incipient normal faulting; and T is σ_v, or state of incipient overthrust faulting.

The term *incipient* does not mean that in the stress states under consideration faulting is imminent, but only that, were the stress character further emphasized, it would eventually result in the types of faults indicated. Since two principal stress directions are always more or less horizontal, according to Anderson, it must evidently be possible to draw stress trajectories in a given area. Furthermore, it should be possible to determine which of the three types of stress states is prevalent in the area. A fundamental task of geophysics is thus the characterization of the tectonic stress field in any one area.

Direct Manifestations of Underground Stresses

Direct measurement would be the most obvious method for ascertaining the stress field. However, stresses cannot be measured directly, but only the displacement which is the result of a stress, so that any direct measurement is to some extent an indirect one.

Generally, so-called stress meters are extensometers; such instruments have been developed for measuring stresses in mine walls (Hast, 1958; May, 1960; Panek, 1961; Emery, 1962; Utter, 1962; Obert, 1962a; Obert et al., 1962; Morgan and Panek, 1963; Merrill, 1963, 1964). The subject has been reviewed by Terzaghi (1962) and by Griswold (1963). Stresses measured by one of these instruments near a mine wall must be extrapolated to yield the regional tectonic stresses. The connection of regional stresses with those around a mine opening is part of the subject of rock mechanics (Isaacson, 1962; Mohr, 1964; Sprang and Kahler, 1963; Judd, 1964; Jaecklin, 1965). Specific calculations of stresses around certain types of openings have been made by a number of investigators (Merrill and Peterson, 1961; Osterwald, 1961; Adler, 1962; Hiramatsu and Oka, 1962; Obert, 1962b; Fara and Wright, 1963; Menzel, 1965). In principle, the various relations they have obtained can be used to ascertain the regional stress state from the stress values near the mine wall. Conversely, the stresses near the mine opening can be calculated if the regional stress state is known.

Apart from direct measurements, the prevailing regional stresses may be estimated on the basis of some phenomena in which the stresses manifest themselves more or less directly. In mines, rock bursts, which are a direct effect of underground stresses, are well known (Milne and White, 1958; Cook, 1963). However, the mechanics of such bursts are so complex, that the theory explaining them is not well enough developed for use in establishing the prevailing regional stress.

Another direct manifestation of the regional tectonic stresses occurs in oil wells, particularly during hydraulic fracturing operations. Artificial increase of the fluid pressure in a well eventually fractures the formation, and the pattern of the fracturing depends partly on the prevailing regional stresses. In fact, these stresses can be calculated from the bottom-hole pressure logs of wells during fracturing operations (Pulpan and Scheidegger, 1965).

It would seem, therefore, that the various direct manifestations of underground stresses are evidence of their anisotropy.

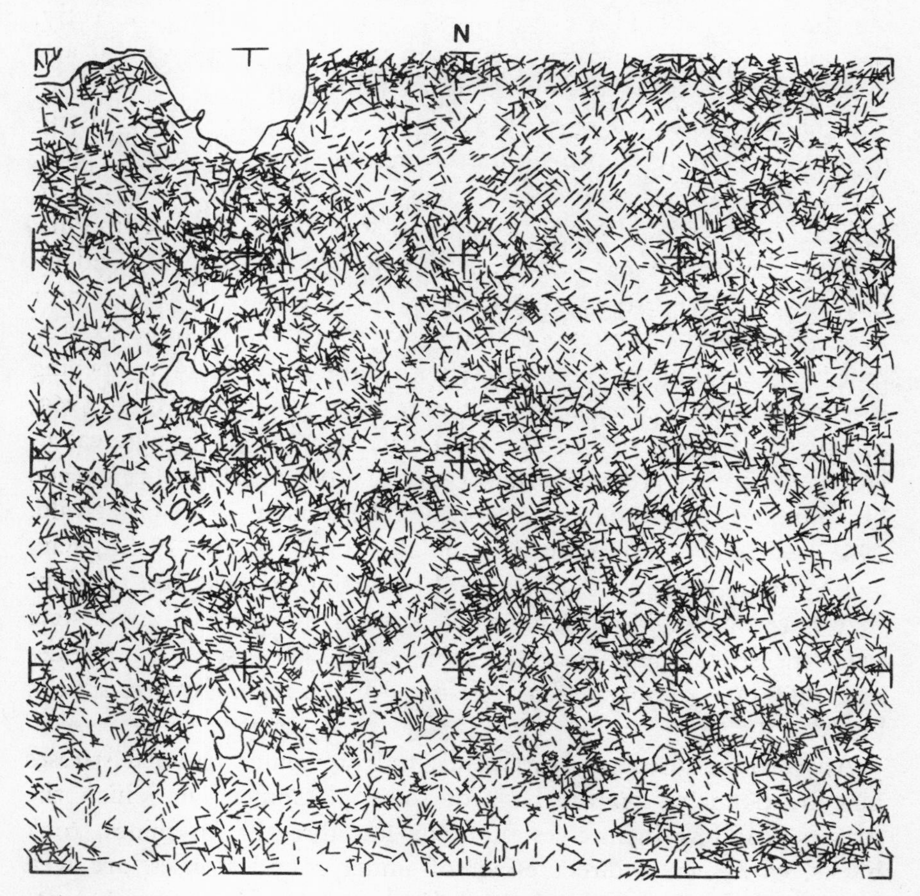

Fig. 1. Example of a fault pattern in northwestern Alberta. After Haman (1964).

Faulting Patterns on Earth's Surface

Anderson (1942) postulates two potential conjugate fault surfaces in a given tectonic stress system. These surfaces should show potential traces on the earth's surface. The actual traces of conjugate fault systems observed in many places (Fig. 1) may thus be interpreted as the surfaces of the tectonic stress tensor prevailing at the time these faults occurred.

Lensen (1958) was probably the first to make use of this concept, by introducing the term *isallo stress lines.* The angle between the traces of a conjugate fault system, which always points towards the direction of the largest compression, is termed *compressional angle* γ (Fig. 2); the isolines of equal angles Lensen characterized as *isallo stress lines.* He noted that if $\gamma = 180°$, the stress state producing the faults is one of incipient reverse faulting; if $\gamma = 90°$, one of incipient transcurrent faulting; and if $\gamma = 0°$, one of incipient normal faulting (in Anderson's terminology). A corresponding characterization of the stress field is given for intermediate angles. Thus, for each possible angle γ there is an associated angle of inclination of the stress axes toward the vertical (Scheidegger, 1963a; Fig. 3). Isallo stress lines can therefore be used to deduce the orientation of the principal stress axes of the stress system which resulted in a given conjugate fault system. The method has been developed and applied to many areas and has yielded results consistent with those obtained from geologic inter-

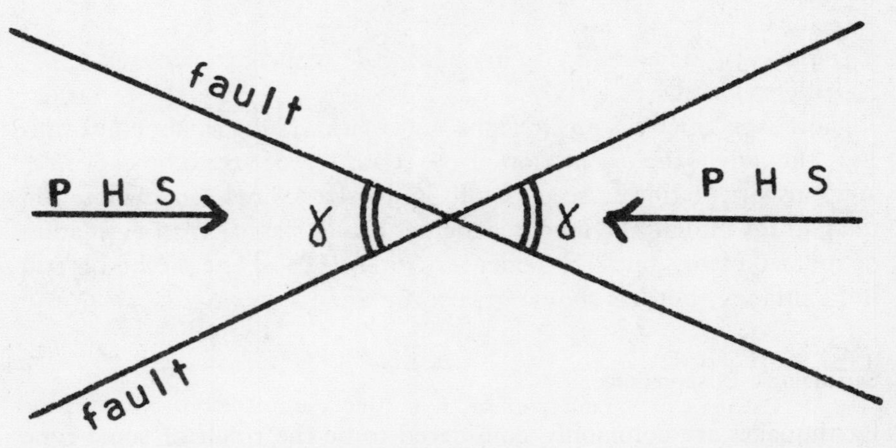

Fig. 2. Compressional angle γ in a fault system. PHS, principal horizontal stress.

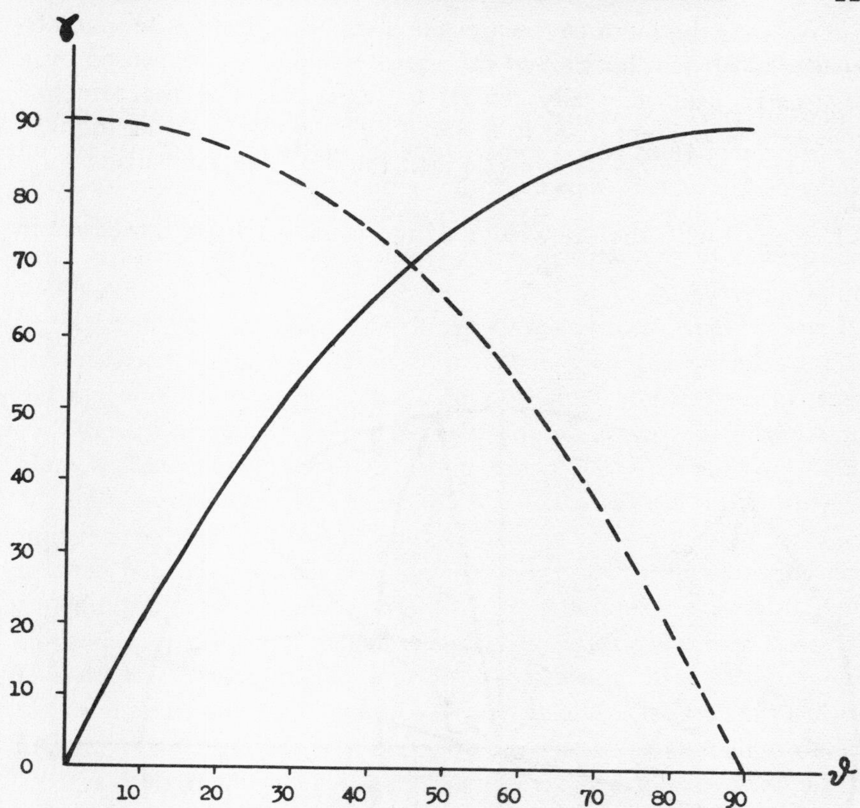

Fig. 3. Relation between γ and ϑ. Note that for a given angle γ there are two angles ϑ (ϑ and $90 - \vartheta$); hence the two lines in the figure. After Scheidegger (1963a).

pretation of the fault systems in question (Lu and Scheidegger, 1965; Scheidegger and Lu, 1965).

Moreover, since subsurface inhomogeneities (such as nonconformities) that affect the orientation of the stress tensor are to be expected, and this orientation is seen in the isallo stress lines, an exploration method for finding the inhomogeneities can be based on an evaluation of the isallo stress lines (Scheidegger, 1964, 1966). The method is still in its infancy, but has shown great promise.

Seismologic Observations

Earthquakes are commonly considered to be the result of some type of faulting process, representing the response, through failure, of the material near the focus to the prevailing stresses. Study at seismic observatories of the direction of the initial ground motion caused by a large earthquake generally reveals the orientation of the principal stress axes (the direction of *P*, *B*, and *T*) giving rise to the earthquake (Fig. 4). Fara (1964) has catalogued the available fault-plane solutions.

Equipped with the knowledge of the principal stress directions in

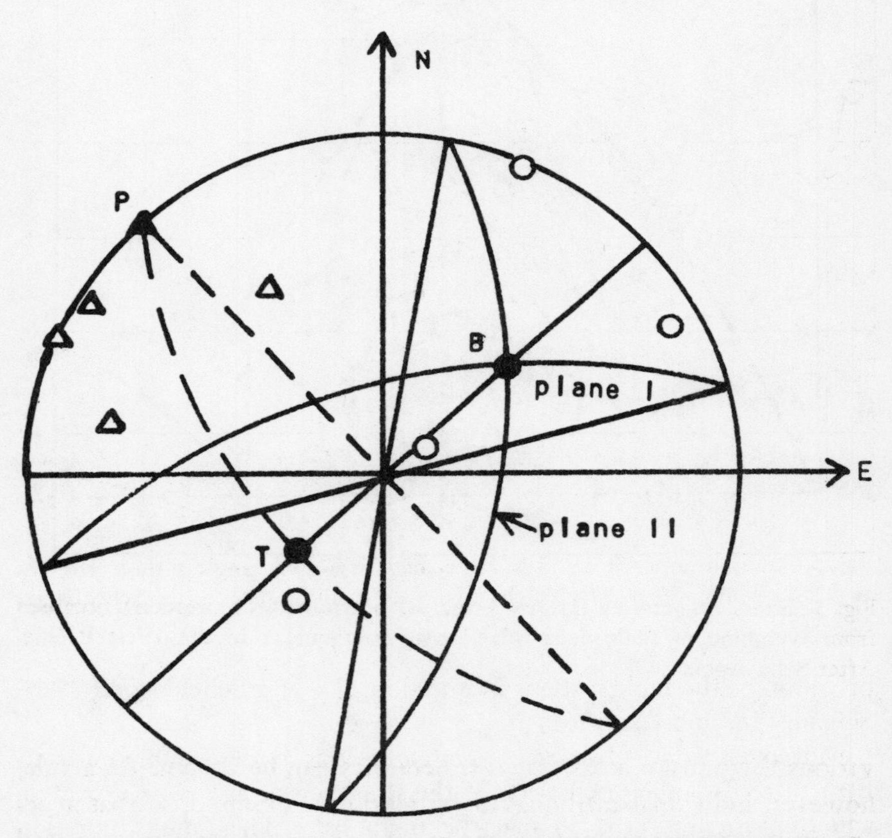

Fig. 4. Schematic example of a fault-plane solution of an earthquake. Interior of large circle, stereographic projection of lower half of a sphere around focus of earthquake (focal sphere). Small circles and triangles–points of intersection (in stereographic projection) of seismic rays, with focal sphere leaving focus toward earthquake observatories; circles–outward initial motion on seismic ray; triangles–inward initial motion. Planes I and II separate quadrants with outward and inward motion of focal sphere (*fault planes*): *P*, *T* and *B*–intersection points of principal stress directions with focal sphere.

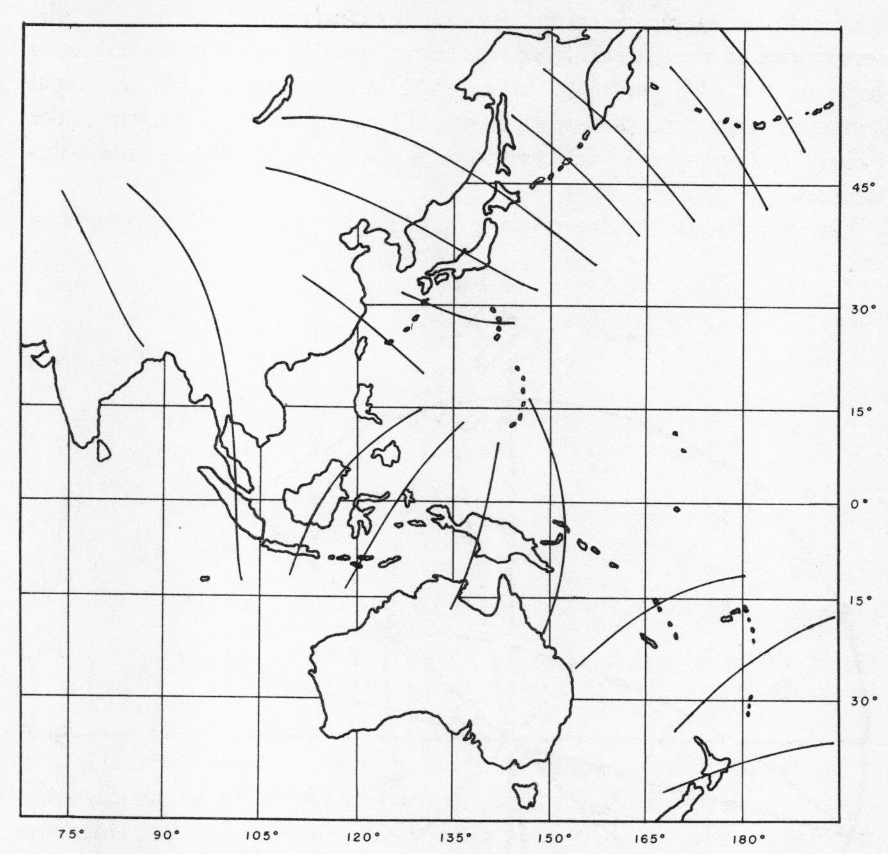

Fig. 5. Stress trajectories (largest principal horizontal compression) obtained from averaging of fault-plane solutions of earthquakes in the West Pacific. After Scheidegger (1965).

various earthquake areas, stress trajectories can be drawn. As a rule, however, individual earthquakes are randomly oriented, so that average positions must be used for the axes of groups of closely associated earthquakes. The averaging of axes posed a certain problem, and a special algorithm had to be developed. It then became possible to draw regional stress trajectories for many areas (Fig. 5). Generally, the picture in most areas is consistent with that obtained by the evaluation of geologic field observations on recent orogenic activity (Scheidegger, 1965; Fig. 6).

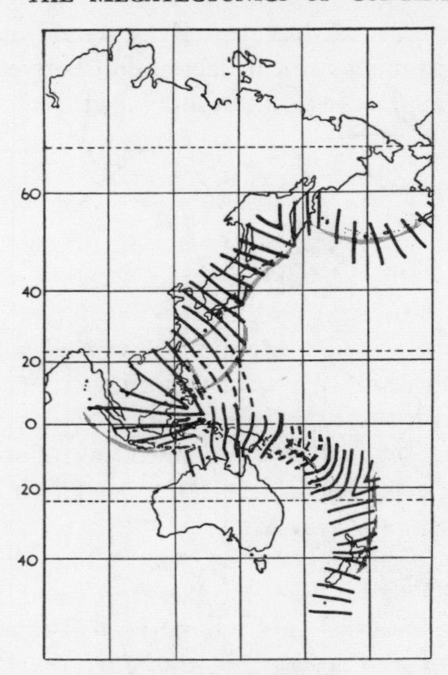

Fig. 6. Stress trajectories (principal horizontal compressions) in the West Pacific, from geologic data. After Lensen (1960).

Study of recent orogenic activity seems to indicate that, with regard to dependence on time, the world's tectonic stress field has remained fairly stationary for the last 60 million years, i.e., since the occurrence of the last major orogenetic activity. However, there is one exception to this general rule; the earthquakes in Switzerland always create compressions in the German area, indicating a *T* mechanism in the N-S direction (Scheidegger, 1967). This differs from the general view of the orogeny of the Alps, with its common inference of a N-S compression. There are geologic indications that a major change of the tectonic pattern in the Alpine region may have occurred at the end of the Pliocene (Fairbridge, 1966). This would be consistent with seismologic observations.

With regard to the dependence of the tectonic stress field on depth, general conclusions on the basis of available evidence are difficult, since the occurrence of deep-focus earthquakes is relatively rare and confined to a few regions. A study of the vicinity of Honshu Island failed to reveal any significant differences in the stress systems causing deep-focus and shallow earthquakes. This would argue against any hypothesis which assumes radical decoupling between the crust and the upper mantle, at least in and around Japan.

Geotectonic Manifestations

Orogeny is clearly the result of the action of tectonic stresses. Of particular interest is the recent orogenic activity, since it might be expected that it is connected with the prevalent underground stresses now present. Indeed, the manifestations discussed here are consistent with the field observations of recent tectonic activity, of which three different types of features have been distinguished, presumably corresponding to the three standard stress states postulated by Anderson.

The best-known geotectonic features are the high mountain ranges of the world. Mountains are essentially linear features, occurring in chains rather than as single, random peaks. The "recent" mountain ranges, built up during the last 100 million years or so, reach around the world in two great belts: the Alpine-Himalayan belt, comprising the Atlas Mountains, the Apennines, the Alps, the Dalmatian mountains (Dinaric Alps), the mountains of Turkey and Persia, the Himalayas, and the mountains of Burma and Indonesia; and the circum-Pacific belt, reaching all the way around the Pacific. Field observations show that most mountains have been formed by a process called *crustal shortening*, i.e., by a contraction of the earth's uppermost parts. The shortening caused the formation of big thrust sheets which were pushed over each other. Anderson's term for the stress state causing this type of mountain range is *incipient overthrust faulting*.

In certain regions, substantial shearing (transcurrent) motions in the earth's crust have been observed. The fracture zones in the ocean bottom off the coast of California (Fig. 7), the Southern Alps in New Zealand, and some of the mountains in Anatolia are examples of this phenomenon. Pavoni (1961) suggests a great prevalence of transcurrent faulting. In such areas, the stress state might be expected to be one of incipient transcurrent faulting.

Finally, in some parts of the world, there are indications that the prevalent stresses must be tensional. Thus, atop the mid-Atlantic Ridge, a rift runs parallel to and on top of the ridge (Fig. 8). On land, there are rift valleys in East Africa, in Palestine, and in the upper Rhine area. Naturally, true tensions cannot exist deep below the

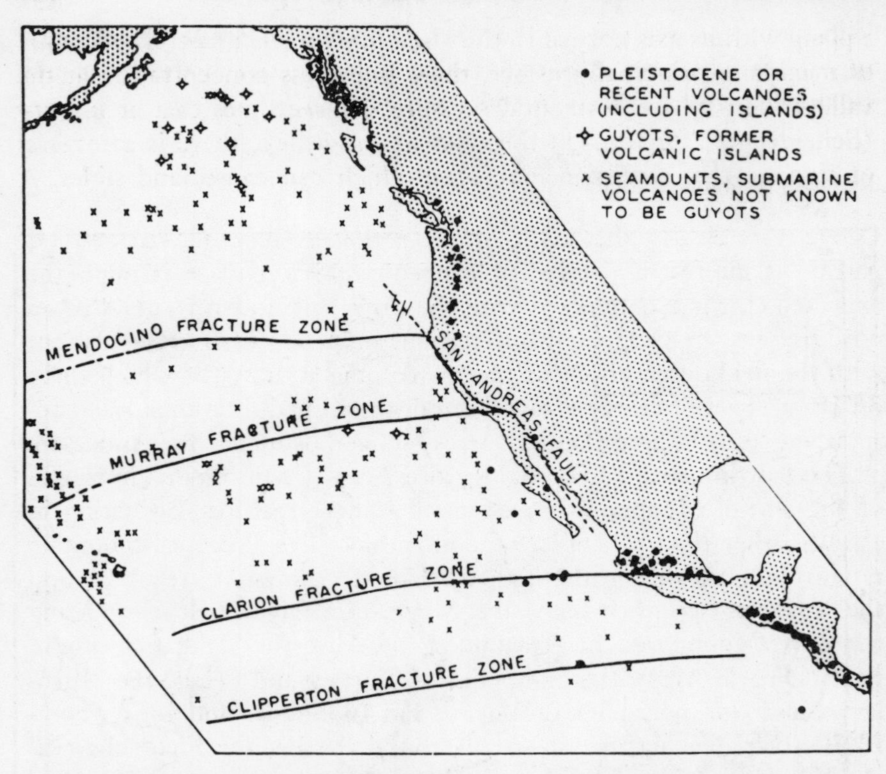

Fig. 7. Menard's (1955) fracture zones in the Pacific, indicating shear fractures.

earth's surface, but rather these rifts represent a relief of pressure. In the areas indicated, there might be a state of incipient normal faulting, with or without concurrent shearing motion.

Geologic evidence thus suggests that the prevalent type of stress in recent times varies from region to region but is fairly uniform within any one tectonic area.

Surface Manifestations

Tectonic stresses may also exert direct manifestations on the earth's surface. Any surface irregularity constitutes an irregularity in the boundary conditions of the tectonic stress field, which will be modified from the state it would be in were the surface of the earth smooth. The effect of some simple surface features on the stress field have been calculated. If a semicircular cross-section of a valley is cut into a plane with its axis normal to the direction (in the undisturbed state) of maximum principal pressure, there is a stress concentration at the valley floor, where horizontal pressure is three times that at infinity (Scheidegger, 1963c). On the sides of the valley, there is a reversal of sign, so that a tension may arise which can cause land slides. A

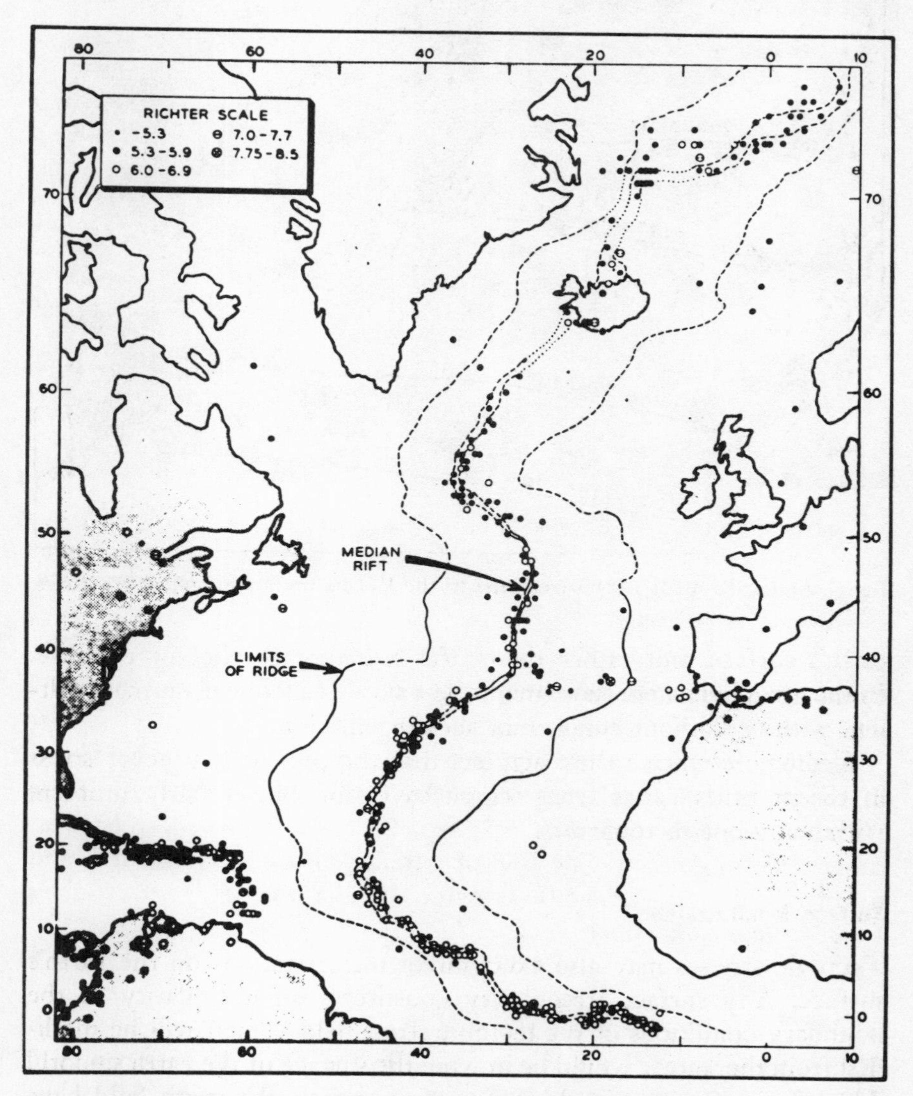

Fig. 8. Mid-Atlantic Rift, indicating tensional effects. After Elmendorf and Heezen (1957).

Fig. 9. Stress-induced weathering of a rock wall (Gisliflue, Aargau, Switzerland). Looking upward, the sawtooth-shaped decay pattern corresponds to (more or less vertical) Mohr-Anderson surfaces in the prevailing stress state. (Gerber and Scheidegger, 1965.)

mountain range is the reverse of a valley. A circular cross-section of a range may show a stress reversal at the crest.

Another instance of the action of tectonic stresses may be found in the so-called weathering of rocky walls in mountainous areas (Gerber and Scheidegger, 1965; Fig. 9). The course of the stress trajectories in the neighborhood of a rock wall can be estimated, and the observed recession can be shown to be in accordance with Mohr's (1938) theory of failure.

Conclusion

The existence of a tectonic stress field in the earth, i.e., a stress field which cannot be explained by the action of gravity (weight of rocks) alone is firmly established. The present stress field, as inferred from actual measurements and from contemporary effects (earthquakes) is in most areas consistent with that inferred from the investigation of recent orogenetic activity. The one exception is in central Europe and can be explained by postulating a very recent change in orogenetic activity there.

The character, predominantly compression, tension, and shear, of the present stress field is uniform over large areas but differs fundamentally from one large region to another. Any tenable geotectonic hypothesis must be able to explain fundamentally different stress fields in widely separated regions of the world.

With regard to the depth dependence of the tectonic stress field, the available evidence fails to support any hypotheses of decoupling between crust and mantle, i.e., the stress field seems to be independent of depth. The evidence for this is scanty, however.

Obviously, the character of the stress field in any one region may have changed with time throughout geologic history, but definite statements about conditions in the distant past are difficult. Nevertheless, the available evidence seems to indicate that the character of the stress field has remained fairly static for the last 60 million years in most of the world's areas except in Europe, where a change in tectonic activity, and therefore in the stress patterns, may have occurred at the end of the Pliocene.

References

Adler, L.: Engineering analysis of geologic pressures and design effects. *Trans. Soc. Min. Eng.* 223:358–362, 1962.

Anderson, E. M.: *The Dynamics of Faulting and Dyke Formation with Applications to Britain.* Edinburgh, Oliver & Boyd, Ltd., 1942.

Cook, N. G. W.: The seismic location of rock bursts. In C. Fairhurst, ed., *Rock Mechanics.* New York, Pergamon Press, 1963, pp. 493–516 (Proceedings of the Fifth Rock Mechanics Symposium, University of Minnesota, 1962).

Elmendorf, C. H., and B. C. Heezen: Oceanographic information for engineering submarine cables. *Bell System Tech. J.* 36:1047–1093, 1957.

Emery, C. L.: The measurement of strains in mine rocks. In *Mining Research: An International Symposium*, B. Clark, ed. New York, Pergamon Press, 1962, vol. 2, pp. 541–553.

Fairbridge, R.: Personal communication, 1966.

Fara, H. D.: A new catalogue of earthquake fault-plane solutions. *Bull. Seism. Soc. Am.* 54:1491–1517, 1964.

Fara, H. D., and F. D. Wright: Plastic and elastic stresses around a circular shaft in a hydrostatic stress field. *Trans. AIME* 226:319–320, Sept., 1963.

Gerber, E. K., and A. E. Scheidegger: Probleme der wandrückwitterung, im besonderen die ausbildung mohrscher bruchflächen. *Rock Mech. & Eng. Geol.*, Suppl. 2:80–87, 1965.

Griswold, G. B.: How to measure rock pressures—new tools and proved techniques and mine design. *Eng. Mining J.* 164:90–95, 1963.

Haman, P. J.: Geomechanics applied to fracture analysis on aerial photographs. West Can. *Res. Pub. Geology and Related Sciences* (Calgary) 2, No. 2, 1964.

Hast, N.: The measurement of rock pressure in mines. *Sveriges Geol. Undersök.* 183, 1958.

Hiramatsu, Y., and Y. Oka: Stress around a shaft or level excavated in ground with a three-dimensional stress state. *Mem. Fac. Eng. Kyoto Univ.* 24:56–76, 1962.

Isaacson, E. de St. Q.: *Rock Pressure in Mines,* 2d ed. London, Mining Publications, Ltd., 1962.

Jaecklin, F. P.: Beitrag zur felsmechanik. *Schweiz. Bauztg.* 83: H. 15, 27, 1965.

Judd, W. R.: Rock stress, rock mechanics and research. In *State of Stress in the Earth's Crust,* W. R. Judd, ed. New York, Elsevier Publishing Co., 1964, pp. 5–53 (International Conference on State of Stress in the Earth's Crust, Santa Monica, Calif., 1963).

Lensen, G. J.: Measurement of compression and tension: Some applications, New Zealand *J. Geol. Geophys.* 1:565–570, 1958.

Lensen, G. J.: Principal horizontal stress directions as an aid to the study of crystal deformation. *Pub. Dom. Obs. Ottawa* 24:389–397, 1960.

Lu, P. H., and A. E. Scheidegger: An intensive local application of Lensen's isallo stress theory to the Sturgeon Lake south area of Alberta. *Bull. Can. Petrol. Geol.* 13:389–396, 1965.

May, A. N.: Instruments to measure the stress conditions existing in the rocks surrounding underground openings. *Trans. Can. Inst. Mining Met.* 63:497–500, 1960.

Menard, H. W.: Deformation of the northeastern Pacific Basin and the west coast of North America. *Bull. Geol. Soc. Am.* 66:1149–1198, 1955.

Menzel, W.: Die Entwicklung von Näherungsformeln zur Spannungsberechnung um isolierte Grubenbaue mit quadratischem, rechteckigem oder trapezförmigem Querschnitt. *Bergakademie* 17:151–159, 1965.

Merrill, R. H.: *In situ* determination of stress by relief technique. *Rand Corp. Memo* RM-3583, pp. 1–36, 1963.

Merrill, R. H., and J. R. Peterson: *Deformation of a Borehole in Rock.* U.S. Bur. Mines Rep. Invest., No. 5881, 1961.

Merrill, R. H., et al.: *Stress Determinations by Flatjack and Borehole Deformation Methods.* U.S. Bur. Mines Rep. Invest., No. 6400, 1964.

Milne, W. F., and W. R. H. White: A seismic investigation of the mine "pumps" in the Crowsnest Pass coal field. *Pub. Dom. Obs. Ottawa* 51: 678–685, 1958.

Mohr, F.: *Gebirgsmechanik.* Goslar, Hübner, 1964.

Mohr, O.: *Abhandlungen aus dem Gebiete der Technichen Mechanik,* 3d ed. Berlin, W. Ernst, 1928.

Morgan, T. A., and L. A. Panek: A method for determining stress in rocks. U.S. Bur. Mines Rep. Invest., No. 6312, 1963.

Obert, L.: *In situ* determination of stress in rocks. *Mining Eng.* 14, No. 8:51–58, 1962a.

Obert, L.: Effects of stress relief and other changes in stress on the physical properties of rock. U.S. Bur. Mines Rep. Invest., No. 6053, 1962b.

Obert, L., et. al.: *Borehole Deformation Gage for Determining the Stress in Mine Rock.* U.S. Bur. Mines Rep. Invest., No. 6053, 1962.

Osterwald, F. W.: *Deformation and Stress Distribution Around Coal Mine Workings in Sunnyside No. 1 Mine, Utah.* U.S. Geol. Surv. Profess. Paper No. 424-C:349–353, 1961.

Panek, L. A.: Measurement of rock pressures with a hydraulic cell. *Trans. AIME* 220:287–290, 1961.

Panek, L. A., and J. A. Stock: Development of a rock stress monitoring station based on the flat slot method of measurement. U.S. Bur. Mines Rep. Invest., No. 6537, 1964.

Pavoni, N.: Faltung durch Horizontalverschiebung. *Eclogiae Geol. Helv.* 54:515–534, 1961.

Pulpan, H., and A. E. Scheidegger: Calculations of tectonic stress from hydraulic well-fracturing data. *J. Inst. Petrol.* 51:169–176, 1965.

Scheidegger, A. E.: Geometrical significance of isallo stress. *New Zealand J. Geol. Geophys.* 6:221–227, 1963a.

Scheidegger, A. E.: *Principles of Geodynamics,* 2d ed. New York, Academic Press, 1963b.

Scheidegger, A. E.: On the tectonic stress in the vicinity of a valley and a mountain range. *Trans. Roy. Soc. Victoria* 75:141–145, 1963c.

Scheidegger, A. E.: On the use of stress values as an exploration tool. *Pure & Appl. Geoph.* 59:38–44, 1964.

Scheidegger, A. E.: Grosstektonische Bedeutung von Erdbebenherdmechnismen. Z. *Geophys.* 31:300–312, 1965.

Scheidegger, A. E.: Isallo stress prospecting. Z. *Geophys.* 32:183–199, 1966.

Scheidegger, A. E.: The tectonic stress in the vicinity of the Alps. Z. *Geophys.* 33:167–181, 1967.

Scheidegger, A. E., and P. H. I. Lu: Beeinflussung der Spannungen im Gestein durch oberflächennahe Inhomogeneitäten. *Rock Mech. Eng. Geol.* 3:93–102, 1965.

Sprang, J., and F. Kahler: Gebirgsdruckbegriffe. *Felsmech. Eng. Geol.* 1:245–249, 1963.

Terzaghi, K.: Measurements of stress in rocks. *Geotechnique* 12:105–124, 1962.

Utter, S.: Stress determination around an underground mine opening. In *Proceedings, International Symposium on Mining Research, Missouri, 1961,* G. B. Clark, ed. New York, Pergamon Press, 1962, vol. 2, pp. 569–582.

11

Biostratigraphy, Magnetic Stratigraphy, and Sea Floor Spreading *

RICHARD K. OLSSON

Department of Geology, Rutgers University
New Brunswick, New Jersey

The model of sea floor spreading as interpreted by the magnetic anomaly pattern (Vine and Matthews, 1963) of ocean ridges implies that the layer of sediment overlying igneous basement has an age not older than that of the basement rock. The model also implies that a sedimentary facies of increasing age in a distal direction from the ridge axis lies on the igneous basement and that, where preserved, a sedimentary section of older and older geologic age will be encountered at increasing distances from the ridge axis (Fig. 1). The proof of sea floor spreading therefore lies in documenting an age relationship among the igneous basement, the magnetic anomaly, and the overlying sediment. The biostratigraphy of fossil plankton not only has provided much data supporting sea floor spreading but major advances in biostratigraphy have developed as a result of the intensive study of geomagnetics and paleontology.

The distribution of paleontologically determined ages on older sediments either outcropping on the sea floor or penetrated in piston cores is consistent with a spreading sea floor. Riedel (1967) analyzed

* Data from Leg V of the Deep Sea Drilling Project were obtained during this writer's participation in that cruise. The Deep Sea Drilling Project is supported by National Science Foundation and is managed by Scripps Institute of Oceanography. Thanks are given to Dr. Raymond C. Murray for suggesting this paper, for reading the manuscript, and for offering helpful suggestions. Thanks are also given to Dr. Steven K. Fox for reading of and comment on the manuscript. This work has been partially supported by the Research Council of Rutgers University.

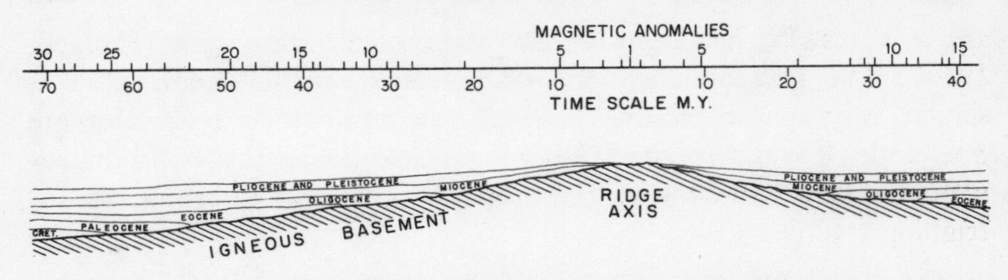

Fig. 1. Sea floor spreading model showing distribution of sedimentary strata in relation to age in a distal direction from ridge axis. Sedimentary thicknesses greatly exaggerated. Magnetic anomaly scale and time scale after Heirtzler et al. (1968).

fossil radiolarians from samples taken west of the East Pacific Rise in an area bounded by latitude 0° and 20°N and longitude 115° and 150°W. By plotting the paleontologic data obtained from his samples he showed a pattern of more western occurrences of the geologically older samples. A similar pattern was shown in the equatorial Pacific and southern Pacific north of 30°S of increasing age (Pliocene to lower Miocene) away from the axis of the East Pacific Rise (Burckle and co-workers, 1967). Further, the paleontologic ages based on planktonic foraminifera were found to be consistent with spreading rates calculated by Pitman and Heirtzler (1966) and Vine (1966) on the basis of magnetic profiles of the southern part of the East Pacific Rise. The age distribution of piston core samples in the North and South Atlantic was also seen to be consistent with sea floor spreading (Funnell and Smith, 1968). A model of the opening of the Atlantic Ocean based on a geometrical reconstruction, the stages of which were dated on the paleontological data, was formulated by them. The model predicted the age of the ocean crust at any particular location. They made provisional use in their model of the magnetic reversal pattern in which some associated absolute dates were estimated. The Deep Sea Drilling Project would be testing such predictions in that the minimum actual age of the ocean floor at a drill site would be determined by the oldest fossil remains encountered. The ages of sediment samples taken above basement during Leg III of the Deep Sea Drilling Project which drilled 5 sites on a traverse in the South Atlantic across the Mid-Atlantic Ridge confirmed their predictions and indicated symmetrical spreading with a half rate of 2 cm. a year (Percival, 1969). The paleontological ages of the basal sediments were converted into absolute ages by relating the relative age to dates that had been set into the absolute time scale (Funnell, 1964). The geomagnetic scale of reversals can be compared in a similar way to the relative scale of paleontology and the absolute time scale. It is the comparison of these three scales that is producing major advances in biostratigraphical concepts and is producing intriguing results.

The Geomagnetic Scale

The work of Cox and co-workers (1963, 1964) provided a wealth of data from measurements on lava flows that confirmed the validity of reversals of the geomagnetic field and allowed them to construct a geomagnetic polarity history of the Pleistocene. Their Pleistocene geomagnetic scale was first used by Vine and Wilson (1965) in their analysis of the Juan de Fuca Ridge to account for the observed symmetrical anomaly patterns of the ridge. Subsequently, Pitman and Heirtzler (1966) and Vine (1966) proposed detailed time scales for the geomagnetic polarity history back more than 10 million years. Detailed mapping of magnetic patterns of the sea floor in the North Pacific, South Pacific, South Atlantic, and Indian Oceans (Heirtzler and co-workers, 1968) showed the magnetic pattern to be the same in each of these ocean areas. Extrapolating from well dated paleomagnetic events they derived a time scale for the sequence of magnetic polarity events predicted by a basic spreading model. Using the South Atlantic profile as a standard they identified 32 anomalies in a scale which reaches into the upper Cretaceous to a point approximately 80 million years ago. Thus a new scale, a geomagnetic scale, for analyzing geologic time has emerged. It was a logical step to compare the geomagnetic scale to one based on biostratigraphy.

Biostratigraphic Zonation, Geomagnetics, and Datums

Over the last decade increased emphasis has been placed on the use of planktonic microfossils in ocean research and in the long distance correlations of marine strata. Detailed biostratigraphic zonations of the calcareous plankton, foraminifer and coccolith, have made it possible to achieve stratigraphic resolution on the order of a half-million to a million years within the Cenozoic and the Cretaceous. The retrieval of continuously cored samples in the Deep Sea Drilling

Project is making it possible to extend this biostratigraphy into the deep sea and to develop biostratigraphic frameworks for the other groups of fossil plankton such as radiolarians, diatoms, silicoflagellates, and dinoflagellates.

Biostratigraphic zonation of fossil plankton is based on assemblages of species, points of species origin, stages in the evolution of species, and points of extinction of species. Much emphasis has been placed on evolution of species as reliable datums or reference planes in time. For example, the evolution of the Holocene planktonic foraminifer *Orbulina* was seen by Blow (1956) to occur over a short stratigraphic interval in the lower Miocene in Venezuela. Subsequently the *Orbulina* datum has been widely recognized in many Miocene rock sections of the world in both northern and southern hemispheres. Other useful evolutionary datums such as the *Pseudohastigerina* datum (Berggren, Olsson, and Reyment, 1967) and the *Globigerinoides* datum (Bolli, 1957) can be used to recognize epoch boundaries; in this case Paleocene-Eocene and Oligocene-Miocene, respectively.

The points of extinction of species have not generally been considered reliable by paleontologists for correlation because it is believed portions of a population might survive longer in some places than in others. The extinction of planktonic groups such as the foraminifera and the coccoliths at the close of the Mesozoic Era appears to be a simultaneous event. But a simultaneous point of extinction of individual species within epochs, for instance, has lacked substantiation even though extinction levels have been used in intercontinental correlations and studies of Pleistocene glacial history in deep sea cores. The comparison of biostratigraphic data and paleomagnetic data has provided valuable insight into this problem.

Harrison and Funnell (1964) first reported that they had found evidence for a geomagnetic polarity reversal in an equatorial Pacific core that coincided with the extinction of a radiolarian species, *Pterocanium prismatium*. A study of Pleistocene radiolarian biostratigraphy and magnetic stratigraphy (Hays, 1965; Opdyke and co-workers, 1966) on cores in the Southern Ocean indicated a coincidence or near coincidence of faunal changes with geomagnetic reversals; two reversals and two faunal changes were found to be closely associated. Later Watkins and Goodell (1967) confirmed that the upper extinction horizon of Opdyke and co-workers (1966) was closely correlated with the Brunhes-Matuyama magnetic epoch boundary 0.7 million years ago. In the 9 deep sea sediment cores they studied from the Pacific-Antarctic Basin the faunal extinction was found to be virtually synchronous and not related to sedimentary factors. Additional studies in the Antarctic Ocean (Hays and Opdyke, 1967) extended into the Pliocene and showed further faunal changes associated with reversals. In an evolutionary study (Berggren and co-workers, 1967) the development of the planktonic foraminiferal species *Globorotalia truncatulinoides* from the ancestral species *Globorotalia tosaensis* was shown to occur within the Olduvai normal event of the reversed Matuyama magnetic epoch and was correlated with the Pliocene-Pleistocene boundary of the stratotype section of Calabria, Italy. A biostratigraphic analysis on tropical east Pacific cores of various groups of planktonic fossils detailed the evolution and extinction of certain species in relation to the geomagnetic reversal sequences of the Pliocene and Pleistocene. This work (Hays and co-workers, 1969) showed that seven equatorial foraminiferal species, two radiolarian species, and two diatom species became extinct near reversals. Further study by Hays (1969) of long cores with continuous or nearly continuous sedimentation from tropical, high northern, and high southern latitudes allowed him to define a paleomagnetically controlled radiolarian stratigraphy ranging from the Recent to the upper Miocene. He established that during the last 3 million years at least 9 species of radiolaria became extinct; seven of them disappeared within a few 10's of centimeters of a magnetic reversal. Application of regression analysis indicated a very low probability of this close a correlation occurring by chance.

There has been considerable speculation on a possible causal relation between times of a magnetic reversal and the extinction of a species (Uffen, 1963; Black, 1967; Waddington, 1967; Harrison, 1968; and Hays and co-workers, 1969) but no satisfactory explanation is apparent. Regardless of the ultimate explanation of these associations the biostratigraphic and geomagnetic studies show the extinction level of a species of plankton to be a highly reliable datum over very broad areas of the oceans. The boundary planes of magnetic reversals have provided the reference horizons with which to analyse the extinction of a species in time. The results show the extinction of a species to be synchronous throughout its total habitat and, in fact, they show extinction datums to be more precise than evolutionary datums because it is more difficult to recognize a precise level during a gradual evolutionary structural change. The true chronostratigraphic relationships of certain fossil species has been established by these studies and, importantly, they have established absolute dates that have been adjusted to the biostratigraphic scale (Opdyke and co-workers, 1966; Berggren and co-workers, 1967; and Hays and co-workers 1969) for the Pliocene and the Pleistocene.

Datum Biostratigraphy in Correlation and Sea Floor Spread

The usefulness of datum biostratigraphy can be observed in the correlation of a Pliocene-Pleistocene foraminiferal rich section taken in the northeastern Pacific during Leg V of the Deep Sea Drilling Project with the tropical Pacific sections of Hays and co-workers (Fig. 2). Climatic barriers had prevented most of the diagnostic

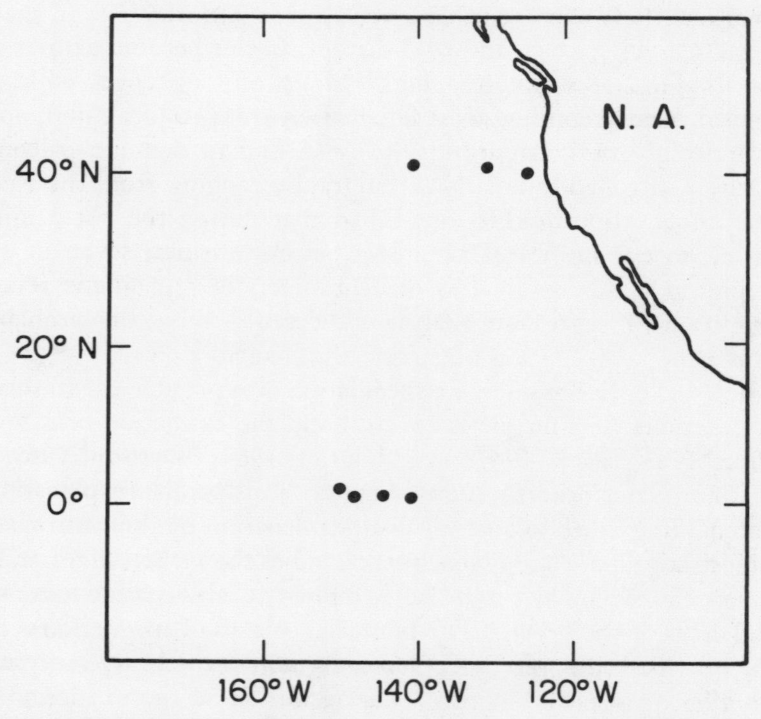

Fig. 2. Location of Pliocene-Pleistocene sections in northeastern Pacific, Leg V, Deep Sea Drilling Project, and in tropical Pacific, Hays et al. (1969).

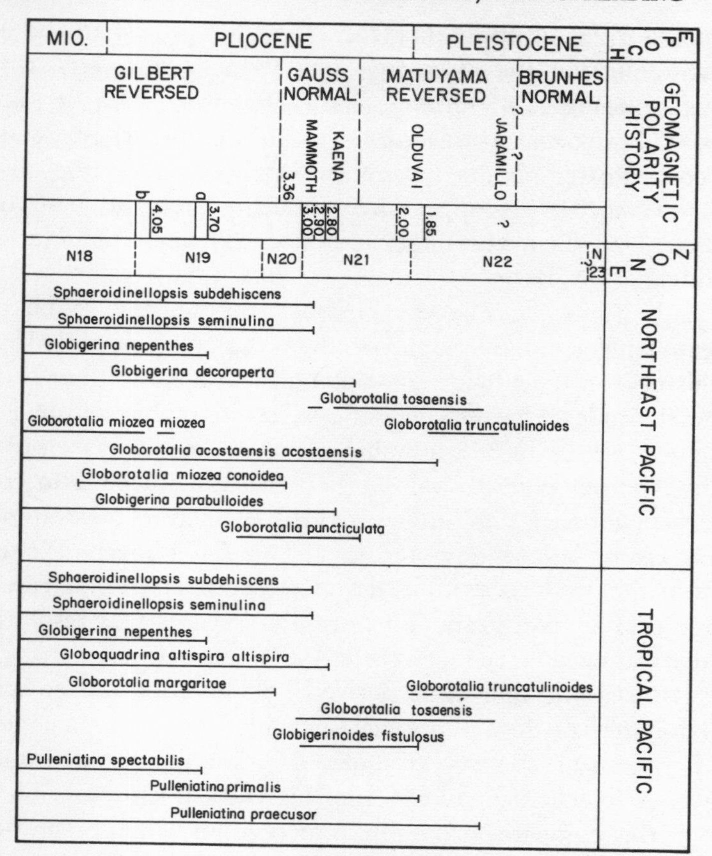

Fig. 3. Comparison of planktonic foraminiferal species in the Pliocene-Pleistocene of the northeastern Pacific and the tropical Pacific sections. ▲ denotes level of discoaster extinction.

tropical zonal species from occurring in the Pliocene-Pleistocene northeast Pacific seas. Figure 3 shows the Pliocene-Pleistocene section of the northeast Pacific and compares it with the tropical Pacific section. The extinction datum of *Globigerina nepenthes* at 3.7 million years at the top of the "a" event of the Gilbert reversed geomagnetic epoch establishes a point of reference between the two sections. The extinction datum of the genus *Sphaeroidinellopsis* at approximately 2.9 million years at the top of the Mammoth event of the Gauss normal epoch is another correlation point. The extinction of discoasters was found to occur in the tropical Pacific cores close to the base of the Olduvai event at approximately 2.0 million years. In the northeast Pacific section the discoaster extinction occurs slightly above the initial appearance of *Globigerina pachyderma.* The first appearance of *Globorotalia truncatulinoides* in the northeast section is related to climatic factors; its evolutionary origin is not observed there. Nevertheless its appearance indicates that the base of the Pleistocene (*i.e.*, the evolutionary origin of *G. truncatulinoides* within the Olduvai event, Berggren and co-workers, 1967) occurs somewhat below and allows the Pliocene-Pleistocene boundary to be placed with reasonable accuracy. Such correlations make it possible to study Earth history on worldwide fundamental levels along lines of longitude and latitude. The meaning of a biostratigraphic datum is clear whereas the use of biostratigraphic zones is not always clear to all workers. Another point of value other than the advances in correlation is the increased resolution provided by datum biostratigraphy. The numerous datums now recognized in the Pliocene-Pleistocene sections of the deep sea sections has increased stratigraphic resolution to within 0.25 million years. These results indicate that resolution of this order can be achieved as well on the older sections.

The geomagnetic time scale derived by Heirtzler and co-workers (1968) for the last 80 million years provided a useful framework to which Berggren (1969) has extrapolated various planktonic foraminiferal datums. A generalized version of his work is shown in Figure 4. Although this comparison is subject to revision as new data become available, it provides a useful table for converting relative paleontological ages into absolute time values. Figure 5 shows the results of drilling of Leg V of the Deep Sea Drilling Project in which paleontologic age determinations were converted to absolute time by using Berggren's extrapolations and then compared to the magnetic anomaly present at each site. It demonstrates general agreement with the predicted age of the igneous basement as determined by the magnetic pattern (Heirtzler and co-workers, 1968). Percival (1969) extrapolated absolute ages in a similar way and was able to demonstrate general agreement with predicted ages of ocean basement across the Mid-Atlantic Ridge in the South Atlantic. Thus a basic framework of reasonable accuracy is available to which the biostratigraphic and geomagnetic data that will be forthcoming from the deep sampling of ocean sediments can be related. Some of the more fundamental aspects of ocean history will be analyzed in the decades ahead through

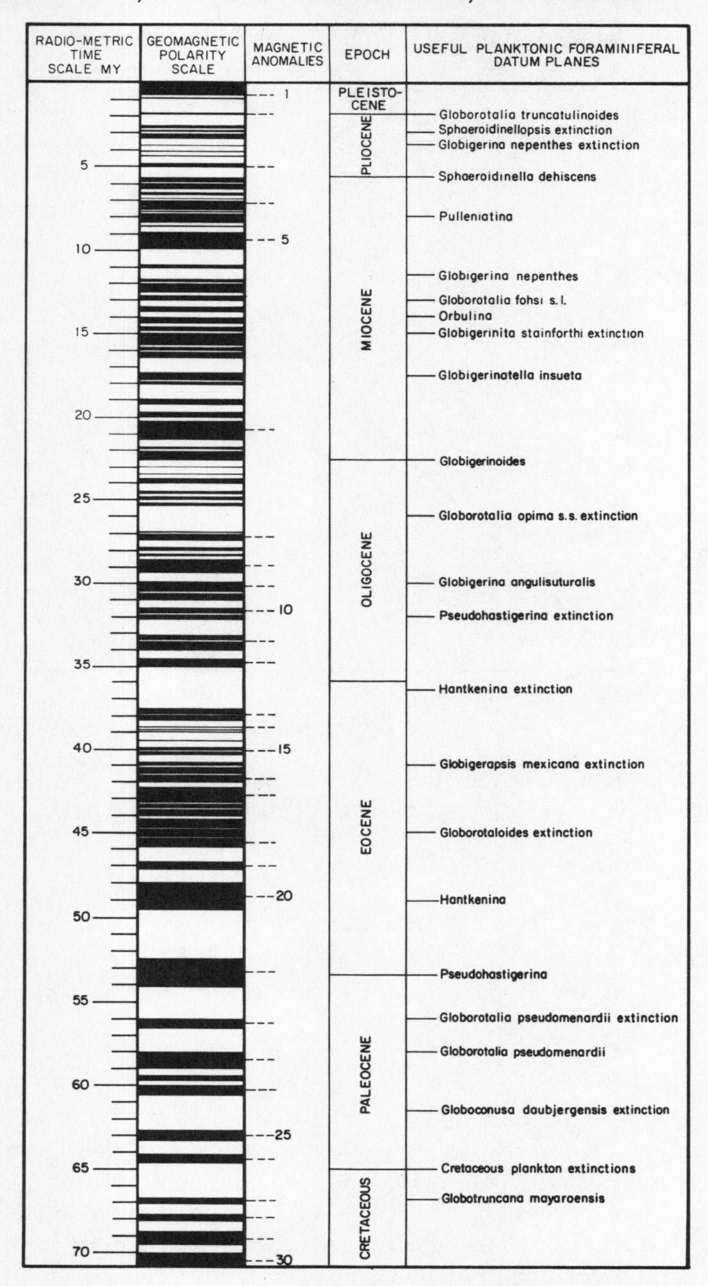

Fig. 4. Geomagnetic scale and time scale for the last 70 million years showing some diagnostic foraminiferal evolutionary and extinction datum planes. Modified after Berggren (1969).

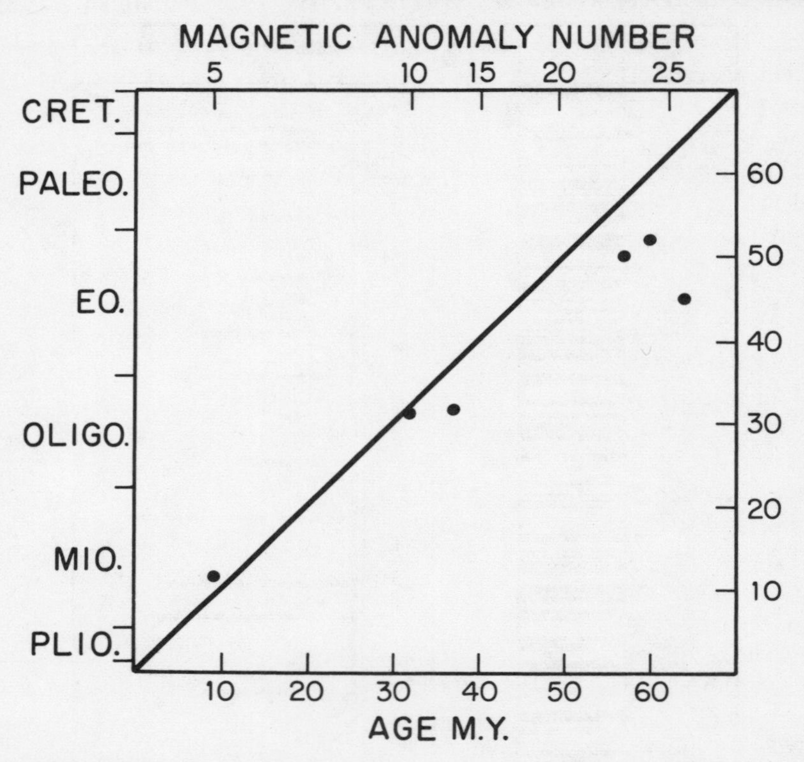

Fig. 5. Comparison of paleontological age of oldest fossil remains in drilling sites, Leg V, Deep Sea Drilling Project, to predicted magnetic anomaly age of igneous basement.

the continued development of datum biostratigraphy and magnetic stratigraphy. The development of this approach has already provided rather precise data supporting the hypothesis of sea floor spreading.

References

Berggren, W. A.: Rates of evolution in some Cenozoic planktonic foraminifera. *Micropaleontology*, vol. 15, no. 2, pp. 351–365, 1969.

Berggren, W. A., R. K. Olsson, and R. A. Reyment: Origin and development of the foraminiferal genus *Pseudohastigerina* Banner and Blow, 1959. *Micropaleontology*, vol. 13, no. 3, pp. 265–288, 1967.

Berggren, W. A., J. D. Phillips, A. Bertels, and D. Wall: Late Pliocene-Pleistocene stratigraphy in deep-sea cores from the south-central North Atlantic. *Nature*, vol. 216, pp. 253–254, 1967.

Black, D. I.: Cosmic ray effects and faunal extinctions at geomagnetic field reversals. *Earth and Planetary Sci. Letters*, vol. 3, pp. 225–236, 1967.

Blow, W. H.: Origin and evolution of the foraminiferal genus *Orbulina* d'Orbigny. *Micropaleontology*, vol. 2, no. 1, pp. 57–70, 1956.

Bolli, H. M.: Planktonic foraminifera from the Oligocene-Miocene Cipero and Lengua formations of Trinidad, B.W.I. *U.S. Nat. Mus., Bull.*, no. 215, pp. 97–124, 1957.

Burckle, L. H., J. Ewing, T. Saito, and R. Leyden: Tertiary sediment from the East Pacific Rise. *Science*, vol. 157, pp. 537–540, 1967.

Cox, A., R. R. Doell, and G. B. Dalrymple: Geomagnetic epochs and Pleistocene geochronology. *Nature*, vol. 198, p. 1049, 1963.

———: Reversals of the earth's magnetic field. *Science*, vol. 144, pp. 1537–1543, 1964.

Funnell, B. M.: The Tertiary Period. In: The Phanerozoic time scale: a symposium. *Geol. Soc. London, Quart. Jour.*, Spec. vol. 120 (5), pp. 179–191, 1964.

Funnell, B. M., and A. G. Smith: Opening of the Atlantic Ocean. *Nature*, vol. 219, pp. 1328–1333, 1968.

Harrison, C. G. A.: Evolutionary process and reversals of the Earth's magnetic field. *Nature*, vol. 217, pp. 46–47, 1968.

Harrison, C. G. A., and B. M. Funnell: Relationship of paleomagnetic reversals and micropaleontology in two late Cenozoic cores from the Pacific Ocean. *Nature*, vol. 204, p. 566, 1964.

Hays, J. D.: Radiolaria and late Tertiary and Quaternary history of Antarctic Seas. *Am. Geophys. Un. Antarctic Res. Ser.*, 5, pp. 125–184, 1965.

———: Paleomagnetically controlled late Cenozoic radiolarian stratigraphy in high and low latitudes. *Abstracts*, Annual Meetings, Geol. Soc. Amer., Atlantic City, 1969.

Hays, J. D., and N. D. Opdyke: Antarctic radiolaria, magnetic reversals and climatic change. *Science*, vol. 158, pp. 1001–1011, 1967.

Hays, J. D., T. Saito, N. D. Opdyke, and L. H. Burckle: Pliocene-Pleistocene sediments of the equatorial Pacific—their paleomagnetic, biostratigraphic, and climatic record. Geol. Soc. Amer., Bull., vol. 80, pp. 1481–1514, 1969.

Heirtzler, J. R., G. O. Dickson, E. M. Herron, W. C. Pitman, and X. Le Pichon: Marine magnetic anomalies, geomagnetic field reversals, and motions of the ocean floor and continents. *Jour. Geophys. Research*, vol. 73, pp. 2119–2136, 1968.

Opdyke, N. D., B. Glass, J. D Hays, and J. Foster: Paleomagnetic study of Antarctic deep sea cores. *Science*, vol. 154 pp. 349–357, 1966.

Percival, S. F., Jr.: Paleontologic evidence for sea floor spreading in the South Atlantic–Leg III (Dakar-Rio de Janiero) of the Deep Sea Drilling Project. *Abstracts*, Annual Meetings, Geol. Soc. Amer., Atlantic City, 1969.

Pitman, W. C., III, and J. R. Heirtzler: Magnetic anomalies over the Pacific-Antarctic ridge. *Science*, vol. 154, p. 1164, 1966.

Riedel, W. R.: Radiolarian evidence consistent with spreading of the Pacific floor. *Science*, vol. 157, pp. 540–542, 1967.

Scientific Staff, Leg V: Deep Sea Drilling Project Leg V. *Geotimes*, Sept., pp. 19–20, 1969.

Uffen, R. J.: Influence of the earth's core on the origin and evolution of life. *Nature*, vol. 198, p. 143, 1963.

Vine, F. J.: Spreading of the ocean floor; new evidence. *Science*, vol. 154, pp. 1405–1415, 1966.

Vine, F. J., and D. H. Matthews: Magnetic anomalies over ocean ridges. *Nature*, vol. 199, pp. 947–949, 1963.

Vine, F. J., and J. T. Wilson: Magnetic anomalies over a young oceanic ridge off Vancouver Island. *Science*, vol. 150, p. 485, 1965.

Waddington, C. J.: Paleomagnetic field reversals and cosmic radiation. *Science*, vol. 158, pp. 913–915, 1967.

Watkins, N. D., and H. G. Goodell: Geomagnetic polarity changes and faunal extinctions in the Southern Ocean. *Science*, vol. 156, pp. 1083–1086, 1967.

12

Science of the Earth

J. TUZO WILSON

Erindale College,
University of Toronto, Toronto, Canada

Contributions of Geology

The precise and accurate description of rocks cropping out on the surface of the earth was among the first branches of science to be developed, and a century ago geology was perhaps the leading science.

Physics had not then developed, and chemistry was far less advanced than it is now. The life sciences were primarily concerned with anatomy and classification. Geologists had their own techniques and these were sound. They were also involved in matters of great public interest: the famous debate on evolution, the exploration of new lands, great gold rushes, and the question of whether certain phenomena were the result of "the flood" or of a continental glaciation. It was natural that geologists should come to regard the study of the earth as a special compartment of science, separate from others, and largely independent of them. At that time this was indeed true. If geologists only learned limited parts of chemistry and biology and little mathematics or physics, the reasons were sensible. To have learned more then would not have greatly helped them.

The task of describing the rocks of the earth's surface has been an enormous one and is not yet completed. It is a valid part of descriptive science or data collecting, but suffers from two severe limitations. The techniques employed cannot be extended to discover much about those greater parts of the earth which constitute its interior and its ocean floors, and the results are so complex that they have not led to any simple explanation or theory of the earth's behavior. Geologic mapping of the earth's surface has been a tremendous task and so absorbing that most geologists have paid scant attention to the discovery of new physical techniques which might partly replace field work and to the development of those precise and valuable theories which enable other branches of science to make daring but valid predictions. Any prospector can testify that predictions about the earth have rarely been accurate. The few geologists, such as Werner, Hutton, Chamberlin, du Toit, Daly, Holmes, Bucher, and van Bemmelmann who have speculated about theories of the earth did not agree among themselves. Most geologists were repelled by this uncertainty and eschewed speculation, calling it "armchair geology." They concentrated upon collecting more data. This they tended to do by improving existing methods or by adding such new methods as fitted conveniently into their accustomed pattern of work rather than by examining wholly new approaches to the study of the earth. Their method of work has always been field mapping with the collection and later laboratory study of rock specimens.

Some of the new techniques which fitted well into this approved pattern have been petrography, chemical analysis, and the crystallographic and X-ray study of minerals. More recently, geologists have studied paleomagnetism and determined the age of rock specimens by radioactive methods. Other techniques they have adopted are the studies under high pressures and temperatures of the physical and mechanical properties of rocks and the chemistry of natural and artificial rock melts. These investigations have contributed to our ideas about the interior of the earth, but by far the greatest contributions have been confined to the earth's surface. Geologists have preferred to map the distribution of accessible rocks and study their properties. They have not played a major part in initiating indirect methods of surveying the ocean floor or the earth's interior. Although they have paid lip service to the study of the whole earth, they have concentrated upon their original field, its surface, a part made complex by boundary conditions with the air and water.

Geophysics, Its Slow but Accelerating Rise

Geophysics has had a very different history of development. Contrary to common opinion, it is not a new subject. One of the earliest of the true scientific treatises was William Gilbert's *De Magnete* (1600); Newton formulated the concept and laws of gravity; the ancients described earthquakes, volcanic eruptions, and other geophysical phenomena. On the other hand, it developed slowly. A century ago, when geologists were describing the surface of the earth with great accuracy and in detail, the slow techniques of geomagnetism, gravity, geodesy, and tidal observation, and the general descriptions of earthquakes and eruptions contributed little to help geology or say much about the nature of the earth. There was no unified science of geophysics, no body of geophysicists, no schools; even the name had not been coined. Geophysics was not studied as an entity but by a succession of physicists who independently attempted to apply the latest discoveries and techniques in physics to the study of various aspects of the earth. This approach has made innovation easy, and geophysicists have been less conservative than geologists in developing completely new techniques.

Physicists who have been responsible for inventing airborne and shipborne magnetometers, gravimeters, methods for measuring heat flow, geophysical prospecting, and the development of the whole subject of seismology which has provided most of our knowledge about the earth's interior. The application of these methods has remained largely in the hands of physicists and engineers who have usually had small knowledge of geology. This has maintained a dichotomy in the study of the earth.

The success of theory in other realms of physics has led physicists to be more optimistic than geologists about the development of theories about the earth. Lacking the detailed knowledge which geologists acquired, indicating that the surface of the earth is complex, physicists have tended to oversimplify, but their theories have differed quite as widely as those produced by geologists. The contradictory views of such physicists as Wegener (1924) and Jeffreys (1959) remind us of this and show that many geophysical theories about the earth, like many geological theories, must be ill-founded.

Dichotomy in the Earth Sciences

The recent, rapid rise of geophysics, as a parallel but largely separate study of the earth from geology, has produced some feelings of insecurity and self-criticism among geologists. Many of them, feeling a need to apply more rigorous methods, have tended to limit their research to those subjects within their accustomed orbit which can be treated precisely. While it is an excellent thing to measure minerals accurately, to apply statistics to studies of sand grains, and do good silicate chemistry, these studies are too limited to solve by themselves the major problem of discovering how the solid earth behaves. The recent efforts of geologists have thus been largely directed toward improving a few methods of data collection, although many other new methods now exist and experience in other sciences shows that, for advancement, sound theory is as necessary as good data.

Geology has increasingly tended to become solely the study of continental rocks, minerals, fossils, and land forms, rather than the study of the earth. Most geophysicists learn little geology and do not appreciate the complexity of the surface rocks. What is needed is a combination of all techniques and ideas into a single science of the study of our planet, but this has not yet been achieved.

It is sad to reflect that in the recent past many geologists were engaged in trying to improve methods of data collection, methods which are not the only ones available and are not necessarily the most useful for solving the problems of the earth; many geophysicists, on the other hand, lacking an appreciation of the earth's true complexity, have spent their time interpreting data in too simple a fashion or in refining physical theories which do not apply in the least to the reality of the earth.

Role of Defense Expenditures

A further practical problem has been the expense and time required to collect information about the earth. More than 150 years have been spent by geologists, surveys, mining and petroleum companies in all the countries of the world to gain our present knowledge of the surface rocks. Their coverage is not yet complete; to obtain a similar knowledge by the more expensive methods of geophysics might have required a very long time, and even greater expense.

Help came unexpectedly, but not from the establishment, if we may so describe the geological surveys of mining and petroleum companies and university departments of geology. It has been the great expenditures for defense and for reasons of national prestige that have speeded up the study of the earth. Naval research, stimulated by the need to detect and to navigate nuclear submarines, has charted the ocean floors, shown them to be quite unlike continents, and revealed the presence of the largest mountain systems on earth—the midocean ridges (Heezen and Ewing, 1963). Disarmament research, with its need to distinguish between artificial explosions and natural earthquakes, has provided worldwide networks of greatly improved seismologic stations and transformed our knowledge of the earth's interior. Space research is now making possible comparisons between the earth and other planets: the moon, Mars, Venus, and Jupiter.

The billions of dollars spent on these investigations, in part done by universities and industry under contract, have provided, with unexpected suddenness, a body of data about the hitherto unknown parts of the earth which is comparable with the bulk of geologic data about its surface which took so much longer to acquire.

Formerly, when science lacked information about so much of the hidden earth, it was not surprising that proper theories could not be framed. Today, with the sudden acquisition of so much new data to constrain and direct speculation, it would be odd if a satisfactory theory of the earth's behavior could not be found.

Ocean Floor Spreading

The theory of ocean floor spreading due to shallow advection in the upper mantle may provide a satisfactory concept [or explanation] of the earth's behavior. If this theory proves to be generally acceptable, as seems to be the case, a revolution will have occurred in earth science which can only be compared with such other major turning points as Harvey's proof of the circulation of the blood, or Becquerel's and Rutherford's discovery of radioactivity which opened the modern era in physics. We may very well be facing such a surge forward in earth science.

It seems safe to predict that, whereas until this present time, it was not considered sound to speculate upon the earth's behavior, hereafter it will not be considered sensible to reject the theory of ocean floor spreading as a precise explanation and cause of geological phenomena.

This is not to claim that continental drift of one special form has been actually proved. The events of the past hundreds of millions of years are history and not subject to repetition or mathematical proof; but the present theory, to which many have contributed, provides such a precise, unique, and satisfying explanation of a vast variety of data it will probably come to be accepted. This meeting may even mark the turning point.

The new theory embodies one form of continental drift. The general arguments for some form of drift have been so extensively reviewed that only some recent, general references seem necessary (Runcorn, 1962; Blackett and co-workers, 1965; Irving, 1964; Elsasser, 1966; Wilson, in press). Many, particularly in the past, have also expressed opinions against it (van Waterschoot van der Gracht, 1928; Beloussov, 1962; Stehli and Helsley, 1963; Axelrod, 1963, 1964).

We are so accustomed to thinking of a single, fixed, and rigid pattern in the earth it is hard to realize that drift and ocean-floor spreading are theories which permit an infinity of different histories. Their acceptance, so far from solving the problems of geology, only open the whole new challenging subject of discovering the earth's past movements in detail.

The theory which now seems so certain is that the ocean floors are spreading by growth of new oceanic crust along the midocean ridges and that they are being reabsorbed again into the deep ocean trenches off mountain and island arcs. Vast, rigid, crustal plates, about 50 km thick, are moving about, with continents embedded in some of them. A proposal for this form of ocean spreading was advanced by Holmes (1931, 1945), and has more recently been advocated with improvements in the light of modern knowledge by Hess (1962) and Dietz (1969).

This particular theory would have remained one of the many general suggestions advanced as hypotheses (Beloussov, 1962, gives a long list of them), had it not been related to other observations which give it a precision quite foreign to the rest. Foremost among these is the suggestion by Vine and Matthews (1963) and Morley and Larochelle (1964) that the ocean floors are magnetically imprinted as they spread. This theory owes its origin to Brunhes' discovery in 1906 that some lava flows are reversely magnetized and his explanation of this as being due to reversals in the earth's magnetic field. These are similar to the quicker reversals of the sun's magnetic field every 11 years. The last 9 reversals of the earth's field have been dated in piles of lava flows by McDougall and Chamalaun (1966) and by Cox and co-workers (1967); Opdyke et al. (1966) have noted them in deep-sea sediments.

Large magnetic anomalies over the ocean floors had been mapped by many expeditions, especially from Columbia, California, and Cambridge universities and U.S.C.G.S. U.S.S.R. These have the form of long strips in the Pacific Ocean (Raff and Mason, 1961; Vacquier, 1965), and of strips parallel with the mid-Atlantic Ridge (Heirtzler and Le Pichon, 1965). The contribution of Vine and Matthews (1963) was to point out that spreading of the ocean floors and formation of basalt in a reversing field could produce such anomalies. We located and interpreted another example of anomalies parallel with a midocean ridge, and recognized the precise location of a new section of it off Vancouver Island (Vine and Wilson, 1965). We showed that if the widths of the anomalies were measured from the crest of the midocean ridge, the ratios so formed equalled the ratios between the times of magnetic reversals. This, we suggested, was compatible with a uniform rate of expansion of 6 cm per year along the ridge off Vancouver Island. The rate and direction of the motion of the San Andreas fault are the same. The magnetic layer, we thought, was 1.7 km thick, and the basalt overlying a nonmagnetic layer might be about 6 km. thick. This is in agreement with the prediction of Hess (1962) that a layer of serpentine beneath basalt might form the oceanic crust.

Since then Pitman and Heirtzler (1966), Vine (1966), and Matthews and Bath have found much more extensive examples of ocean-floor spreading, as indicated by magnetic anomalies. All find that the distance scale from the center of ridges equals the time scale of recent reversals. This relation between distances on the ocean floor and times of magnetic reversals is worldwide, and cannot be due to chance. The only explanation so far offered is that the midocean ridges are spreading at uniform rates and that the sea floor is being imprinted like a magnetic tape as the floor is generated.

Another line of evidence supporting the concept of expansion is the increasing age of cores and islands away from midocean ridge crests (Funnell, 1964; Riedel and Funnell, 1964; Saito and co-workers, 1966; Wilson, 1963). Only one observation among 600 cores studied is contradictory, and this may be due to an error or complication.

In Iceland, which is well studied and the only island lying athwart the midocean ridge crest, Bodvarsson and Walker (1964) found strong geologic evidence that it has been expanding due to the intrusion of vast numbers of dikes.

The rates of spreading of the ocean basins measured by Vine, by Heirtzler, and their colleagues are such that it is unlikely they can be explained by expansion of the earth. Even were the earth expanding, it is hardly possible for physical reasons that it is doing so with such rapidity. Thus, the excess ocean floors must be in process of being absorbed. Coats (1962) has described such a process, in which the crust is being pushed down beneath island arcs and into ocean trenches. Wilson (1965) has discussed the pattern of present motions. This theory also seems able to explain the different suites of lavas found in various geologic environments. The chemistry of lavas on ocean floors and islands has been discussed recently in a series of reports, including those by Eaton and Murata (1960); Yoder and Tilley (1962); Muir and Tilley (1964); Engel and co-workers (1965); Carmichael (1964); Taylor and White (1965); and Green and co-workers (1967). The consensus is that partial melting in the mantle has produced a uniform tholeiite which forms a layer of dikes and flows on the ocean floors.

In some places, an extra supply of lava builds islands. These, unlike the floor proper, often contain much alkali basalt, trachyte, and even soda rhyolite. The field evidence suggests that this is due to differentiation in the tall conduits of the volcanoes, but some laboratory work suggests a separate source in the mantle.

If in the process of ocean-floor spreading the ocean-floor basalts are reabsorbed in trenches, their remelting under different conditions could produce the andesites of island arcs. These are rapidly eroded and the sediments so formed are deposited in eugeosynclines. Subsequent metamorphism and squeezing could produce gneisses and lead to the growth of continents. This process has been discussed by Jacobs and co-workers (1959), Coats (1962), and Taylor and White (1965).

Verhoogen (1954) drew attention to the probability that lavas melt in a layer about 50 to 300 km deep. If the solid crust were pushed back to that depth in the mantle it might be reabsorbed. This layer would correspond with the low-velocity layer recognized by seismologists (Anderson, 1962). A layer near its melting point would be expected to have a low velocity for the transmission of seismic waves and might easily be able to creep. Elsasser (1966), Tozer (1965), and Orowan (1964) have discussed the possibility that shallow advection occurs in this layer, but Runcorn (1965) favors convection of the whole mantle.

Fundamental Difference between Continents and Ocean Basins

The creation and destruction of ocean floors means that oceanic crust is not conserved; this implies that the mechanics and geometry are different from those applicable to continents, which are conserved. On land, mountains are raised and lowered as standing waves, without migrating relative to the crust and without the crust flowing through the mountain. This is Lagrangian mechanics. On the ocean floor, the midocean ridges are moving waves; the crust is continuously generated at the crest and flows away down the sides, passing through the mountain shapes and out of them. This is Eulerian mechanics. In 1965, it was proposed that a new class of transform faults should be common on ocean floors, and shown that their properties, as predicted by simple geometry, fitted those observed in the great fracture zones discovered by Menard (1964), Heezen et al. (1964), and Matthews (1966). In a test, Sykes (1967) has shown that the direction of motion in recent earthquakes supports this concept.

If the mechanics of the behavior of continents and ocean basins are different, it follows that the theories of faulting and the directions of fault motion are also different. This appears to be the case, lending further support to the concept of ocean-floor spreading.

Changing Continental Patterns and Recurrence of Drift

The present system of midocean ridges was first fully recognized by Heezen and Ewing (1963) and by Menard (1964), although many of their ideas were foreshadowed by others, including Holmes (1945), Gutenberg and Richter (1954), and King (1962). The essential form of the present system can be regarded as a nearly complete circumpolar ring through the Southern Ocean with a branch northward in each ocean. It seems to have originated during the Mesozoic era and to have caused the separation and dispersal of the present continents.

A major task in earth science is to establish the preexisting configurations. Wegener (1924) suggested that there was only one continent, Pangaea. Du Toit (1937) proposed two, Laurasia and Gondwanaland. Melville (1966) has advocated four: Gondwanaland, Atlantica, Pacifica, and Angara. Before discussing which view is likely to be correct, it must be recalled that the continents are assumed to have been in motion, intermittently or continuously, and that they would have had different configurations at different times. All the details are as yet far from being established. What we are most certain about is that South America and Africa fitted closely together until mid-Cretaceous time (Stoneley, 1966; Belmonte and others, 1965; Bullard et al., 1965), and that previously they had been joined for several geologic periods to the Falkland Islands, Antarctica, Madagascar, India, Seychelles Islands, Australia, and perhaps part of New Zealand. This formed the continent of Gondwanaland.

In the north, Greenland and much of Eurasia and North America were joined, forming Laurasia or Atlantica. The presence of many volcanic rocks of Cretaceous-Eocene age on the coasts of Baffin Island, Greenland, Scotland, Ireland, and Portugal suggests that the North Atlantic may have opened later than the South Atlantic. On the other hand, the Gulf of Mexico seems to have formed during the Jurassic period. Perhaps related to its opening were the Palisadean volcanics of Upper Triassic to Lower Jurassic age which are found from Florida to Nova Scotia. Thus, the opening of the Atlantic Ocean may have been a complex affair, occurring at different times at different latitudes. Melville (1966) has discussed the possible existence and dispersal of another continent in the Pacific area and the separate existence of Angara, now forming most of Siberia.

It has often been supposed that drifting only started during the Mesozoic, but in view of the great length of earlier geologic time this seems surprising. Drift has probably been recurrent.

Two examples of this follow: one is the proposal that, whereas the opening of the present Atlantic during the Mesozoic had separated what had been a single continent into the present parts, an earlier proto-Atlantic Ocean had closed during the Paleozoic era, bringing earlier fragments together to form the unified continent of Laurasia or Atlantica. I suggested (in 1966 and 1967) that this process fitted the descriptions of the Appalachian region by Kay (1951) and by Schuchert (1923) and Barrell, and caused the transposition of fragments bearing Cambrian and Ordovician faunas across the ocean, accounting for faunas of the European type in eastern Canada and New England, of faunas of the African type in the basement of Florida, and for areas of originally North American rocks in northern Europe and western Africa.

The other example in the Pacific is less clear. Several investigators have contributed evidence. Hess (1962) and Menard (1964) have suggested that the Darwin Rise in the West Pacific is the remnant of a former midocean ridge. Gilluly (1963) pointed out that there was an extraordinary burst of intrusive activity in the Andes and Cordillera during the Jurassic period. Could that have been due to a reversal in direction of motion which caused continental shelves to collapse into mountains?

Menard (1964) and Cook (1962) have proposed that the East Pacific Rise, which now ends in the Gulf of California and the San Andreas fault (Hamilton and Myers, 1966; Wilson, 1965) formerly underlay and tended to expand and uplift the Basin and Range province. If so, it seems plausible, from Eardley's (1962) account, that in pre-Miocene time the shear that ended the East Pacific Rise was not the San Andreas fault, but the Rocky Mountain Trench. Carey (1958), Wise (1963), and Watkins (1965) have all noted evidence for major disturbances and dislocations of the Pacific coast. Melville (1966) has carried the matter further and has proposed that a former continent of Pacifica split to form eastern Asia and western North America. White (1966) has remarked that the Permian faunas of eastern and western British Columbia are quite unlike one another.

Is it not possible that the Tertiary shearing in western North America is a late phase of the coming together of part of Pacifica and the rest of North America and the closing of what had been an ocean through British Columbia and the western United States? Were not the shears of the Cabot, Great Glen, and Brevard faults formed during the late stages of the closing of the proto-Atlantic Ocean? Could the Rocky Mountain Trench and the San Andreas fault be analogous features?

Is it true that as recently as the late Paleozoic era parts of what is now North America were separated by oceans to form parts of four continents (the Maritime Provinces and New England of Europe, central Florida of Africa, and the whole west coast of Asia)? If true, how does it affect all the accounts of the geologic history of North America? If not true, many questions still remain to be answered, as Wise (1963) pointed out.

Conclusions

It is now apparently beginning to be established that the oceans are all young. The oldest fossils found in any core or island which lacks continental structure are uppermost Jurassic marine fossils on Maio in the Cape Verde Islands. No others exceed Cretaceous age. In contrast, the continents are 20 times as old, and all seem to be composite mosaics, as though they had been repeatedly pulled apart and pushed together in different patterns.

The structure of the earth's surface in its own slow fashion does not so much resemble a rigid solid as the top of a boiling and convecting pot of soup. The continents are like froth, the ocean floors like the clear liquid between. Like froth and soup, continents and ocean floors differ in composition, age, structure, and mechanics. The continents, like froth, are in some parts relatively old. They are forever growing and accumulating, and their dominant structures are compressed zones parallel with their margins. The ocean floors are relatively young, forever upwelling like the clear soup. Their dominant structures are tensional along the midocean ridges, with lines of flow and of shearing at right angles to the midocean ridges. The latter are marked by chains of islands and fracture zones.

If, broadly speaking, these ideas are correct, the whole of historic geology needs rewriting, and all the standard textbooks of general and historic geology are to a greater or lesser extent out of date. All books on structural geology will require an extra chapter about the different tectonic behavior of ocean floors, and it is open to question whether a lot of ideas about petrogenesis and economic geology also need revision.

Earth science may be at a turning point. Geology need no longer lack a theory, and the need to integrate geology and geophysics is obvious. What is the point of including such unrelated subjects as mineralogy and paleontology in one science, while other, no more disparate subjects, such as geomagnetism and seismology, are placed in another? Their common bond is that all treat of the earth; they should be combined into one science of the earth. We have a great opportunity.

References

Anderson, D. L.: The plastic layer of the earth's mantle. *Sci. Am.* 207: 52–509, 1962.

Axelrod, D. J.: Fossil floras suggest stable, not drifting continents. *J. Geophys. Res.* 68:3257–3263, 1963; 69:1669–1671, 1964.

Belmonte, Y., P. Hertz, and R. Wenger: The salt basins of the Gabon and the Congo (Brazzaville): A tentative paleographic interpretation. *The Salt Basins around Africa*, D. C. Ion, ed. Institute of Petroleum, London, pp. 55–74, 1965.

Beloussov, V. V.: *Basic Problems in Geotectonics.* New York, McGraw-Hill Book Co., 1962.

Blackett, P. M. S., E. C. Bullard, and S. K. Runcorn, eds.: *A Symposium on Continental Drift.* Phil. Trans. Roy. Soc. London s.A, 258, 1965.

Bodvarsson, G., and G. P. L. Walker: Crustal drift in Iceland. *Geophys. J.* 8:285–300, 1964.

Bullard, E. C., J. E. Everett, and A. G. Smith: The Fit of the Continents Around the Atlantic. In *A Symposium on Continental Drift*, P. M. S. Blackett, E. C. Bullard, and S. K. Runcorn, eds. Phil. Trans. Roy. Soc. London s.A, 258, 1965, pp. 228–251.

Carey, S. W.: The Tectonic Approach to Continental Drift. In *Continental drift: A Symposium . . . , March, 1956.* Hobart, University of Tasmania, Geology Dept., 1958, pp. 177–355.

Carmichael, I. S. E.: The petrology of Thingmuli, a Tertiary volcano in Eastern Iceland. *J. Petrol.* 5:435–460, 1964.

Coats, R. R.: Magma type and crustal structure in the Aleutian arc. *Am. Geophys. Union, Geophys. Mon.* 6:92–108, 1962.

Cook, K. L.: The problem of the mantle-crust mix. *Advan. Geophys.* 9: 295–360, 1962.

Cox, A., G. B. Dalrymple, and R. R. Doell: Reversals of the earth's magnetic field. *Sci. Am.* 216:44–54, 1967.

Dietz, R. S.: Continents and Ocean Basins. In *The Megatectonics of Continents and Oceans*, H. Johnson and B. L. Smith, eds. New Brunswick, Rutgers University Press, 1970, pp. 24–46.

Du Toit, A. L.: *Our Wandering Continents.* Edinburgh, Oliver & Boyd, 1937.

Eardley, A. J.: *Structural Geology of North America*, 2d ed. New York, Harper & Row, 1962.

Eaton, J. P., and K. J. Murata: How volcanoes grow. *Science* 132:925–939, 1960.

Elsasser, W. M.: Thermal structure of the upper mantle and convection. In *Advances in Earth Science*, P. M. Hurley, ed. Cambridge, Mass., Massachusetts Institute of Technology Press, 1966, pp. 461–502.

Engel, C. G., R. L. Fischer, and A. E. J. Engel: Igneous rocks of the Indian Ocean floor. *Science* 150:605–610, 1965.

Funnell, B. M.: Studies in North Atlantic geology. *Geol. Mag.* 101:421–434, 1964.

Gilluly, J.: The tectonic evolution of the western United States. *Quart. J. Geol. Soc. London* 119:133–174, 1963.

Green, T. H., D. H. Green, and A. E. Ringwood: The origin of high-alumina basalts and their relationships to quartz tholeiites and alkali basalts. *Earth and Planetary Sci. Let.* 2:41–51, 1967.

Gutenberg, B., and C. F. Richter: *Seismicity of the Earth*. Princeton, N.J., Princeton University Press, 1954.

Hamilton, W., and W. B. Myers: Cenozoic tectonics of the Western United States. *Geol. Surv. Canada, Paper* 66-14:291–306, 1966.

Heezen, B. C., and M. Ewing: The mid-oceanic ridge. In *The Sea*, M. N. Hill, ed. New York, John Wiley & Sons, Inc., 1963, vol. 3, pp. 388–410.

Heezen, B. C., et al.: Chain and Romanche fracture zones. *Deep-Sea Res.* 11:11–33, 1964.

Heirtzler, J. R., and X. La Pichon: Magnetic anomalies over the mid-Atlantic Ridge. *J. Geophys. Res.* 70:4013–4033, 1965.

Hess, H. H.: History of Ocean Basins. In *Petrologic Studies: A Volume to Honor A. F. Buddington*, A. J. Engel, ed. Geol. Soc. Am., 1962, pp. 599–620.

Hill, M. N., ed.: *The Sea*. New York, John Wiley & Sons, Inc., 1963.

Holmes, A.: *Principles of Physical Geology*. New York, Ronald Press, 1945.

Holmes, A.: Radioactivity and earth movements. *Trans. Geol. Soc. Glasgow* 18:559–607, 1928–1931.

Hurley, P. M., ed.: *Advances in Earth Science*. Cambridge, Mass., Massachusetts Institute of Technology Press, 1966.

Irving, E.: *Paleomagnetism*. New York, John Wiley & Sons, Inc., 1964,

Jacobs, J. A., R. D. Russell, and J. T. Wilson: *Physics and Geology*. New York, McGraw-Hill Book Co., 1959.

Kay, M.: North American geosynclines. *Geol. Soc. Am. Mem.* 48, 1951.

King, L. C.: *The Morphology of the Earth*. Hafner Pubn. Co., New York, 1962, 699ff.

Matthews, D. H.: The Owen fracture zone and the northern end of the Carlsberg Ridge. *Phil. Trans. Roy. Soc. London* s.A, 259:172–186, 1966.

Matthews, D. H., and J. Bath: Formation of magnetic anomaly pattern of mid-Atlantic Ridge. *Geophy. Jour.* 13:349–357, 1967.

McDougall, I., and F. H. Chamalaun: Geomagnetic polarity scale of time. *Nature* 212:1415–1418, 1966.

Melville, R.: Continental drift, Mesozoic continents and the migrations of the angiosperms. *Nature* 211:116–120, 1966.

Menard, H. W.: *Marine Geology of the Pacific*. New York, McGraw-Hill Book Co., Inc., 1964.

Morley, L. W., and A. Larochelle: Paleomagnetism as a means of dating geological events. *Trans. Roy. Soc. Can.* Spec. Pubn. 9:39–51, 1964.

Muir, I. D., and C. E. Tilley: Results from the northern part of the rift zone of the mid-Atlantic Ridge. *J. Petrology* 5:409–433, 1964.

Opdyke, N. D., et al.: Paleomagnetic study of the Antarctic deep-sea cores. *Science* 154:349–357, 1966.

Orowan, E.: Continental drift and the origin of mountains. *Science* 148: 1003–1010, 1964.

Pitman, W. C., III, and J. R. Heirtzler: Magnetic anomalies over the Pacific-Antarctic Ridge. *Science* 154:1162–1164, 1966.

Raff, A. D., and R. G. Mason: Magnetic survey of the west coast of North America, 40°N latitude to 50°N latitude. *Bull. Geol. Soc. Am.* 72:1267–1270, 1961.

Riedel, W. R., and B. M. Funnell: Tertiary sediment cores and microfossils from the Pacific Ocean floor. *Quart. J. Geol. Soc. London* 120: 305–368, 1964.

Runcorn, S. K., ed.: *Continental Drift.* New York, Academic Press, 1962.

Runcorn, S. K.: Changes in the convection pattern in the earth's mantle. In *A Symposium on Continental Drift*, P. M. S. Blackett, E. C. Bullard, and S. K. Runcorn, eds. Phil. Trans. Roy. Soc. London. s.A, 258:228–251, 1965.

Saito, T., M. Ewing, and L. H. Burckle: Tertiary sediment from the Mid-Atlantic Ridge. *Science* 151:1075–1079, 1966.

Schuchert, C.: Sites and nature of North American geosynclines. *Bull. Geol. Soc. Am.* 34:151–229, 1923.

Stehli, F. G., and C. E. Helsley: Paleontologic technique for defining ancient pole positions. *Science* 142:1057–1059, 1963.

Stoneley, R.: The Niger delta region in the light of the theory of continental drift. *Geol. Mag.* 103:385–397, 1966.

Sykes, L. R.: Mechanism of earthquakes and nature of faulting on the mid-oceaning ridges. *J. Geophys. Res.* 72:2131–2153, 1967.

Taylor, S. R., and A. J. R. White: Geochemistry of andesites and the growth of continents. *Nature* 208:271–273, 1965.

Tozer, D. C.: Heat transfer and convection currents. In *A Symposium on Continental Drift*, P. M. S. Blackett, E. C. Bullard, and S. K. Runcorn, eds. Phil. Trans. Roy. Soc. London, s.A, 258:252–271, 1965.

Vacquier, V.: Transcurrent faulting in the ocean floor. In *A Symposium on Continental Drift*, P. M. S. Blackett, E. C. Bullard, and S. K. Runcorn, eds. Phil. Trans. Roy. Soc. London s.A, 258:77–81, 1965.

Verhoogen, J.: Petrological evidence on temperature distribution in the mantle of the earth. *Trans. Am. Geophys. Union* 35:85–92, 1954.

Vine, F. J.: Spreading of the ocean floor: new evidence. *Science* 154: 1405–1415, 1966.

Vine, F. J., and D. H. Matthews: Magnetic anomalies over oceanic ridges. *Nature* 199:947–949, 1963.

Vine, F. J., and J. T. Wilson: Magnetic anomalies over a young oceanic ridge off Vancouver Island. *Science* 150:485–489, 1965.

van Waterschoot van der Gracht, W. A. J. M., ed.: *Theory of Continental Drift, a Symposium.* Amer. Assoc. Petrol. Geologists, Tulsa, Okla., 1928, 240 pp.

Watkins, N. D.: Paleomagnetism of the Columbia Plateaus. *J. Geophys. Res.* 70:1379–1406, 1965.

Wegener, A.: *Origin of Continents and Oceans.* New York, Dutton, 1924.

White, W. H.: Summary of tectonic history, in W. G. Stevenson, ed.: Symposium on the tectonic history and mineral deposits of the Western cordillera. Spec. vol. 8, Can. Inst. Min. Metallurgy, pp. 185–189, 1966.

Wilson, J. T.: Evidence from islands on the spreading of ocean floors. *Nature* 197:536–538, 1963.

Wilson, J. T.: A new class of faults and their bearing on continental drift. *Nature* 207:343–347, 1965.

Wilson, J. T.: Did the Atlantic close and then re-open? *Nature* 211:676–681, 1966.

Wilson, J. T.: On a possible explanation for the crust's growth and movement. *Quart. J. Geol. Soc. London* (in press).

Wise, D. V.: An outrageous hypothesis for the tectonic pattern of the North American cordillera. *Bull. Geol. Soc. Am.* 74:357–362, 1963.

Yoder, H. S., and C. E. Tilley: Origin of basaltic magmas: An experimental study of natural and synthetic rock systems. *J. Petrology* 3: 342–532, 1962.

Index

Acadian orogeny, 77, 84, 103
aeromagnetic surveys, *x–xi*, 3–9, 10, 55–56
Affleck, James, *viii*, *x–xi*, 3–10, 150
Africa, 32, 33, 35, 44, 84, 188; area of, 34(*tab.*); coastal magnetic anomalies of, 91; European fit of, *ix*, 168, 170, 172, 189; Florida and, 262, 263; foldbelt of, 103, 106, 168, 171, 172, 173; North Atlantic subsea features and, 76, 77(*fig.*); plateaus of, 39, 40; rift valleys of, 195, 198, 233; sial of, 168; South American fit of, 17, 18(*fig.*), 106, 164, 169, 262
Alaska earthquake (1964), 28, 130
Aleutian Islands, 28, 129
Alfven, H., 25
Algeria, 105, 172, 174, 175
Algoman unconformity (Karelian unconformity; Kenoran unconformity), 82, 148, 150, 159, 165
alkali metasomatism, 206, 216, 218, 220
Alleghenian orogeny, 77, 84, 103
Alpine-Himalayan belt, 233
Alps, The, *ix*, 105, 122, 168; low velocity channel and, 123; marine sediments of, 170, 171, 207; Mediterranean mountain similarities, 173–74, 175, 177, 178, 182; stress fields of, 232, 233
Ampferer, O., cited, 54, 68
Anderson, E. M., cited, 225, 226, 228, 233
Andes, 28, 29, 32, 40, 106; Jurassic period in, 263; straightness of, 63
andesites, 177; formation of, 28, 260
Angara, 261, 262
anomalies, 55–56, 258; Atlantic anomaly belts, 89–91, 97, 98, 106, 259; basement magnetic, *viii*, *x–xi*, 5–9, 10, 89–91, 99, 106, 147, 148, 150, 151(*fig.*), 153, 165, 241, 243, 248, 259; Bouguer gravity anomaly map, 47–49, 87–89, 117, 122–23, 126, 152(*fig.*), 153(*fig.*), 156, 157; depths of, 150, 151, 152, 153; of heat flow, 201–202, 217–18; Italian gravity anomaly, 119, 126; midcontinental, 155–58; Pacific anomaly belts, 63, 65, 162–63, 259; of pole positions, 22
Antarctica, 32, 33, 34(*tab.*), 262
Antarctic-Africa Ocean, 214(*tab.*)
Antarctic Ocean, 245
Antillean foldbelt, 78(*fig.*), 80, 106
Apennines, The, 179, 233; submarine origin of, 180, 182, 187
Appalachian area, 5, 75–106, 155, 263
Appalachian geosyncline, 77, 78(*fig.*), 80, 99, 262; Africa and, 103, 106;

Appalachian geosyncline (cont.)
anomalies of, 3–4, 5, 88, 89, 97, 98; crustal thickness of, 93; earthquakes and, 86; isotopic age of, 82, 84; sediments of, 101–102; structure, 3–5; termination (Eastward), *ix*, 79, 102
Arabia, 34*n*, 35
Arctic Ocean, 37, 43
Armorican structural system, 9
Asia, 32, 35, 161, 263; plateaus of, 39, 40
Atlantica, 261, 262. *See also* Laurasia
Atlantic Coastal Plain, 80, 99
Atlantic Ocean, 27, 43, 84, 85; anomalies of, 87, 91, 106, 157, 243; coastal displacement, 5–6, 18; closing and reopening of, *viii*, *ix*, 18, 44, 242, 262, 263; greenstones of, 206; mid-ocean rift, 233, 235(*fig.*); Mendocino fracture and, 10, 97. *See also* Mid-Atlantic Ridge; North Atlantic; South Atlantic
Atlas foldbelt, 105, 168, 173
attenuation factors, 113, 114, 130, 132–33; lower and upper mantle differences and, 131, 132, 136, 142
Atwater, Tanya, cited, 198–99
Aubouin, J., cited, 182
Australia, 32, 33, 34(*tab.*), 262
Avalonian tectonic belt, *ix*, 83–84, 102, 103
Axial Valley, *xi*, 194–220, 233, 259
Azema, J., cited, 174, 175

basalt, 27, 28, 30, 177, 182; axial valleys and, *xi*, 197, 200–201, 204, 205–207, 208, 209–13, 215–16, 218, 220, 259, 260
bathymetry, 196–99
Bay of Biscay, 14, 105
Becquerel, Antoine Henri, 257
Berggren, W. A., cited, 244, 246, 248, 249(*fig.*)
Berlage, H. P., cited, 26
Betic Mountains, 168, 173, 174, 187; peridotite of, 175, 177, 183
Bight of Africa, 106
biostratigraphy, *viii*, 19, 241–43, 250; zonation and, 243–48
Blackett, P. M. S., quoted, 12
Blake Plateau, 99, 100(*fig.*), 101
Blumenthal, M., 173*n*
Bonneville Basin, 49, 52, 67, 135
Bouguer gravity anomaly map of the United States (Woollard and Joesting), 47, 87, 117, 121, 122, 126, 152(*fig.*); Greenleaf anomaly and, 156; Mid-Atlantic Ridge and, 157; of Oklahoma, 153(*fig.*)
Brevard fault, 95, 96, 263
Briançonais geanticline, 175, 178
British Association for the Advancement of Science, 12
British Columbia, 263
British Isles, *ix*, 8, 103, 106
Brunhes-Matuyama magnetic epoch boundary, 245, 258
Brunn, J. H., cited, 182–83
Bucher, John Emery, 254
Buena Vista Hills, 54
Bullard, Sir Edward, cited, 74, 105–106
Bullen, K. E., earth interior model of, 115–17, 128, 130

Cabot fault zone, 95–96, 103, 263
Caire, A., cited, 175
Calabria, Italy, 172, 177, 178, 180, 245
Caledonian orogeny, 103, 105, 171
California, 42, 51, 52, 63, 64, 68, 95, 164, 197; Moho under, 119, 121; transcurrent faulting in, 233
California, University of, 259
California Coast Ranges, 52, 55
Cambrian Period, 4, 22, 82, 158, 177; earth expansion since, 20; fossil forms of, *viii*, 262
Cambridge University, 259
Canada, *viii*, 62, 79, 148, 262, 263
Canada Geological Survey, 83
Canadian Shield, *ix*, 42, 76, 78(*fig.*), 102, 148; Archean rock of, 148, 149

(*fig.*), 150, 163–64; earthquakes and, 85, 86; gravity anomalies of, 89; isotopic age determinations, 82
Cape Mendocino, 64, *65*
Cape Verde Islands, 264
carbonates, 19, 38, 180
Carboniferous Period, 78, 79, 95, 96, 170, 174, 178; polar wander in, 18
Carder, D. S., cited, 127
Carey, S. W., cited, 170
Carpathian Mountains, 172
Cascade Mountains, 49, 55, 58, *65*
catastrophism, 25–26
Cenozoic Era, 39, 43, 67, 106, 243; Alpine foldbelt and, 105, 174; Mediterranean in, 170, 178, 180, 189; NRM directions and, 13–15, 22
Central America, 34*n,* 105, 106
Chamberlin, Thomas Chrowder, 254
Charleston earthquake (1886), 86
China, 163, 165
Chinnery, M. A., cited, 128–29
Chugwater formation, 20, 21(*fig.*)
Clarion fracture zone, 63–64, 66
Collinson, David W., *vii, viii,* 12–22
Colorado Plateau, 40(*fig.*), 48, 51, 52, 55
Columbia Plateau, 58
Columbia University, 259
Conrad, V., 147
Conrad discontinuity, *xi,* 50–51, 147–65
continental arch, 158–59, 160, 161
continental drift, *vii,* 12–21, 74, 76, 167–68; Conrad discontinuity and, 147; crust and mantle properties and, *x,* 69, 140–43; distribution and, *ix, x,* 32–37, 41, 141–42, 189, 261–62, 263; East-West fault system and, 164–65; impingements, 40, 44; island arcs and, 30, 32; landmass rotation and, 13–15; opposing-continental tectonics and, 102–106, 164, 168, 172–73, 189; paleomagnetism and, *viii,* 12–22, 47; Precambrian stress patterns and, *viii,* 3-5, 10; process continuity, 16, 17, 262; sea floor spread and, *ix–x,* 27, 28–29, 32, 37, 42–43, 257–58; Tethys and, 168–71
continental shelves, 105, 263; eastern North American, *ix,* 76, 77(*fig.*), 79, 86, 90, 96–97, 98–102
continental slopes, 97, 98–102
continents: process of formation by accretion, *ix,* 25, 26, 28–32, 37, 39, 43, 99, 102, 260, 264; anomaly trends and, 5, 10, 87; Conrad layer of, 147, 148; distribution of, *ix, x,* 32–37, 41, 141–42, 189, 261–62, 263; mantle properties and, 124, 140–43; Moho depths of, 117; ocean basins and, 24–44, 171, 256, 261, 264; oceanic fault interrelations, 63–67; ocean rises and, 38–39, 40, 42, 160, 161–63, 261; permanency of, *xi,* 10; polar wander and, *viii,* 3, 10, 13, 14–15, 16, 17(*fig.*); transatlantic fit of, 74, 75(*fig.*), 103–106, 164, 262, 263. *See also specific continents*
convection, *ix,* 27–44, 202, 140–41
Cordilleran foldbelt, 93, 263
core of the earth, *x,* 27, 116, 142, 223; growth of, 35–36; motion in, 21, 113–14, 130–32, 138
Corsica, 123, 172, 178, 187
Cramer, C. H., cited, 213
Creer, K. M., cited, 17
Cretaceous Period, 42, 66, 100(*fig.*), 162–63, 243, 262, 264; Mediterranean, 170, 175, 176(*fig.*), 178, 180, 181
Crittenden, M. D., Jr., cited, 49
crust of the earth, 253–54, 257; breaking strength, 136; Conrad layer, 147–65; deformations of, *x,* 47–69, 114, 121–22, 123, 124–25, 136, 223, 233, 237; density, 223, 224; diapirs and, 186, 187, 188; fragmentation, 99; Great Basin spread rate, 58–62; heat flow anomalies and, 218; mantle convection and, *x,* 128, 140–43;

crust of the earth (cont.)
the Moho and, 117, 118, 120, 154; ocean basin level and, 29; oceanic generation of, *vii, viii,* 194, 195, 196, 214, 218, 258, 259, 260, 261; oceanic crust permeability, 219–20; oceanic structure of, 125–26, 170, 171(*fig.*), 177, 188, 205–207, 242, 259, 260, 261; plate analogy, *vii,* 202–203, 214–15, 258; shortening of, 233; sial and, *viii–ix,* 24, 25, 26, 29, 40–41, 120; spin axis and, *xi,* 10; stress fields and, 225, 228–29, 233, 234, 236(*fig.*), 237; thickness variations, 49–50, 51, 54, 63, 67, 91–93, 115, 117, 120, 121, 122–23, 126, 141, 154; velocity variations, 50–51, 52, 53(*fig.*), 100(*fig.*), 115, 118, 119–21, 122, 126, 128
crust-mantle mix, 49, 52, 123, 260
Curie temperature, 4, 6, 20, 56, 150
Cyprus, as ocean-floor model, *xi,* 207–208, 209, 212, 213(*fig.*), 215–16, 220

Dalmatian Mountains (Dinaric Alps; Dinarides), 172, 233
Daly, Reginald, 254
Damaran episode, 84
Darwin Rise, 37, 162–63, 165, 262–63
Deep Sea Drilling Project, 241*n*, 242, 243–44, 246, 248, 250(*fig.*)
Deffeyes, Kenneth S., *viii, xi,* 194–220
densities, distribution of, 116–117, 121, 131, 156, 170–71, 186–87, 223; earth mantle, 132, 163–64. *See also* gravity; isotasy
Devonian Period, 18, 65, 95, 174
diapirism, 183-88, 189
Dietz, Robert S., *vii, viii–ix,* 24–44, 258
Dinaric Alps (Dalmatian Mountains; Dinarides), 172, 233
discontinuities, 126–128, 132, 141, 142–43. *See also* Conrad discontinuity; Mohorovičić (Moho) discontinuity
Donn, W. L., cited, 26
Drake, C. L., cited, 106
Durand-Delga, M., 173*n*
DuToit, A. L., 254, 261

earthquake, 113–14, 134, 141–42, 255, 257; aftershocks, 139–40; deep-focus, 138, 232–33; energy rate-depth function for, 136, 137(*fig.*), 138; epicenter patterns, 84–86, 115, 128–29; inner earth core and, 116, 130–32, 138; midocean ridges and, 203–204, 209, 213, 261; sea floor spread and, 28, 261; stress fields and, 224, 229–33, 237. *See also* faults
East Africa, 195, 198, 233
East Pacific Ocean, 5, 10, 63, 245, 263
East Pacific Rise, 40, 159, 160, 161–62, 263; Cypress and, 208; fossil forms of, 242; heat flow, 203(*fig.*)
echo-sounder profiles, 198–99
elastic wave velocities, 113, 114, 130; attenuation factors and, 132–33, 142
Elba, 172, 178, 179(*fig.*), 181, 187
Elder, J. W., cited, 203, 213
Elsasser, W. M., cited, 202–203
Eocene epoch, 97, 100(*fig.*), 175, 176, 178, 262; fossil forms, 244
epeirogenesis, 39–40, 41(*fig.*), 42
erosion, *ix,* 26, 43, 50, 171, 189; of andesites, 260; Appalachian, 93; continental slope, 100; crust density and, 61; Gondwanaland and Laurasia and, 37–38; of Precambrian shields, 42, 164; rate of, 39, 41–42, 164; subaerial, 180
Escher, B. G., 25
eugeosynclines, 30, 31(*fig.*), 260
Euler, Leonhard, 261
Eurasia, 33, 35, 172, 262; area of, 34(*tab.*); orogenies of, *ix,* 168, 170–71, 172, 233, 237; rotation hypothesis and, 169, 170, 189
Europe, 32, 103, 104(*fig.*), 105, 127, 262, 263; African fit, *ix,* 168, 170, 172, 189; Atlantic continental shelf and, 106; coastal magnetic anom-

alies of, 91; Conrad discontinuity and, 159; North Atlantic subsea features and, 76, 77(*fig.*); stress fields and, *x*, 232, 233, 237
Ewing, M., cited, 261

Fallot, P., 173*n*, 175–76
Fara, H. D., cited, 230
faults, 54–55, 78(*fig.*), 103, 230(*fig.*), 231(*fig.*); continental and oceanic, compared, 64–67, 261; East-West, 153, 154–55, 163, 164–65, 198(*fig.*); friction and, 139–40; incipient, 225–26, 227(*fig.*), 233, 234; isallo stress lines, 228–29; landmass rotation and, 56–63, 68–69, 93–98; midocean ridge axial valleys and, *xi*, 198–99, 200, 202, 203–204, 207, 209, 213, 219. *See also* fracture
Fennoscandia, mantle viscosity, 49, 67, 135, 136, 137
Finland Shield, 150, 161
Flemish Cap, 77(*fig.*), 98, 99
Florida, 84, 95, 99, 100(*fig.*); Africa and, 262, 263; magnetic anomaly belts, 90, 91, 98
Flysch, 29; Mediterranean, 175, 176, 178–80, 183
foldbelts, *see* geosynclines; orogenesis; *and see specific mountain systems*
Fortieth Parallel, 95, 97
fossil forms, *viii*, 38, 177, 243–48, 262; paleomagnetic dating and, *viii*, 19, 241–43, 244–46, 247–50, 263–64
fracture, 47, 74, 75, 135, 136–37, 264; basement relief in, 3–4; crust plate analogy and, 202–203; East-West, 5–7, 10, 68; Pacific zones of, 63, 68–69, 263; North Atlantic, 93–98; sea floor spread and, 27, 219, 261; spin axis and, 9–10; stress fields and, 225–26, 227. *See also* faults; rifts
France, 123, 175
Franciscan formation, California, 42
Front Range, 47, 55, 65
Funnell, B. M., cited, 244

gabbro, 147, 148, 181, 183, 209; Vourinos and, 182, 184(*fig.*)
Galapagos Rise, 214(*tab.*)
Gass, I. G., cited, 207
Gauss normal epoch, Mammoth event of the, 247
geologic time, geomagnetic scale of, *xi*, 241–43, 248, 249(*fig.*), 250
geomagnetic field, 12–13, 15, 20–22
geomagnetism, 255, 264
geophysics, *viii*, 147–65, 254–55, 256, 264
georheology, 113–43
geosynclines, *ix*, 31(fig.), 39, 148, 199; axial valleys of, *xi*, 194–220; basement relief in, 3–4; Conrad discontinuity and, *xi*, 165; eastern North American, 77, 78(*fig.*), 79–80, 101–102; gravity anomalies and, 88, 89, 152, 153; isotopic age determination for, 80–84; opposing-continent comparisons, 103, 104(*fig.*), 105–106, 168, 171, 172–75; sial formation and, 28, 29–30, 32, 42. *See also specific foldbelts*
Gibraltar, 172, 173, 174, 175
Gilbert, William, 254
Gilliland, William, *viii*
Gilluly, James, *vii*, *x*, 47–69, 263
Girdler, E. W., cited, 37
Gondwanaland, 16(*fig.*), 32–37, 39, 261; area of, 33, 34(*tab.*), 35, 36, 262; collision of, 44; sial of, 37
Gorda Ridge, 197, 198(*fig.*), 205 (*fig.*), 208; escarpment, 64, 65, 199 (*fig.*)
Grand Banks, 88(*fig.*), 79, 98, 99; faults, 95, 96; earthquake of 1929, 96
gravity, 67, 223, 254, 255; anomalies, *x*, 47–49, 87–89, 97, 119, 122–23, 147, 148, 150, 151–53, 155–58, 165, 185, 188–89, 223–37; crustal thickness data and, 93, 117, 121; Mediterranean slides and, 177, 178, 179, 180, 187, 189; midocean ridges and,

gravity (cont.)
204; ophiolite emplacement and, 185, 187
Great Basin and Range province, 49, 50–51, 195; faults, 54, 57, 58–62, 68–69; orogenic belts of, 62–63, 263
Greater Antilles, 62, 106
Great Glen fault, 103, 263
Green, L., cited, 32
Greenleaf anomaly, 155–58, 164
greenstones, 206, 208, 213, 218; chemical exchange and, 216, 217(*fig.*), 220
Grenville tectonic belt, *ix*, 76, 78(*fig.*), 89; accretion of, 102; dating and, 82–83, 84, 103, 105; Greenleaf anomaly and, 156
Griscom, Andrew, quoted, 88*n*
Gulf of Alaska, 37
Gulf of California, 66, 263
Gulf Coastal Plain, 79, 80
Gulf of Mexico, 76, 80, 91, 98, 156, 163, 262; salt diapirs of, 185
Gutenberg, Beno, 130

Hack, John, 195
Harrison, C. G. A., cited, 244
Harvey, William, 257
Hawaii, 63, 163
heat, 4, 56, 254, 255; Cyprus hydrothermal alteration, 208; deep-focus earthquakes and, 138; deforming stresses and, 134–35; diapirism and, 183, 187–88; earth interior, 128; magnetite deposit depth and, 150; mantle convection and, *x*, 28, 37, 42, 140, 141–42, 201–202, 260; midocean ridge axes and, 201–203, 206, 213–20; NRM deviation and, 20; of radioactive decay, 27; tetrahedral hypothesis and, 32; tholeiite metamorphism and, 206; vibration conversion to, 132; water convection in rock and, 202, 203, 213–20; western United States heat-flow data, 51, 53
heat flow, 201, 203
Heezen, B. C., 194, 261
Heirtzler, J. R., cited, 242, 243, 248, 259, 260
Hercynian episode, *ix*, 168, 170, 178, 180
Herring-Nabarro creep, 135
Hess, H. H., 194, 206, 215, 259, 262
Himalayan Overlap, 34(*tab.*)
Himalayas, 62, 182, 233
Hobbs, William Herbert, 54, 93
Holmes, Arthur, 254, 258
Holocene, 244
Honshu Island, Japan, 232–33
Howard, L. N., cited, 142
Hudsonian (Penokean) orogeny, 82
Hungarian Depression (Pannonian Basin), 172
Hutton, James, 167

Iceland, 33, 159, 161, 259
India, 33, 35, 40, 161, 262; area of, 34(*tab.*)
Indian Ocean, 27, 35, 38, 43, 206, 214(*tab.*), 243
Indonesia, 233
International Geological Congress (1964), 173
International Tectonic Map of Europe, 172–73
isallo stress lines, 228–29
island arcs, 62–63, 118, 263–64; sea floor spreading and, *vii*, 208, 258, 259, 260; sial formation and, 28–29, 30, 32
isostasy, 24, 40, 43, 223; Gondwanaland-Laurasia hypothesis and, 33–34, 35; gravity anomalies and, 47–49, 88, 158; midocean ridges and, 204, 213; seismological data on, 49–50, 67, 117–18, 121
isotopic age provinces, 80–84, 103, 105
Italy, 168, 172, 175, 178–79, 187, 245; Moho under, 119–20, 121
Ivrea, Italy, 119–20, 121, 126

Japan, 13–14, 15(*fig.*), 28, 29, 32, 62; stress fields, 232–33
Jeffreys, Harold, 53, 130, 255
Joesting, H. R., cited, 47–49
Johnson, Helgi, *viii*
JOIDES core, 199, 204, 208
Juan de Fuca Ridge, 197(*fig.*), 243
Jurassic Period, 16, 42, 262, 263; Mediterranean in, 168, 175, 176(*fig.*), 178, 181; paleomagnetic data on, 16

Kabyle massifs, 174, 175, 177
Kay, Marshall, 199
Kelvin seamounts, 90, 96
King, Philip B., *vii*, *ix*, 74–106
Knopoff, Leon, *vii–viii*, *x*, 67*n*, 113–43, 188

Lagrange, Joseph Louis, 261
Lake Superior, *ix*, *xi*, 82; gravity anomalies, 88, 155, 160–61
Lamont Geological Observatory, 100
landmass rotation, 13–14, 15(*fig.*), 47, 74, 75, 165; eastern North America, 93–98; Eurasian, 169, 189; European, 168; Japan, 15; western United States, *x*, 56–63, 68–69
Laplace, Pierre Simon, marquis de, 43
Larochelle, A., cited, 258
Laurasia, 32, 37, 39, 44, 261; area, 33, 34(*tab.*), 35, 36, 262
lava, 13, 181, 182, 183; in axial valleys, 200, 201, 204; earth magnetic field reversals and, 258; Greenleaf anomaly and, 155, 156, 157; islands of, 260; magma generation and, 24, 260
Lehmann, I., cited, 116, 126
Lensen, G. J., cited, 228
Lesser Antilles, 62
Liguria, 178, 179(*fig.*)
Ligurian Sea, 179, 180
Love waves, 124, 125(*fig.*), 131
Ludwig, W. J., cited, 28
Lyons, Paul R., *viii*, *xi*, 147–65

McGinnis, L. D., cited, 155
Madagascar, 262
mafic volcanics, 51, 52, 93, 98, 99; chemical exchanges and, 220; Mediterranean, 183–87, 189
magmas, 24, 29, 52, 183; basaltic, 208–209
magnetics, 47, 68, 255; anomalies, *x–xi*, 3–10, 51, 89–91, 98, 99, 147, 148, 150, 151(*fig.*), 163, 196, 201, 241, 243, 244–48, 258; definition of regional structures by, 3–10, 55–56; magnetic interruption, 5–7, 96–97; midocean ridge profiles, *viii*, 199–201, 211–12, 241–43; NRM measurement in, 12–13, 20; sea floor regeneration and, 196
magnetic polarity history, *viii*, 243–46, 258–59
Malaga nappe, 174, 175
Malagasy, 33, 34
mantle, 33, 115–16, 237; Appalachian, 93; chemical composition of, 120–21, 127, 128, 141, 142, 143; convection process of, *vii*, *ix*, *x*, 24, 27–28, 29, 36–37, 39, 41–42, 43, 128, 136, 138, 140–43, 202–203, 257, 260; crust deformations and, *x*, 51–56, 68, 121–22, 136; Cyprus outcrop, 207–208; deep-focus earthquakes and, 138, 233; density, 132, 163–64; differentation degree of, 52, 54, 181; domal rises in, 183–88, 189; epeirogeny and, 40, 41(*fig.*), 42; homogeneity postulate, 128–29, 140; lower mantle viscosity, *x*, 135–36, 137, 138, 141, 142; low velocity channels, 117, 123, 126, 127, 128, 129, 133, 137; midocean rises and, 37, 38–39, 42, 43, 202–203, 206–207; the Moho and, 117, 118, 160–61; oceanic basin structure of, 126, 127(*fig.*), 206–207; sea floor spread and, *ix*, 27–29, 157–58, 260; surface feature influence on, 121–22, 123, 124–25; transition zones, 115, 116, 123; up-

mantle (cont.)
per and lower mantle contrasts, *x*, 115–16, 117, 120–21, 126–29, 131–32, 133, 135–38, 141, 142; velocities (Western U.S.), 52, 53(*fig.*), 118; viscosity (Western U.S.), 49, 67, 135; water derivation from, 26
maria, 24–25, 29
Matthews, D. H., 199, 241, 258, 259
Maxwell, John C., *vii*, *ix–x*, 167–89
Mediterranean Sea, 35, 122, 123, 124 (*fig.*), 167–89; basin structure, *ix–x*, 126, 127(*fig.*)
Melville, R., cited, 261, 262, 263
Menard, H. W., cited, 162, 197, 204, 215, 261, 262
Mendocino fracture, 5, 6, 7, 64–65, 66, 68; Atlantic fracture zone and, 10, 97, 163, 164, 165
Mesozoic Era, 80, 96, 106, 189; Alpine foldbelt and, 105, 174; Blake Plateau and, 99, 100(*fig.*), 101; Cordilleran foldbelt and, 93; Florida cover, 91; fossil forms, 244; Gondwanaland and Laurasia in, 32, 33, 35, 37, 43, 44, 261, 262; NRM directions and, 13–15; ocean floor rock of, *ix*, 43, 168, 182
metasomatism, 206, 216, 218, 220
Mexico, 62, 63, 66, 82, 105, 106
micropaleontology, *xi*
Mid-Atlantic Ridge, 38, 91, 164, 242, 259; Conrad discontinuity and, 157, 159, 160; Cyprus and, 208; greenstones of, 216, 217(*fig.*); layer thicknesses, 204; rift of, 233, 235 (*fig.*); sea floor production rate, 214(*tab.*)
midocean ridges, *vii*, 161–63, 164, 257, 258, 264; axial valleys of, *xi*, 194–220, 233, 259; Eulerian mechanics and, 261; extension and basalt supply, *xi*, 208–13, 259; "misplaced," 40, 42; P*n* velocities under, 118; rise system, 37, 38–39, 43; sediment dating, *viii*, 241–43, 259. *See also specific ridges*
mines, stress measurement in, 226–27
Minnesota River Valley, 82
Miocene Epoch, 58, 60, 62, 68, 100 (*fig.*), 242; fossil forms of, 244, 245; Mediterranean in, *x*, 175–76, 177, 180, 189
Mississippian Period, 56, 65
Mohorovičić (Moho) discontinuity, 51, 53, 67, 117–18, 260; Cyprus and, 207; Greenleaf anomaly and, 156, 157, 160–61; locations of, 119–21, 154, 163, 188
Mohole, 162(*fig.*), 163
molasse (detritus), *ix*, 41
moon, the, 24–25, 29, 257; accretion of, 26; earth rotation and, 150, 164
Moores, E., cited, 183, 207
Morgan, W. J., cited, 201, 202, 211, 214
Morley, L. W., cited, 258
Morocco, 105, 106, 172, 176(*fig.*), 177
motion, earthquake fault-plane solution, 230(*fig.*), 261; free oscillation, 130–32
Mudie, J. D., cited, 198–99
Murray fracture zone, 64, 66, 67, 69

Nafe, J. E., cited, 106
nappes, 174, 175–76, 179, 187
National Geographic Society, 34
National Science Foundation, 194*n*
natural remanent magnetization (NRM), 12–14, 19–20
Nayudu, Y. R., cited, 183*n*
Neogene, 39, 180
Nevada, 49, 52, 65; Moho under, 119, 121
Nevada Nuclear Testing Site, 119
Newark Group, 79
New England, *viii*, 79, 95, 262, 263; Spain and, 105
New England (Kelvin) seamounts, 90, 96

Newfoundland, *ix*, 79, 82, 106; Appalachian foldbelt extension through, 102, 103; faults, 95, 96, 103
New Madrid earthquake (1811–1812), 86
Newton, Sir Isaac, 254
Newtonian viscous fluids, 133, 134, 135, 136
New York City, 85, 101(*fig.*)
New Zealand, 34, 35, 207, 233, 262
North America, 32, 33, 37, 44, 262, 263; area of, 34(*tab.*); continental arch of, 158–59, 160, 161; continental shelf, *ix*, 76, 77(*fig.*), 79, 86, 90, 96–97, 98–102; eastern tectonics of, *ix*, 74–106, 170; eastern and western differences of crust and mantle, 118, 121, 127; midocean rise and, 40; peneplanation rate of, 39; western tectonics of, *x*, 52–69, 86, 88, 93, 97, 170
North Atlantic Ocean: Bullard reconstruction, 74, 75(*fig.*); core samples, 242; Drake and Nafe reconstruction, 106; opening of, 18, 262; opposing continental fit, 74, 75(*fig.*), 102–106, 262; sea floor production rate, 214(*tab.*); subsea features of, 76, 77(*fig.*), 233, 235(*fig.*), 259
nuclear explosion, 118, 119, 126–27

ocean basins, 24–44, 47, 93, 261, 263–64; compared with the maria, 24–25; crustal structure, 125–26, 188, 261; depth, 43–44, 80; Mediterranean, *ix–x*, 168, 169, 176, 188–89; North Atlantic, 77(*fig.*), 78(*fig.*), 99, 102; P*n* velocities under, 118; sial distribution, *viii–ix*, 27–28; theories of origin, 25–29
ocean floor, 24, 254; Atlantic magnetic anomaly trends, 91, 106, 259; basalt dominance in, 205–207, 209, 215–16; Conrad layer and, 147, 148; continental fault interrelations, 63–67, 68–69; core drilling, *viii*, *xi*, 162, 163, 199, 205, 208, 241–42, 243–44, 245, 246, 248, 250(*fig.*), 259, 263–64; crust permeability, 219–20; Cyprus as model of, 207–208, 220; Grand Banks, 79; magnetic polarity history and, *xi*, 242, 243, 248, 250 (*fig.*), 258–59; mantle properties, 124; Moho depth below, 117; naval research on, 256–57; new crust production rate, 214–15, 218, 259, 260; ridge axial valleys, *xi*, 194–220, 259; sedimentation rate, 39; spread of, *vii–xi*, 27–29, 32, 39, 42–43, 44, 91, 106, 157, 194–220, 241, 242(*fig.*), 250, 257–60, 261; transcurrent faults, 233
ocean level, peneplain and, 38, 43
Ohio, 82, 155; magnetic interruptions in, 5
Olduvai event, 245, 248
Oligocene Epoch, 62, 100(*fig.*), 175, 179, 180; fossil forms of, 244
olistostromes, 176, 180
Olsson, Richard K., *viii*, *xi*, 241–50
Olympic Mountains, 49, 57–58
Olympic-Wallowa lineament, 58
ophiolites, 168, 177, 178, 179–80, 259; of Cyprus, 207; emplacement, 183–87, 188; stratigraphy and, 181–83, 189
Ordovician Period, 148, 159, 177; fossil forms of, *viii*, 262
Oregon, 51, 55, 57, 63, 197; rotation hypothesis, 62
Oregon State University, 194*n*
ores, 148, 150, 153, 159
orocline, 62–63
orogenesis, 30–32, 39, 63, 68, 261; Appalachian, 77, 93; Canadian Shield, 76, 82, 164; circum-Pacific orogeny, 233; diapirism and, 183–88, 189; earthquakes and, 139, 232; Eurasian, *ix*, 168, 170–71, 172, 233; European, 103, 172, 173, 232, 237; stress fields and, 223, 231–34, 237, 263. *See also* geosynclines

Osburn fault, 58
Ostenso, N. A., cited, 204
Ouachita foldbelt, 79–80, 89, 98, 153

Pacifica (continent), 261, 263
Pacific-Antarctic Ocean, 214(*tab.*), 245
Pacific-Cocos Ridge, 214(*tab.*)
Pacific Coast, 51, 52, 97, 118, 137–38, 263
Pacific-Nazca Ridge, 214(*tab.*)
Pacific Ocean, 37, 43, 127, 234(*fig.*); anomaly trends in, 5, 10, 163, 243; basin origins, 25; circum-Pacific orogeny, 233; crust thickness, 63; fossil forms, 245, 246–48; greenstones of, 206; midocean rises, 38, 40, 159, 160, 161–62, 203(*fig.*), 206, 208, 214(*tab.*), 242, 262–63; West Pacific stress trajectories, 231(*fig.*), 232(*fig.*)
Pakiser, L. C., quoted, 93
Paleocene Epoch, 99, 100(*fig.*), 244
paleoclimatic data, *viii*, 13, 18–20, 22
paleolatitudes, *viii*, 13, 16, 18–20, 22
paleomagnetism, *viii*, 241, 254; continental drift theory and, *viii*, 12–22, 62, 169, 170; data for: South America and Africa, 17; middle and late Paleozoic, 18; North and South Atlantic, 18; Chugwater formation, Wyoming, 20, 21. *See also* magnetics
paleontology, 241, 242, 250(*fig.*), 259, 264
Paleozoic Era, 33, 35, 36, 41, 66, 68, 83, 98, 106; Cabot fault zone and, 95–96; Canadian Shield, 164; foldbelts of, *ix*, 4, 5, 77, 79, 81(*fig.*), 84, 86, 103, 105; Mediterranean and, 168, 172, 174, 175, 177, 189; midocean rise in, 39, 43, 91; polar wander in, 18; proto-Atlantic in, *viii*, 44, 263
Pangaea, 35, 43, 261
Panthalassa, 36(*fig.*), 39, 43
Pelagonian massif, 182
Pennsylvania, 88, 97; magnetic intensity data for, 3, 4(*fig*), 5, 9
Pennsylvanian Period, 38
Penokean (Hudsonian) orogeny, 82
peridotite, 27, 28, 206–207; Mediterranean, 175, 177, 181, 182, 183, 187
Permian Period, 174, 263; continental drift in, 17, 18, 169, 170
petrogenesis, *vii*, 264. *See also* rock formation
pillow basalt, 205, 212, 220
pillow lavas, 181, 182; in axial valleys, 200, 201, 209, 212
Plafker, G., 28
platform areas, 78(*fig.*), 80
plate tectonics, *xii*, 202, 258
Pleistocene Epoch: axial valley sediment of, 199; fossil forms, 244, 245, 246, 247, 248; geomagnetic scale of, 243; ice of, 196, 244; mantle differentiation in, 52
Pliocene Epoch, 177, 232, 237; fossil forms, 242, 245, 246, 247, 248
Plio-Pleistocene, 245, 246, 247(*fig.*), 248; paleomagnetic pole positions, 14(*fig.*)
plutons, 62, 63; Mediterranean, 168, 183, 187
P*n*, 118, 119–21, 126; defined, 117
polar wander, 3, 10; continuity of, 16, 17; paleomagnetic data on, *viii*, 13, 14(*fig.*), 15, 16, 17(*fig.*), 20
potassium-argon dating, 196
Precambrian Era, 78, 88, 97, 103, 105, 154; Algoman unconformity in, 148–50; dating uncertainties in, 22, 38; rock classifications of, 148, 149 (*fig.*); shields, 42, 76, 78(*fig.*); stress patterns, *viii*, *x–xi*, 3–5, 7–10, 55–56, 65, 83

quartz-diorite line, in Great Basin, 62, 63, 68
Quaternary Period, 177, 178

radiometric dating, 163–64, 178, 254
Rayleigh waves, 122–23, 136; spheroidal vibration and, 131
red beds, *viii*, 19
Red Sea, 14
regional stresses, *x*, 226–27, 233–34, 237; earthquake trajectories, 229–31, 232
Rif Mountains, 168, 173, 174, 177, 183, 187
rifts, 32, 156, 233, 234; rise system and, 27, 28, 38, 39, 43, 44
Rift Valleys of East Africa, 195, 198, 233
rock formation, *vii*, 253–54, 256; allochthonous Mediterranean deposits, 178–81; Archean, *xi*, 83, 148, 149(*fig.*), 150, 152, 153, 159, 163–64; climate and, 19; clock minerals, 148, 163–64; Conrad discontinuity and, 147, 148, 163–64; cratonic cover, 38, 39, 42; density contrasts and, 88*n*, 117, 151; earth age time gap in, 26–27; earth expansion and, 19–20; Easter island, 162; fracture strength, 136; of the Great Basin province, 58, 65, 68–69; isotopic age provinces of, 80–84; Mediterranean nappes, 174, 175–76, 179; metamorphic, 42, 76, 78(*fig.*), 82, 84, 148, 177, 178, 180, 181, 187, 206, 216, 218, 260; NRM measurement and, 12–13; ocean floor (sima), 28, 43, 50–51, 120, 180, 205–207, 208, 213, 259; plutons, 76, 78(*fig.*), 79, 82, 84, 175, 183, 187; Precambrian classifications, 148, 149(*fig.*); rock bursts, 226–27; of San Giorgio, 180; sialic, 25–27, 120, 168 (*See also* sial); water convection and, 202, 203, 213–220; weathering, 236(*fig.*). *See also specific varieties*
Rocky Mountains, 50, 52, 55, 66, 118
Rocky Mountain Trench, 63, 263
rotation of the earth, 135–36, 150, 164, 196, 223
Royal Society Symposium on Continental Drift, 12
Runcorn, S. K., cited, 260
Rutgers University, *vii*, *viii*
Rutherford, Ernest, Baron, 257

Sahara Shield, 103, 105
St. Augustine, Florida, 100(*fig.*)
salt diapirs, 185, 186(*fig.*), 187
San Andreas fault, 54, 56, 57, 64–65, 66–67, 68–69, 95; East Pacific Rise and, 263
Sardinia, Italy, 123
satellite navigation, 197, 204, 257
scarps, 64, 65, 199, 206
Scheidegger, Adrian E., *viii*, *x*, 223–37
schists, 175, 177, 178
Scotland, 103, 105, 159, 262
Scripps Institute of Oceanography, 198, 241*n*
Sea floor spreading, *vii*, *viii*, *ix*, 27–44, 140–42, 194–96, 201, 209–220, 241–250, 257–264
seamounts, 77(*fig.*), 90(*fig.*), 97, 162, 183*n*
sediments, 13, 20, 88*n*; dating of, *xi*, 241–43, 245, 248, 258; eastern North America continental shelf and, 99, 100(*fig.*), 101–102; gravity anomalies and, 151, 152; Mediterranean, 168, 170, 174–75, 178–79, 189; in midocean ridge axial valleys, 199, 204, 208–13, 218, 220, 241; ophiolites amongst, 181–82, 183, 188; peneplain rate and, 39, 164; sodium-calcium ratios in, 216
seismology, *vii–viii*, 47, 113–43, 255, 257, 260, 264; Conrad discontinuity and, 147, 154; continental margin surveys, 99, 100(*fig.*), 101(*fig.*); crustal thickness data from, 49–51, 53–55, 67, 92, 93, 113, 117, 122–23, 154; crustal velocity data from, *x*, 51–53, 113, 118, 119–21; on earthquake epicenter patterns, 84–86,

seismology (cont.)
138–40; magnetic anomaly belts and, 89, 96–97; mantle homogeneity postulate and, 128–29, 140, 142, 143; midocean ridges and, 203–204; ophiolites and, 188; regional stress trajectories and, 229–31, 232 (*fig.*); zero-frequency, 114. *See also* earthquakes; faults.
serpentine, 206, 209, 215–16, 259. *See also* ophiolites
serpentinite, 181
serpentinization, 215–16
Seychelles, 34, 262
shale diapirs, 185
shear (S) waves, 116, 132, 237, 263; Conrad discontinuity velocity, 147; diapir, 188; low velocity channels, 117, 123, 124(*fig*), 125, 126, 140, 141; Q values for, 132–33; transcurrent faulting and, 233, 234(*fig.*)
sheeted complex, 207–208, 209
sial, *viii–ix*, 25, 120; convection process and, *ix*, 24, 27–28, 29, 37; cosmic source hypothesis, 26; dating of, 27; Great Basin Conrad discontinuity and, 50–51; Mediterranean, 168, 170, 174, 175, 178, 188, 189; plateau uplift and, 40; water ratio, 43
Siberia, 37, 262
Sicily, 168, 172, 175, 177, 178, 180, 188
Sierra Nevada, 48, 49, 56, 65; boundary fault of, 54, 57; crustal velocity in, 51
Silurian Period, 17, 18, 38, 177
sima, 28, 43, 50–51, 120
Smith, Bennett L., *viii–xii*
solids: deformation of, 133–35; stress fields in, 224–25
South America, 33, 38, 40–41, 63, 84, 105, 130, 214; African fit of, 17, 18(*fig.*), 106, 164, 169, 262
South Atlantic Ocean, 214(*tab.*), 243, 261; core samples, 242; opening of, 18, 262
Southern Hemisphere, 32, 33, 37
South Pacific, 242, 243, 244, 261
Spain, *ix*, 172, 173, 174, 176(*fig.*); rotation of, 14, 105, 106
species extinction, 244; magnetic reversals and, *xi*, 243–50
steady-state concept, ocean ridges and, *xi*, 195–96, 212
Steinman, G., cited, 182
strain rates, axial valley, 210(*fig.*), 211, 212, 220
stratigraphy, 223, 260; fossil forms and, *viii*, 241–43, 248, 250(*fig.*); Mediterranean, 168, 172, 173–77, 181–83
stress fields, *viii*, *x*, 52–55, 67–68, 223–37, 263; deformation of solids and, 133–35
stress patterns, *x*, 67, 98, 224–26; Precambrian, *viii*, *x–xi*, 3–5, 7–10, 55–56; surface, 228–29, 234–36; underground, 226–27, 229–34, 237
structures, *ix*, 76, 77(*fig.*); basement, 3–4, 42; continental, 26, 37–38, 63, 264; crustal velocities and, 50–51, 118, 123; diastrophism and, 27, 39–40; earth rotation and, 135–36, 150, 164, 196, 223; gravity and, 47, 88; magnetic interruptions and, 5–7; marine, 24, 27–28, 63, 196, 260, 264; peneplain, 38, 39; sial formation and, 28–29, 43; stress fields and, 234–36
surface waves, 122–23, 124, 126, 131
Sykes, L. R., cited, 203, 204
symmetry, 114–15, 211, 224, 243; Mediterranean, 172–73, 177, 178, 187, 188

Taconian orogeny, 77, 84, 103
tectonism, 264; convection in, 140–43; earthquake model, 139–40; in eastern North America, *ix*, 74–106; Mediterranean, 167–89; midocean ridge model, 194–220, 261; Pacific Ocean, 160, 161–63; plate analogy

in, *vii*, 202–203, 214–15, 258; stress fields in, *viii*, *x*, 223–37; in western United States, *x*, 52–69. *See also* structures
Tell Mountains, 173, 174, 175, 177, 187
Tertiary Period, 63, 68, 80, 168, 263; Blake Plateau and, 99; mantle differentiation in, 52; NRM directions of, 13, 21
Tethys, *ix*, 168–71, 174, 177, 188
Texas, 64, 82, 151–52, 185
tholeiites, 28, 205, 206, 213; metamorphism of, 216, 260
Tibetan arcs, 62
Tibetan Plateau, 34*n*, 35, 40
Toksoz, M. N., cited, 128–29
transform faults, 202–203
Transverse Ranges, 55, 57, 64, 66, 68
travel-time interpretation, 113, 115, 118
trenches: diapirs and, 187, 188, 189; earth mantle and, 42, 138; eastern North American continental shelf and, 101–102; marine, *vii*, 27, 28, 29, 39, 44, 89, 101–102, 175, 176, 258, 260
Triassic Period, 78(*fig.*), 79, 96, 97, 262; Appalachian dikes, 94; Chugwater formation of, 20; Mediterranean, 170, 174, 175, 178, 179
Trois-Rivières earthquake of 1663, 86
Troodos Massif, Cyprus, 207–208
Tulsa, University of, 154
turbidity currents, 96, 199
Tyrrhenian Sea, 172, 177–81, 187

uniformitarianism, 194, 195–96
United Kingdom, *see* British Isles
United States, 38, 39, 40; earthquake energy rate-depth function and, 137–38; P*n* velocities in, 118, 119 (*fig.*), 121, 126; western and eastern differences of crust and mantle, *x*, 47–69, 118, 127; western United States, 40–66, 195, 263
United States Office of Naval Research, 194*n*
Urey, Harold, 26

Van Bemmelen, R., 26
Van Hilten, D., cited, 169–70
Variscan orogeny, 103
Vela Uniform program, 49
velocities, 113–14, 126–28, 188; axial valley, 209, 212, 213; Conrad layer, 147, 148, 154, 156, 157; dispersion measurement, 122–23; earth mantle, *x*, 119–21, 123, 127, 129, 132, 133, 157, 260; inner core-outer core boundary and, 130; Tyrrhenian crust, 177; western United States, 52, 53(*fig.*), 118
Verhoogen, J., cited, 260
Vine, F. J., cited, 199, 201, 206, 207, 211, 212, 258, 259, 260
Vine-Matthews magnetic model, 199, 241, 258, 259
Virginia, 96
viscosity, *x*, 133–36, 140–43; decay, 213; earthquake aftershocks and, 140; lower mantle, *x*, 135–36, 137, 138, 141; stress fields and, 224–25; velocity dispersion measurement and, 122; western United States, 49, 67, 135
Vogt, P. R., cited, 204
volcanism, 67, 255, 262; eastern North America, 77(*fig.*), 78(*fig.*), 79, 91, 93, 99; Greenleaf anomaly and, 155, 156, 157; magnetic trends and, 56, 91; mantle differentiation and, 52, 260; Mediterranean, 178, 182, 183–87
volcanoes, 63, 115, 260
Vourinos, 179(*fig.*), 182, 184(*fig.*)

Walker Lane, 56, 57
Wasatch front, 49
Wegener, Alfred, 167, 168, 170, 255; Pangaea and, 35, 261
Werner, Abraham Gottlob, 167, 254

West Pacific Ocean, 231(*fig.*), 232 (*fig.*)
Wilson, J. Tuzo, *vii, viii, ix, xi,* 203, 253–64
Woollard, G. P., cited, 47–49
Wyoming, 20, 21(*fig.*), 152(*fig.*)

Yaquina (vessel), 194*n*

Zachos, K., cited, 182–83
Zero-frequency seismology, 114
Zeeman effect, 131
Ziest, I., quoted, 93

This book was set in Janson Linotype and printed by offset on P & S Wove manufactured by P. H. Glatfelter Co., Spring Grove, Pa. Composed, printed and bound by Quinn & Boden Company, Inc., Rahway, New Jersey.